Karst Hydrogeology and Human Activities

Impacts, Consequences and Implications

Edited by

David Drew
Trinity College Dublin, Department of Geography, Ireland

Heinz Hötzl
Karlsruhe University, Department of Applied Geology, Germany

A.A.BALKEMA/ROTTERDAM/BROOKFIELD/1999

Cover photograph: The Katavothoe Kanatas (swallow hole) in the Polje of Tripolis. It swallows the untreated waste water of the city of Tripolis, Peloponnesus, Greece. Photograph by Hötzl, 1986.

Published by
A.A. Balkema, P.O. Box 1675, 3000 BR Rotterdam, Netherlands
Fax: +31.10.4135947; E-mail: balkema@balkema.nl; Internet site: http://www.balkema.nl

A.A. Balkema Publishers, Old Post Road, Brookfield, VT 05036-9704, USA
Fax: 802.276.3837; E-mail: info@ashgate.com

ISBN 90 5410 463 5 hardbound edition
ISBN 90 5410 464 3 paper edition (IAH member)

Contents

Preface

The increasing pollution of our environment as deliberate or unintentional conse-quences of human activities has to a great extent spoiled sensitive karst ecosystems. This has led, for example, to distortions of landscapes, to soil erosion, to increased solution processes and to input of contaminants into the underground system.

Some of the various impairments to karst ecosystems which damage natural re-sources may threaten directly directly the life of man, animals and plants and their living space.

Large parts of our earth are covered by karst areas and in many regions the karst aquifers are the only available water resources for drinking water. Even large cities in Europe, Asia and America obtain their drinking water from karst water resources.

These humanly induced changes to karst are a serious challenge to every hydro-geologist. We have to recognise that the resulting problems cannot be solved just by technical measures. It is necessary also to change of human behaviour and perception in respect of the natural resources involved. In order to preserve the quantity and quality of karst groundwater the efforts of the whole of society are required, but though an increasing awareness of environmental problems can be observed, there is still a lack of awareness of the specific problems of the vulnerable karst systems.

The IAH Karst Commission decided in 1993 to contribute to an increasing aware-ness of the vulnerability of our karst systems and karst groundwater resources by publishing examples of the human impacts on karst groundwater. A great number of colleagues put their experience at our disposal. With introductory chapters by main co-ordinators and succeeding case studies, we hope to present a convincing depiction of the environmental problems of karst aquifers; not only for experts, but also for politicians, managers, administrators, planners and the whole interested public.

Our sincere thanks are due to the co-ordinators and authors for their contributions as well as for their patience with the delay in printing due to editorial and technical reasons. We are very much obliged to William White for reviewing the manuscript and for his many constructive suggestions. We appreciate also the support of the IAH and especially the overall editor, Ian Simmers, as well as of the publisher A.A. Balkema in the Netherlands. We owe special thanks to all those persons who helped us in editorial work. We greatly appreciate the support of B. Reichert in the

XI

different phases as well as the help by V. Busche, U. Schneider and R. Ohlenbusch in handling the files on the computer and Sheila McMorrow for her cartographic work.

David Drew and Heinz Hötzl

List of authors

Bartolomé Andreo Navarro, Dpt. de Geologia, Facultad de Ciencias, Universidad de Malaga, E-29071 Malaga, Spain.

Jose Miguel Andreu, Dept of Environmental Sciences, University of Alicante, Spain.

Philippe Audra, Groupe de valorisation de l'environnement, CNRS Université de Nice-Sophia-Antipolis, 98 Boulevard Edouard Herriot, BP 209, F-06204 Nice Cédex, France.

William Back, US Geological Survey, 431 National Center, Reston, VA 22092, USA.

Bozidar Biondic, Institut za Geoloska, Istrazivanja, Sachsova 2, 10000 Zagreb, Croatia.

Ranko Biondic, Institut za Geoloska, Istrazivanja, Sachsova 2, 10000 Zagreb, Croatia.

Paolo Bono, Dip. Scienze della Terra, Universita La Sapienza, Piazza A. Moro 5, I-00185 Roma, Italy.

Srecko Bozicevic, Institut za Geoloska, Istrazivanja, Sachsova 2, 10000 Zagreb, Croatia.

David B. Bredenkamp, Dept. of Water Affairs and Forestry, Private Bag X313, Pretoria, Republic of South Africa.

Francisco Carrasco Cantos, Dept. de Geologia, Facultad de Ciencias, Universidad de Malaga, E-29071 Malaga, Spain.

Pierre Chauve, Universite Franche-Compte, Structural and Applied Geology, Place Leclerc, F-25030 Besancon-Cedex, France.

Catherine Coxon, Environmental Sciences Unit, Trinity College, Dublin 2, Ireland.

David Drew, Department of Geography, Trinity College. Dublin 2, Ireland.

Franjo Dukaric, Institut za Geoloska, Istrazivanja, Sachsova 2, 10000 Zagreb, Croatia.

Juan José Durán-Valsero, Inst. Tecnologico Geo. Minero de Espana, Rio Rosos 23, E-28003 Madrid, Spain.

Matthias Eiswirth, Dept. of Applied Geology, University of Karlsruhe, Kaiserstr. 12, D-76128 Karlsruhe, Germany.

Mark Ellaway, Dept. of Geography and Environmental Studies, University of Melbourne, Parkville, Victoria, 3052, Australia.

Antonio Estevez, Dept of Environmental Sciences,University of Alicante, Spain.

Maria Dolores Fidelibus, Ist. Geologica Applicata e Geotec., Politecnico di Bari, Via Orabona 4, I-70125 Bari, Italy.

Brian Finlayson, Dept. of Geography and Environmental Studies, University of Melbourne, Parkville, Victoria, 3052, Australia.

David Gillieson, Dept. of Geography and Oceanography, University College, University of New South Wales, Canberra, A.C.T. 2601, Australia.

Radisav Golubovic, Petnica Science Center, 14000 Valjevo, Yugoslavia.

Wolfgang F. Grimmelmann, Geological Survey of Lower Saxony, Postfach 51 01 53, D-30631 Hannover, Germany.

John Gunn, Geographical and Environmental Sciences Dept., University Queensgate, Huddersfield HD1 3DH, UK.

Jose Gutiérrez Diaz, Centro de Hidrología y Calidad de las Aguas, Instituto Nacional de Recursos Hidráulicos, P.O. Box 23, General Peraza 19210, Ciudad de La Habana, Cuba.

George R. Hallberg, Environmental Research, University Hygienic Laboratory, University of Iowa, Iowa City, Iowa, USA.

Paul Hardwick, Geographical and Environmental Sciences Dept., University Queensgate, Huddersfield HD1 3DH, UK.

John W. Hess, Water Resources Center, Desert Research Institute, P.O.BOX 19040, Las Vegas, NV 89732-0040, USA.

Steve Hobbs, Geographical and Environmental Sciences Dept., University Queensgate, Huddersfield HD1 3DH, UK.

Heinz Hötzl, Dept. of Applied Geology, University of Karlsruhe, Kaiserstr. 12, D-76128 Karlsruhe, Germany.

Ian Houshold, Dept. of Environment and Land management, University College, University of New South Wales, Canberra, A.C.T. 2601, Australia.

Arie Issar, Water Resources Center, Ben Gurion University of Negev, J. Blaustein Inst. for Desert Research, Sede Boker Campus, Israel, 84990.

Georg Jentsch, Hydrosond, Geologisches Büro, Nordstr. 24, D-77694 Kehl, Germany.

Kenneth S. Johnson, Oklahoma Geological Survey, 100 E, Boyd, Room N 131, Norman, OK 73019-0628, USA.

Werner Käss, Mühlematten 5, D-79224 Umkirch, Germany.

Janja Kogovšek, Inst. za Raziskovanje Krasa ZRC Sazu, Titov trg 2, SI-6230 Postojna, Slovenia.

Vladimir S. Kovalevsky, Inst. of Water Problems, Academy of Sciences, Novo Basmannaj 10, 107078 Moscow, Russia.

Andrej Kranjc, Inst. za Raziskovanje Krasa ZRC Sazu, Titov trg 2, SI-6230 Postojna, Slovenia.

Bernd Krauthausen, Hydrosond, Geologisches Büro, Nordstr. 24, D-77694 Kehl, Germany.

Neven Kresic, Law Engineering and Environmental Services, 112 Townpark Drive, Kennesaw, GA 30144, USA.

Philip LaMoreaux, P.O.BOX 2310, Tuscaloosa, AL 35403, USA.

Zisheng Liao, Changchun College of Geology, Peoples Republic of China.

Robert D. Libra, Iowa Department of Natural Resources-Geological Survey, Iowa City, Iowa, USA.

Xueyu Lin, Changchun College of Geology, Peoples Republic of China.

Cristina Liñán Baena, Dpt. de Geologia, Facultad de Ciencias, Universidad de Malaga, E-29071 Malaga, Spain.

L.F. Molerio Léon, Centro de Hidrología y Calidad de las Aguas, Instituto Nacional de Recursos Hidráulicos, P.O. Box 23, General Peraza 19210, Ciudad de La Habana, Cuba.

Jacques Mudry, Universite Franche-Compte, Structural and Applied Geology, Place Leclerc, F-25030 Besancon-Cedex, France.

Manfred Nahold, Gruppe Umwelt und Technik, Leonfelderstr. 18, A-4040 Linz/Urfahr, Austria.

Petar Papic, University of Belgrade, School of Mining and Geology, 21000 Belgrade, Yugoslavia.

Antonio Pulido Bosch, Dpto. Geodinamica, Facultad de Ciencias, Avta. Fuentennera s/n, E-18071 Granada, Spain.

Sakir Simsek, Hacettepe University, Int. Research and Application Center for Karst Water Resources, 06532 Beytepe Ankara, Turkey.

Linda D. Slattery, Ohio Environmental Protection Agency; Division of Emergency and Remedial Response, 1800 WaterMark Drive, Columbus, Ohio 43216-1049, USA.

Luigi Tulipano, Engineering Faculty of Rome, University La Sapienza, Via Endosiana 20, I-00184 Roma, Italy.

Andrzej Tyc, Department of Geomorphology, University of Silesia, ul. Bedzinska 60, 41-200 Sosnowiec, Poland.

Iñaki Vadillo Pérez, Dpt. de Geologia, Facultad de Ciencias, Universidad de Malaga, E-29071 Malaga, Spain.

John Webb, Department of Geology, Latrobe University, Bundoora, Victoria, 3083, Australia.

Steven R.H. Worthington, Dept. of Geography, McMaster University, Hamilton, Ontario L8S 4K1, Canada.

Daoxian Yuan, Institute of Karst Geology, 40 Qixing Road, Guilin, Guangxi 541004, Peoples Republic of China.

Hans Zojer, Institute of Hydrogeology and Geothermics, Joanneum Research; Elsiabethstr. 16/II, A-8010 Graz, Austria.

Part 1: Karst waters and human activities: An overview

CHAPTER 1

Introduction

DAVID DREW
Department of Geography, Trinity College, Dublin, Ireland

1.1 THE PURPOSE OF THE BOOK

A concern for the effects that human activities have upon the environment is now widespread, in large part of course, motivated by the adverse effects that environmental degradation may have upon the quality of human life. One particular hydrogeological environment, that commonly termed karstic, is the concern of this book for three main reasons:

1. Karstic terrains (underlain by highly soluble rocks) are highly sensitive and vulnerable to imposed stress. Karsts usually lack resilience in the face of such stresses – in this instance those generated by human processes.

2. The response of karst and karst waters to such pressures is highly distinctive and sets the karst environment apart from other terrains irrespective of location or geological type.

3. Although only some 7-12% of the Earth's land surface is underlain by karstic rocks (depending on the definition of karst chosen) karst regions provide water supplies for up to a quarter of the world's population (Ford 1990) including more than 100 million people in China alone and including significant proportion of the populations of many countries of Europe (Fig. 1.1). Also many karsts are areas of outstanding scenic value and are major tourist attractions thereby increasing the intensity of human impact on their geological and hydrological systems.

Therefore, an understanding of the basis for the distinctiveness of the karst system and its workings in the context of human activities is likely to be of value to a variety of professionals whose work may at times involve them in aspects of karst and karst waters. Such people would include hydrogeologists who lack specialist training in karstic hydrology and engineers, particularly geotechnical engineers operating in karst terrain. Equally, environmental planners, legislators and other policy makers and decision takers who function occasionally or frequently in a karstic milieu would benefit from a knowledge of the fundamentals of human impact on karsts. It is to these people rather than to the professional karst hydrogeologist and geomorphologist that this book is addressed.

1.2   CHARACTERISTICS OF THE KARSTIC SYSTEM

The term *karst* is used to describe an area of limestone or other highly soluble rock, in which the landforms are of dominantly solutional origin and in which the drainage is underground in solutionally enlarged fissures and conduits (caves). In this book the emphasis is on karst terrains developed in limestone and dolomite rocks with the exception of Chapter 6.7 which is concerned with impacts in a halite (sodium chloride) karst region of the USA. However:

'All terrains in which carbonate, sulphate or chloride rocks crop out or are in the shallow subsurface are probably some type of karst (Johnson & Quinlan 1995).'

Another useful definition of non-evaporite karst is provided by Quinlan et al. (1995):

'...any terrane in which carbonate bedrock (limestone and/or dolomite and/or chalk) is exposed, is present beneath soil and/or regolith, and/or may be partially or totally capped by other sedimentary rock. Sinkholes, springs, caves, integrated conduits (dissolutionally enlarged joints (fissures) and/or bedding planes) may be present in a carbonate terrane but may not be obvious. The presence at the ground surface of one, some, or all of these features is not necessary for a carbonate terrane to be defined as a karst. ...Some karst terranes are more extensively developed than others.'

Carbonate rocks are widespread as may be seen from Figure 1.1 with Europe having the largest percentage of carbonate rock outcrop relative to its area of any continent (Fig. 1.2).

The degree to which the rocks have been karstified varies greatly from place to

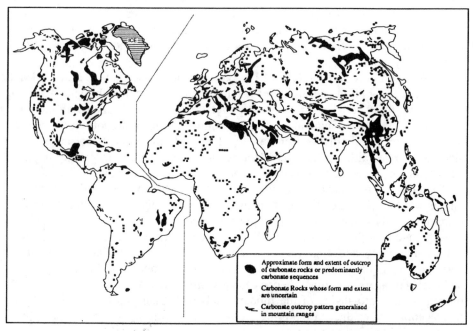

Figure 1.1. Major outcrops of carbonate rocks, most of which exhibit karstification at least to some extent (after Ford & Williams 1989).

Figure 1.2. Major outcrops of karstified carbonate rocks in Europe (after Bakalowitz & Biondic in EC COST 65).

place depending upon how much the fissures in the rock have been enlarged by the solutional action of acidified rainwater and the extent to which the underground drainage system has become organised and integrated into efficient conduits for the collection, transport and ultimately discharge of recharge waters. In general, pure, highly soluble carbonate rocks with well developed secondary permeability karstify best. In some areas the karstified rocks may be overlain with non-carbonate strata or unconsolidated deposits and this is termed a covered or mantled karst.

Old karstic landforms, surface and underground, which have been infilled by subsequent deposits, often have no surface expression and do not function hydrologically. They are called palaeokarsts. Palaeokarst may be reactivated to some extent if environmental conditions change.

Prohic (1989) remarks that what distinguishes karst systems from non-karst systems is the existence of solutionally enlarged underground spaces in which the groundwater is stored and moves and the occurrence of complex and deep interconnections between surface and underground which are much more apparent than in conventional hydrogeology. He also identifies the characteristics of many karsts that render its groundwater particularly liable to contamination (Fig. 1.3):

1. There is often little or no soil or cover which means poor filtration, poor prepurification, rapid infiltration (this is not explicitly a karst water factor but an incidental aggravating factor).

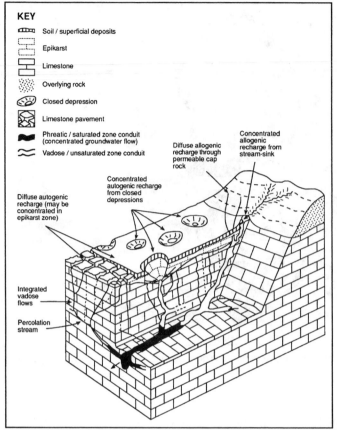

Figure 1.3. Illustration of the various modes of recharge, flow and discharge in a karst area (adapted from Gunn 1985).

2. Flow is turbulent and conduit dominated and there are no fine grained aquifers and so only limited self-purification.

3. High flow velocities allow transit times that may be too short for microorganisms to die off, particularly in shallow groundwater systems.

4. Large numbers of interconnected fissures mean that pollution inputs are possible from the surface almost anywhere.

Prohic (1989) also observes that it is unfortunate that enclosed depressions such as dolines are such a common landform in most karst areas because they are at one and the same time, tempting sites for waste disposal (and for leachates etc. to flow down into) and also the major recharge input zones to ground water.

For many people a karst region is visualised as a spectacularly rocky and barren terrain with disappearing rivers and many caves (Photos 1.1, 1.2). This is a true picture of some exceptionally well developed karst regions but many degrees of karstification can exist from slight to extreme. The absence of obvious karst features such as those mentioned above does not necessarily mean that karst processes are not operative in an area.

Photo 1.1. Karst landscape in the Yorkshire Dales, England, with limestone pavement and impressive karren formations (photo: Drew 1972).

Photo 1.2. The 'small natural bridge' of Rakov Škocjan, Slovenia, forming a collapsed part of the underground connection from the Cerkniško Polje to the Planinsko Polje (photo: Hötzl 1982).

All of the above statements concerning karst hydrogeology are generalisations but the extent to which they apply to any given area is much more difficult to ascertain. It is often said that the only predicable thing about karsts are their unpredictability, and indeed the site specific nature of karsts is a major problem as is apparent from the examples presented in this book.

Much fuller descriptions of the character of karst landforms and hydrology may be found in Ford & Williams (1989) in White (1988) and in LaMoreaux et al. (1984). A good summary of the nature of karst groundwater systems with particular emphasis on their vulnerability is also given in the pan-European project report COST action 65 (EC-COST 65 1995).

1.3   HUMAN IMPACT ON KARST

As remarked in Section 1.1, in recent decades there has been a shift from an attitude towards the environment that centred upon exploitation of its resources and containment of those aspects of the environment that threaten human beings, to a more holistic approach towards living in and intelligently managing the resources of our planet. Environmental impact assessments for proposed projects are now the norm and concern over possible long term environmental impacts caused by human activity such as changes in climate is increasingly a part of the political agenda.

This changed perspective has been reflected in attitudes towards karstic environments. For example, a symposium focused on karst water resources is held in Turkey every five years and the papers delivered are an indication of priorities in karst investigation. At the 1985 meeting (Günay & Johnson 1985) only one of 46 papers was concerned with human impacts. At the 1990 symposium (Günay et al. 1990) some 5% of presentations were oriented towards human impacts but at the 1995 meeting, which was explicitly entitled *Karst Waters and Environmental Impacts* (Günay & Johnson 1997) 15% of contributions dealt with human-environmental problems in karsts.

A similar change of emphasis may be seen in the series of symposia held by the Sinkholes Institute in the USA at approximately two year intervals and concerned with sinkholes in karst terrains Beck (1984, 1989, 1993, 1995), Beck & Wilson (1987), Beck & Stephenson (1997). The thrust of the contributions to these symposia has altered from the hazards posed by karst terrains to investigations of anthropogenic processes in karsts.

In recent years a series of publications have dealt with human impacts on karsts although in many cases the focus has been upon changes caused to soils, vegetation, landscapes and caves rather than to water in karst regions. Examples of these studies include the publications of the International Geographical Union study group on Human-Karst Interaction: *Mans Impact on Karst* (International Geographical Union 1987), *Resource Management in Limestone Landscapes* (Gillieson & Smith 1989), *Environmental Changes in Karst Areas* (Sauro et al. 1991). A special supplement of the journal *Catena* entitled *Karst Terrains: Environmental Changes and Human Impact* (Williams 1993) contains a number of case studies of degradation of karst areas together with a summary of the character of karst aquifers and their susceptibility to pollution (Smith 1993). Similarly an issue of *Environmental Geology* (Ford 1993) had as its theme *Environmental Change in Karst Areas*. *Environmental Effects on Karst Terrains* was the theme of an issue of *Acta Geographica Szegediensis* (Barany-Kevei 1995) with an eastern European perspective. Similarly a special issue of the journal *Acta Carsologica* dealt with the topic: *Man on Karst* (Kranjc 1995) as did the symposium held in Poland in 1996 *Karst-Fractured Aquifers – Vulnerability and Sustainability* (Rozkowski et al. 1996). Although concerned largely with caves, the Australian book: *Caves: Processes Development, Management* (Gillieson 1996) also explicitly addresses more general problems of conservation of karsts.

Although comparatively few of the publications mentioned above are primarily concerned with impacts on karst waters they do illustrate the complexity of chains of cause and effect in karst terrains.

Figure 1.4 is adapted from Williams (1993a) and summarises the consequences of various human activities on the karst system. The impacts upon karst waters have been emphasised. As Williams notes, modifications to vegetation can provoke soil erosion, but unlike non-karstic areas, the eroded soil does not accumulate in river valleys but rather in the numerous enclosed depressions (for example *dolines*) that characterise many karst regions. Dolines are points at which concentrated recharge of a karst aquifer takes place and therefore modifications to dolines alter the ground-water hydrology of the area. Even if dolines are not apparent in a karst region solutionally enlarged fissures have the same hydrological function. Changes in the hydrology of a karst are often the synthesis or summary of all the other environmental changes made in the karst.

A summary chart of human impacts on a single component of the karst environ-

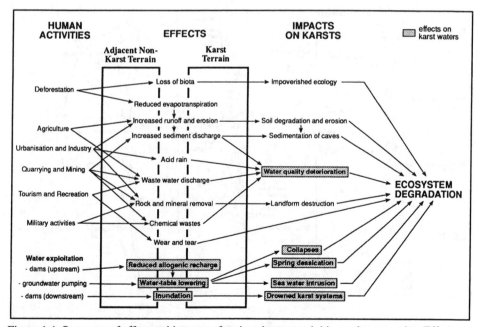

Figure 1.4. Summary of effect and impacts of various human activities on karst terrains. Effects on karst waters are shaded (after Williams 1993a).

ment, caves, is shown in Figure 1.5 (Marker & Gamble 1987) again showing the complexity of the inter-relationships, whilst an even more restrictive summary chart (Fig. 1.6) details the impact of just agricultural activities only on caves (Hardwick & Gunn in Williams 1993).

An alternative overview of the effects of impacts (disturbances) on karst is that given by Lichon (1993) that considers aspects of the karst system separately rather than in terms of the processes that cause changes (Table 1.1).

The importance of understanding the workings of the karst system in order to appreciate the likely consequences of human actions is well expressed by Yuan (1988) in the context of Chinese karst.

'[there are] many environmental problems in karst regions, these include drought, floods, deforestation, surface collapse, ground water pollution, scenic despoilation.

...new problems occur following the treatment of old ones or problems appear in neighbouring regions ...the reason [is] ignorance of the karst environmental systems, lack of proper environmental planning and effective management.

...karst has been regarded as a fragile environment by environmental scientists ...system with a low capacity, difficult to restore if once disturbed. Karst environmental problems in China are a cause of concern because of their extent, population pressure and rich natural resources.'

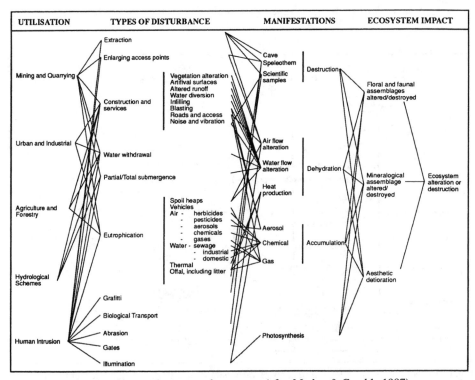

Figure 1.5. Summary of human impacts on karst caves (after Marker & Gamble 1987).

surface changes	impacts on cave geoecosystems — geomorphology microscale	geomorphology macroscale	work environment	micro climate	ecology	agricultural operations which may produce this type of change
allogenic recharge / autogenic recharge						
direct changes to water quality	speleothem growth & development / contamination of inwashed & in-situ sediments		gas hazards / explosion hazards		hypogean fauna / stream fauna	fertilisers (inc. irritation) agrichemicals slurry / silage leaks petroleum products septc tank leaks
indirect changes to water quality	speleothem growth & development				hypogean fauna / stream fauna	field drainage land drainage (open ditching) ploughing afforestation practices
direct changes to water quantity	speleothem growth & development	boulder choke stability	flood hazard	humidity airflow air/water temperature	hypogean fauna / stream fauna trogloxenes	field drainage land drainage (open ditching) irrigation water extraction
indirect changes to water quantity	speleothem growth & development		flood hazard	humidity airflow air/water temperature	hypogean fauna / stream fauna trogloxenes	afforestation practices ploughing crop changes
indirect changes in sediment loads	speleothem growth & development / burial of speleothems, speleogens & in-situ sediments	inpilling of cave passages & blockage of cave entrances	flood hazard	humidity airflow temperature	hypogean fauna / stream fauna - habitat loss trogloxenes - access lost	field drainage land drainage (open ditching) afforestation practices ploughing construction (roads, buildings)
direct alteration of cave entrances	speleothem growth & development	blockage of cave entrance	loss of access to cave health hazards: e. g. Leptospirosis (from rats/livestock)	humidity airflow temperature	trogloxenes - access lost	waste disposal deliberate infilling capping for water supply
indirect alteration of cave entrances	speleothem growth & development	blockage of cave entrance	loss of access to cave	humidity airflow temperature	trogloxenes - access lost	afforestation practices ploughing construction (roads, tracks, buildings) heavy mechanisation

Figure 1.6. Modelling the potential impacts of agricultural operations on cave geoecosystems: An impact matrix (after Hardwick & Gunn 1993).

Table 1.1. Effects of surface disturbances on karst (after Lichon 1993).

Aspect	Natural karst	Disturbed karst
Landscape	Slow karst development	Rapid subsidence and collapse
Soil quality	High organic content	Depleted, erodible residue
Soil quantity	Slow accumulation, uneven distribution	Rapidly lost
Soil structure	Heterogeneous, stable	Disintegrates, erodes
Caves	Stable karst processes	Fill with debris and sediment
Speleothems	Stable growth	Re-solution
Cave fauna	Fragile habitats kept	Loss of habitat
Water quality	Good, hard	Turbid, bacteria, contaminants
Vegetation	Many communities and niches	Diversity lost, weeds invade

One of the earliest and most prescient expositions of the singularities of karst hydrogeology from the perspective of resource management was made by Legrand (1984, p. 191) when he stated:

'Changes in hydrological balances are not unique to karst regions, but karst regions are more sensitive than other regions ...An evaluation of the environmental problems of karst regions must continuously focus on certain key features and prin-

ciples of karst hydrology, so that problems which are unique to karst regions may be better understood.'

It is apparent that there are problems inherent in coping with karst terrains but it is also easy for humans to exacerbate these problems by mis-perception and/or mis-management of the environment. There is an unfortunate dichotomy between the value of the groundwater resource in karst and the ease with which it can be degraded.

1.4   THE LAYOUT OF THE BOOK

The book is divided into three parts. Part 1 provides an overview of the characteristics of karst terrains in both a scientific and a broader environmental context. It also presents (Chapter 2) a summary of the historical relationship between human activities and karstic terrains thus demonstrating the often precarious balance between successful exploitation of karst waters and the adverse consequences of mismanagement. The examples given, from the Americas from Asia and from Europe, emphasise that human interaction with karst waters has existed from the time of the earliest civilisations to the present day. However, it is only in comparatively recent times that humans have impacted severely and extensively on the karst system.

The core topic of the book is treated in Part 2 which examines the consequences of major groups of human activities on karst waters. Chapter 4 deals with agriculturally driven impacts, Chapter 5 with industrial-urban impacts, Chapter 6 with the effects of extractive industries, and Chapter 7 with the effects of karst groundwater exploitation. In each chapter a general overview and summary of the impacts is followed by a series of short case-studies or specific examples designed to amplify the general overview by reference to varied locales or to particular processes.

Part 3 attempts to view the human impacts on karst groundwater within a broader legislative and societal framework. The legal and technical aspects of management of karstic environments are examined with particular reference to conditions prevailing in Europe and in North America with emphasis on the means that may be adopted to minimise adverse impacts upon karst waters.

Finally, future trends that are likely to significantly impact upon the interaction between human activity and karst water resources are examined. These include environmental modifications such as short to medium term climatic changes and also wholly anthropogenic changes such as the probable large increases in population and hence in water demand in areas of the world reliant on karst groundwater. Also regarded as likely to be an important stress is the inevitable future increase in tourist visitors to many of the worlds more scenic karst terrains as this is also likely greatly to stress water resources in these areas.

CHAPTER 2

The historical perspective

PHILIP E. LAMOREAUX (co-ordinator)
Tuscaloosa, USA

2.1 OVERVIEW OF THE HISTORY OF KARST HYDROGEOLOGICAL STUDIES (Philip E. LaMoreaux)

Man was intrigued with karst, initially perhaps particularly with caves, long before the word karst came into use. In pre-historic times caves provided humans with a living space, water supply, and protection. In the cave area of southern France, in the Pyrenees and in northern Spain that were outside the influence of the massive continental Pleistocene glaciers, the Palaeolithic Period of man's evolution was when remarkable cave drawings illustrating an amazing capability of early man to duplicate pictures of animals and hunting activities were created. He had chosen these caves as living areas, and his exploratory drive led him deep underground to sources of water. These were man's first attempts to explore and understand karst and use this natural phenomenon to enhance his living conditions and safety.

During Greek times, caves, underground rivers, and springs were a part of Greek mythology, and the subject of much discussion by Greek and Roman philosophers. The earliest hydrologic concepts of the hydrologic cycle, water source, occurrence, and quality were related to a karst setting.

An early example of man's use of karst features in the Mediterranean area is King Hezekiah's Tunnel dating from 1000 BC in the city of David (Jerusalem), Jebusites (Canaanites) enlarged the underground solution cavity system beneath the city to provide water from the Gihon Spring, a spring probably used by early man hundreds or even thousands of years before the Canaanites.

In 701 BC, Hezekiah brought waters from the Gihon into the city in preparation for Sennacherib's siege, an act that succeeded (2 King 20:20; 2 Chronicles 32:2 – 4,30). Sennacherib failed to capture Jerusalem. This early development of a karst spring is described in Section 2.2.4 and in detail in an article by Gill (1994).

In ancient Greece we can observe the earliest development of karst terrains. Crouch (1993), describes the methods and techniques of the urbanisation of the ancient Greek world. These were all karst settings and provide a documentation that early civilisation felt the profound impact of karst on all aspects of urbanisation, development, society, transportation, and water supply in a karst area.

2.2   HISTORICAL EXAMPLES OF HUMAN ACTIVITIES AND KARST ENVIRONMENTS

2.2.1   *General comments*

The interaction between human societies and karstic environments has existed from the earliest times as remarked earlier in this chapter and also in Chapter 1. Whether coincidentally or otherwise a large number of significant early civilisations were not simply 'hydraulic civilisations', dependent on the effective management of water resources, but were '*karstic* hydraulic civilisations', located in important karst regions of the world. Some more detailed examples of such interactions in historical times drawn from various parts of the world, are given below.

2.2.2   *The Yucatan Peninsula, Mexico* (William Back)

The Yucatan Peninsula provides an excellent example of societal consequences of human impact on an environmentally sensitive, highly karstified plain. Management of water resources has long had a major influence on the cultural and economic development of the Yucatan. Two critical objectives of groundwater management in the Yucatan are to develop regional groundwater supplies for the increased demand resulting from an expanding population and a governmental programme to promote tourism based on the Mayan archaeological sites (Photos 2.1 and 2.2) and excellent Caribbean beaches (see also Chapters 3 and 5) and to control groundwater pollution in a chemically sensitive system made critically vulnerable by climate and geologic conditions.

The Yucatan Peninsula in south-eastern Mexico projects northward between the Caribbean Sea on the east and the Gulf of Mexico on the north and west. It contains the states of Campeche on the west, Quintana Roo on the Caribbean Sea, and Yucatan in the north-central portion. Mérida is the capital of Yucatan and principal city of the peninsula. The northern third of the peninsula, 150 km wide, is an almost level karst plain underlain by nearly horizontal Tertiary formations consisting mainly of limestone, marl and gypsum. It ranges in elevation from sea level along the coast to slightly more than 30 m above sea level in the interior. Bare, fluted limestone is exposed over large areas and generally it is pitted and scarred by solution depressions and small ridges. The rough surface and many sink holes make travel across country very difficult despite the apparent flatness. This northern coastal plain is bounded on the south by a range of low hills, the Sierrita de Ticul or Puuc Hills, with an elevation of 150 m above sea level.

The great degree of karstification has caused environmental and ecological problems similar to those in karst areas of other parts of the world. Structure, topography and presence or absence of other geologic formations play an important role in the development of hydrogeological conditions in a karst area because these, along with the climatic factors, control the permeability and the soil formation which strongly influence the type of water management required for any area. For example, a pure limestone, such as that of the Yucatan, can dissolve and leave essentially no residue; and also, owing to the lack of other geologic formations that would produce clay or sand, the soil is almost non-existent. In addition, the permeability developed from the solution of the limestone remains high with the concomitant porosity in the form of

Photo 2.1. Maya pyramid at Uxmal, Yucatan, Mexico (photo: Drew 1982).

Photo 2.2. Cenote at the famous Maya site of Chichenitza, Yucatan, Mexico (photo: Drew 1982).

open channels in the Yucatan because other sediments are not available to decrease the permeability and porosity by filling the solution channels. If the limestone above the water table had lower permeability in the Yucatan as a result of containing sediments in the solution channels the head would be higher and the fresh water-salt-water interface would be appreciable farther below sea level. In addition, the presence or absence of sediments is a major control on the development of surface drainage and river systems. No rivers have developed in the Yucatan because the absence of sediments permits the rainfall to infiltrate rapidly to the water table.

In the Yucatan Peninsula these factors have combined with the regional geologic and physiographic setting to produce long-term problems of water management. Although the rainfall is high – as much as $1500 \, mm{\cdot}a^{-1}$ – it is seasonal and requires storage for use during the extended dry season. The type of storage usually provided in non-karstified areas by rivers and aquifers to maintain base flow of streams, virtually does not exist in the Yucatan. There, the extreme permeability of the limestone causes rapid infiltration of rainfall and nearly simultaneous discharge of groundwater to the ocean. The limestone is so permeable and movement through the system so rapid that not enough head is developed in the groundwater body to provide adequate storage in the aquifer.

Not only does the high permeability decrease the amount of fresh water available, it also makes the aquifer particularly susceptible to contamination by domestic and municipal sewage, farmyard wastes, and the natural decomposition of the abundant vegetation in the warm, humid environment. If the limestone were overlain by less permeable sediments, they could serve as a filtering system. This would tend to purify the water by decreasing the flow rate and providing longer residence time, thus permitting the decomposition of the organic contaminants. In addition, the warm climate is particularly conducive to the growth of bacteria in polluted water. Widespread pollution has generated a host of endemic water-borne diseases.

Encroachment of salt water from the ocean that surrounds the peninsula on three sides is another serious problem. The large hydraulic conductivity along with the lack of high heads permits an extensive body of salt water to underlie the entire northern third of the peninsula.

The only naturally-occurring bodies of water in the Yucatan are those openings formed by solution of limestone or by tectonic activity. The most numerous bodies of water are the cenotes (from the Mayan word d'zonot) or sinkholes; sources of water also occur in other solution features such as caves and shallow, clay-lined depressions called aguadas. Tectonically controlled bodies of water occur along the east coast as bays in an elongated, faulted and fractured depression. Also, the groundwater level in this zone is shallow and occasionally comes to the surface resulting in a series of small lagoons, called 'sabanitas'.

Such a setting as the Yucatan should provide abundant supplies of water; however, factors of climate and hydrogeology have combined to form a hydrologic system in which fresh water is scarce and whose chemical environment decreases even that restricted supply. These factors include:
– Extensive karstification that permits rapid infiltration to prevent formation of rivers, rapid discharge that precludes development of enough head to provide adequate storage, also low head permits seawater to permeate the aquifer.
– Uneven seasonal distribution of rainfall.

– Long-term pollution and contamination from organic and inorganic sources.

The northern part of the Yucatan, no doubt, presented an environmental challenge to the first people who came to the area, and it has continued to do so ever since. The paucity of fertile soil, the scarcity of potable water and shortage of other natural resources make this one of the most inhospitable regions in which a sophisticated civilisation has ever developed. The small plots near the family settlements that grew beans, squash and corn were often constructed by transporting the soil from areas of accumulation, sometimes underground, to where it was more convenient to plant the crops. Nonetheless, plentiful small game of both birds and animals, accessibility to the oceans, easily quarried limestone for building materials, and the knowledge of maize cultivation combined to permit a livelihood.

At times it was quite pleasant, requiring a minimum amount of effort to provide for the family and thus permitting leisure time to be used, in lieu of paying taxes, for the construction of public buildings, public roads, and public water sources. The need for water is constant and its uneven distribution, both seasonally and spatially caused both the priests and the common people to spend time and effort in providing reasonably ample and secure supplies.

Because of the absence of rivers and with the exception of a few brackish lakes along the east coast, the Yucatan would be a waterless plain if it were not for the cenotes and caves containing groundwater and the scattered aguadas intermittently containing rainfall perched above the water table. The Maya were one of the few early civilisations to utilise a groundwater supply extensively. Many of the early cities developed around cenotes. Had it not been for the groundwater supply, northwestern Yucatan would have been largely uninhabitable.

As with many early people, the Maya practised water management through religion and simple engineering tasks. Water management involves some human activity to make the spatial and temporal occurrence compatible with the spatial and temporal need for the water. The religion and much of the culture was water-oriented and the Mayan priests prayed and performed rituals and sacrificed to Chac, the water god, for assistance in water management – primarily to decrease the severity of droughts. Chac is represented with a large elephant-like nose in the Maya glyphs and codices (the books of picture writing that tell much of the history and culture) and is sculptured on many buildings and painted in many murals. His 'T' shaped eyes, full of tears, symbolically represent the rain. In the Dresden Codex, Chac is seated on a coiled snake enclosing a reservoir of water from which he dips water to sprinkle on the earth in the form of rain.

The hydraulic works of the Maya are among the first engineering efforts to control water in the Americas. In a few places small aqueducts and canals were constructed to carry the water from the hills to the areas of settlement. The Mayas constructed aguadas (reservoirs) and cultuns (cisterns) of both large and small scale to store water supplies, particularly in the southern part of the peninsula. Aguadas are natural shallow depressions some of which the Mayas lined with stones or impermeable clay. Cisterns or wells in the limestone were built in the bottom of the aguadas to furnish water during the season of no rain when the aguada itself was dry. The first people to the Yucatan migrated from the south and crossed the Puuc Hill region to reach the northern coastal plain. In this region the caves and caverns were the sources of water, and the Maya examined every cave they found in the hope it would lead to a permanent source of water. The Maya have been speleologists for the past 2700-3000 years. They developed water supplies from the many caves and caverns by constructing steps and ladders from the entrance to the deep water level and water was carried out in clay jars or lifted by rope and bucket from the land surface. As with primitive religions, the art and ceremonies are either to honour or appease the gods, or to imitate them. The Maya honoured Chac with the paintings and sculpture; they appeased him with offering of jewellery, art objects and human sacrifices in sacred cenotes.

The three great accomplishments of the Maya civilisation were the design and construction of temple centres, Photos 2.1 and 2.2, the development of an accurate perpetual calendar and, independently of the Hindus, the use of the concept of zero. Water is a dominant theme in all these accomplishments. For example, aqueous symbols and creatures, such as turtles, conch shells, and water jugs are common architectural embellishments. The Maya had an understanding of the concept of 'water year'. Their impressive calendar, based on knowledge of astronomy, permitted the priests to foretell the beginning and ending of the rainy season, so that maize could be planted and harvested at the best possible time.

Like the Mayas, the Spanish conquistadors faced serious water problems when they arrived in 1517. The long period of time required for the conquest of the Yucatan, almost twenty years, was due in large part to the terrain and notably to the tremendous difficulties of obtaining adequate water supplies. The Spaniards came from a land of rivers and lakes and did not know how to cope with water supplies that were to be obtained only from below the surface. During the conquest, much of the warfare was of guerrilla type, and for defence, the Mayas would burn their houses and fill the wells with stone. Carrying water jars, they would take their wives and children into the hills where they obtained water from caves. The Spaniards, often wounded and fatigued from travelling across the difficult terrain would suffer greatly during the three or four days it often took to clean and repair the wells. During and after the conquest, the Spaniards constructed wells of their own, some of which are still in operation today. The construction of wells by digging with pickaxe and shovel and using a bucket to remove the broken rocks continued through the many centuries up to modern times.

Some of the wells were equipped with large wooden wheels with attached buckets for lifting the water. A burro or ox was used to rotate a wheel that was geared to the vertical wheel containing the buckets. Introduction, by the Spaniards, of these wells was practically a revolution in the agricultural system because it permitted irrigation and the development of a cattle industry. Construction of wells made water available throughout the Yucatan coastal plain and permitted development of numerous haciendas owned by wealthy landholders which were worked by the Maya descendants, somewhat similar to serfdom. This form of society based on economic development of cattle and henequen led to the centuries-long oppression of the Indians and was largely overthrown by the Mayan revolt between 1847 and 1855.

The first public water supply was organised in the village of Tekax on 22 November 1825, and resulted in the construction of a public well. The water from public wells was lifted by rope and bucket. In the late 19th century, increases in population resulting primarily from the world demand for sisal, prompted the digging of thousands of wells. The first mechanical pumps, driven by steam engines, were introduced in 1865. In 1880 the first windmill was installed in the patio of a private home in the city of Mérida. This was a most important historical event in the use of water in the Yucatan, because the windmill became an extremely popular way to obtain the water. Wells equipped with windmills became so numerous in Mérida that by 1930 more than 20,000 windmills existed within the city limits, and Mérida was known as the 'city of windmills'. These have now been abandoned because of the pollution resulting from the use of a sewage disposal well adjacent to each windmill. For many years much of the city was supplied by horse drawn buggies carrying a water barrel from which the people obtained their domestic water supplies. In 1946 planning began for the modern water supply, and Mérida today has a city well field with electric pumps and a distribution system for the potable water.

2.2.3 *China* (Yuan Daoxian)

China is a country of great size and with many varieties of karst. The history of Chinese understanding, utilisation and modification of karst started in the distant past.

There are three major types of karst in mainland China: subtropical humid karst in south China, semiarid karst in north China, and high mountain karst in the west. Karst hydrogeology and other karst features have linked with social and economic development in many respects. In north China, more than 50 big karst springs each with a minimum discharge of more than 1 $m^3 \cdot s^{-1}$ are the major source for local water supply. However, the Cambrian-Ordovician karst aquifers are overlain by Carboniferous coal measures which are the main energy source in China. The potential for conflicting resource-exploitation environmental impacts given such a hydrogeological setting are apparent. In south China, the long-term intensive karstification and the strong Cenozoic uplift mean that the region is characterised by thin soils and a shortage of surface water, even though there are 2836 underground streams with a total length of 13,919 km, and total minimum discharge of 1482 $m^3 \cdot s^{-1}$. All these hydrogeological features and relevant karst phenomena have been long investigated by the Chinese people.

Although the subtropical karst in south China is far more spectacular than the North (Photos 2.3 and 2.4), the development of knowledge on karst hydrogeology began in northern China, because the middle reach of Yellow River Basin was the cradle of ancient Chinese culture. Later on, following the unification of the whole territory into one country in the Qin Dynasty (221 BC) and the shifting of the economic and cultural centre, karst studies expanded gradually to all over China. From ancient writings we can see that the investigation and records of knowledge on karst hydrogeology by ancient Chinese scholars before the Qin and Han Dynasties (221 BC-220 AD) were concentrated in north and northwest China. In the time of Three Kingdoms (220-280 AD) and Sui Dynasty (581-618 AD), the main focus of attention was southeast China. In the Tang Dynasty (618-907 AD), which is considered to be the most prosperous period in ancient China's science and culture, and the Song Dynasty (960-1279 AD), the spectacular subtropical karst in Guangdong and Guangxi Provinces was investigated in depth. From the Ming Dynasty (1368-1648 AD) through the Qing Dynasty (1583-1911 AD), karst researches were extended to the Yunnan-Guizhou Plateau, the most extensive, most colourful and most difficult karstland to exploit in China.

Karst phenomena appear in quite a few ancient Chinese books. For example, „The Mountain Scripture‚‚‚ written during the period of Warring States (475-221 BC) to the early Western Han Dynasty (206 BC-8 AD), records many karst phenomena relating to caves, sinking streams and subterranean rivers. It notes that: '*there is much water and many caves in the south mountain, the water flows into the cave in the spring, flows out of the cave in the summer, and stops in the winter*'. Moreover, it described the arid karstic environment, saying: '*in the mountainous area of Yingmu, most of the valleys below are underlain by green rock* (an ancient expression to designate limestone still used today by local farmers) *without water*'.

The extensive tropical karst peak-forest landforms were noticed by the ancient Chinese long ago. An ancient map, unearthed in the No. 3 Han Dynasty Tomb of Mawangdui, Changsha in 1973, vividly shows the peak-cluster landforms of the Mount Jiuyi in Ningyuan County, Hunan Province (Fig. 2.1). The map was drawn in 168 BC and is the oldest classical colour map on silk in China, even in the world. Undoubtedly, the map is the oldest karst geomorphic map in the world. On the map, the stone peaks are represented by columns and the peak-cluster depressions by whorl symbols.

The first materia medica works of China, written during the Western Han Dynasty (or in the first century), the 'Compendium of Materia Medica' gave names for 'stalactite' and some other speleothems and described their medicinal functions (Fig. 2.2). There are many karst springs in North China which are important sources for irrigation and potable water. It has been found that many geographical names are those of springs in the unearthed inscriptions on bones or tortoise shells of the Shang Dynasty (16th-11th century BC). The development and usage of many springs also has a long history. For example, the Jinci Spring in Shanxi Province was recorded in

Figure 2.1. Part of an ancient map of Mount Jiuyi, Hunan Province, China (after Cultural Relic Publishing House).

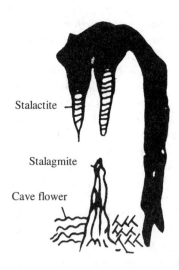

Stalactite

Stalagmite

Cave flower

Figure 2.2. Plate from the 'Compendium of Materia Medica' showing cave calcite deposits.

'The Mountain Scripture' for a long time. According to 'The Water Scripture' (written from the 4th to 5th century), the Jinci Spring was extensively used for irrigation in 453 BC of the period of Warring States.

The Three Kingdoms was a period of warring but also of fruitful political, military and diplomatic strategic thinking which is still used today at home and abroad. The most interesting event of karst research in the Period was an ambitious cave expedition ordered by Sun Quan, the emperor (222-252 AD) of Wu Kingdom, who ruled the middle and lower reaches of Yangtze River. The cave entrance is on the Dongting Hill on an island in Taihu Lake, Jiangsu Province. It was recorded, that the expedition finished after getting into the cave for 10 km in 17 days. The motive of the cave exploration was to verify a legend that the cave was connected with Zhucheng County, Shandong Province in the north, and Changsha County, Hunan Province in the south – obviously with military considerations in mind. The most representative karst publication in this period is the 'note on Jingzhou' written by Sheng Hongzhi at the time between 420-479 AD, which described many karst phenomena in the middle reach of Yangtze River, including the exploitation of cave water for irrigation, the impact of cave collapse, and even the chimney effect of a wind cave in Yidu county as saying 'the cave blows out in Summer, and blows in during Winter'.

The Tang and Song Dynasties were flourishing periods of ancient Chinese Literature. Many classic poems handed down to this day were written at that time. Poets and painters came to Guilin, Liuzhou in Guangxi province and found the unique tower karst landforms a source of creativity. Moreover, Buddhism and Taoism were spreading over China at the promotion of central government. Both religions used caves as places for cultivation and temple building. Accordingly, karst investigation made remarkable progress. Many scholars made contributions in this field, including Yuan Jie (Tang Dynasty, around 764 AD), Liu Zongyuan (773-819 AD), Shen Kuo (1031-1095 AD) and Fan Chengda (1126-1173 AD). Yuan Jie was the governor of Daozhou in southern Hunan province, and made extensive investigation of landforms and caves in the region. His most interesting finding is about cave micro meteorology saying: 'the cave is full of fog in Spring, cold in the Summer, clear in the

Autumn, and warm in Winter'. Liu Zongyuan was the governor of Liuzhou. His trips and field observations covered karst and caves of Liuzhou, Guilin and southern Hunan Province. In his works, he tried to explain the contribution of fluvial action to karst formation. Besides describing some underground streams, he first introduced a word equivalent to the present-day term 'flowstone'. Moreover, he also noticed the deforestation problems in limestone areas. For instance, he described Fish Hill (now in downtown Liuzhou) as 'all rocky, without big tree and grass'. These kinds of description increased later in ancient Chinese literature, and are useful for a better understanding of the process of rock desertification today in the karst regions of southwest China. Shen Kuo was a great scientist of the Song Dynasty. He had not only linked the formation of speleothems with the activities of groundwater, but was also far-sighted enough to use the term 'brine water' in this respect – long before the development of modern chemistry. Fang Chengda was the most important karst scientist before Xu Xiake. Besides his explanation of speleothem formation mentioned before, he wrote a monograph 'on the Taihu Stone' to explain its origin. This is a sub-soil and sub aqueous dissolution feature used for Bonsai. It was first mined by the Taihu Lake in Jiangsu province as an annual tribute to the King of Song Dynasty. A special bureau was established in Suzhou Jiangsu to collect this kind of tribute. Later this collection of tributes spread to Guangdong, Hunan and Hubei Provinces. Fang Chengda criticised the malpractice of over – quarrying limestone on Dongting Hill as 'stripping the skin' ('Note on Jiangsu'). Moreover, his 'Note on caves around Guilin' is the first book of speleology in China (1172 AD).

Xu Xiake (1587-1641 AD) (Fig. 2.3) was a pioneer of studies on karst landforms and caves and was engaged in geographical investigation for about 30 years. 'Travel of Xu Xiake', about 600,000 words long, is one of the most important results of his scientific investigation, and also very valuable literature in studying history of karstology and speleology. Xu Xiake travelled about 5000 km and accurately and carefully recorded the distribution and characteristics of the tropical karst and subtropical karst landforms, as well as various types of karst phenomena from 1636 to 1640 in South China, including the Guizhou and Yunnan Plateau. He surveyed about 100 caves and noted down not only their structures, shapes, directions and the various

Figure 2.3. Portrait of Xu Xiake.

Photo 2.3. Tropical karst with tower karst and peak cluster from Guilin, China (photo: Hötzl 1988).

Photo 2.4. Deep karren formation from the Stone Forest, Yunnan, China (photo: Hötzl 1988).

types of speleothems, but also the exploitation of water resources, e.g. for irrigation and for water mills for refining rice (the Suiyuan Cave, Lanxi county, Zhejiang province). After Xu Xiake, in the Qing Dynasty, more investigation were made on karst in south China, especially the Yunnan-Guizhou plateau, including many interesting observations, e.g. about karst eco-hydrogeology, Cheng Ding (around 1672 AD) wrote: 'The stone peak is bare and without any grass, the thickness of soil on the ricefield is less than 30 cm. If there is no rain for 5 days, the farmland will be dry, and the seedlings of rice will die'. This is quite similar to the present-day situation in this region.

2.2.4 *The ancient biblical world* (Arie Issar)

Perhaps in few other places of the world is the relation between the evolution and development of civilisation and water so intimately related as in the ancient biblical world. Owing to the arid climate and the extent of limestone rocks over the area, it should be expected that the occurrence of ground water in limestone that issues as springs would play an important role in this development. Documents describing these occurrences include the clay tablets of the Sumerians, the Bible and the Koran, and the results from archaeological studies as well as paleoclimatic studies (Issar 1990).

O.E. Meinzer, generally thought of in the United States as the 'Father of Ground Water' made the statement that in places the Bible reads like a water supply paper. References in the Bible to springs, many of which issue from limestone rocks, are related to locations that can be traced today (Photos 2.5 and 2.6).

The setting is the region of the Fertile Crescent. The geography is the Middle-East from the Sahara Desert in the west to the Plateau in the Gulf of Persia in the east. Much of the history was obtained while Arie S. Issar worked in this area as a hydrogeologist. Much of the history related has already been published in the book, *Water shall flow from the rock* (Issar 1990), a summary of the hydrogeology and climate change in the lands of the Bible. The Fertile Crescent, as used herein, mainly includes Lebanon, Syria, Palestine, Mesopotamia, and Egypt. The area is widely underlain by marls, clays, limestones, chalks, and dolostones. It is these rocks that are some of the most prolific aquifers in the area. It is an area where the rains come as winter rains between the months of November and March. The coastal areas are underlain by massive deposits of windblown sand brought in by the sea waves, and more recent, alluvial aeolian sands, underlain by marls, clays, and sandstones. The Fertile Crescent, due to its situation on the fringe of the desert, is also a region where geology, structure, and lithology play a very dominant role in deciding the geomorphology as well as the availability of water resources.

Perhaps the oldest references to ground water in karst are related to 'The Eyes of Tiamat'. The archaic source of the term 'Ayin', or 'Ayun' or 'Eye', which is Hebrew or Arabic may mean fountain as well as spring and is related to the appearance of springs. They were 'The Eyes of Tiamat', the fountains of the deep. The springs may be associated with 'The Eyes of Tiamat' and in turn to the two large rivers, the Tigris and the Euphrates, which emerge from large springs, most of them karst springs or limestone solution outlets along the foothills of the Zagros Mountains on the border between Iran, Turkey, and Iraq. These springs in many instances form large circular pools of water out from which a stream flows. In many cases the spring emerges from large caverns. An AsSyrian Stele is a pictorial description of Salmanassar III's visit to such a cave at the source of the Tigris. This diagram, or relief, shows cattle sacrificed at the source while workmen are busy cutting commemorative reliefs on the rock (Reade 1983).

The oldest agricultural remnants of civilisation where the earliest irrigated agriculture was found is at Jericho in the Jordan Rift Valley about 10 km north of the present northern end of the Dead Sea (Kenyon 1957). It is the lowest town, elevation-wise, in the world, at about 300 m below sea level. Jericho is in the arid valley of the Jordan River and its perennial water supply is dependent on a large karst spring. One can claim that indeed dependence on irrigated agriculture may have started here. Jericho is a oasis, the life there depends on the spring called in Arabic 'Ein-el-Sultan'. The spring is fed from the rains falling and infiltrating into the permeable limestone mountains of Judea and Samarea. Its flow is about 10 million m^3/year. This, therefore then, would be one of the oldest known uses of a karst spring for domestic and agricultural purposes.

Perhaps amongst the most interesting springs are those at Ayun Musa mentioned in the Biblical story of the Exodus. From Exodus 15:22, 23 it is quoted: 'So Moses brought Israel from the Sea of Reeds and they went out unto the wilderness. And found no water. And when they came to Marah, they could not drink the waters, because they were bitter; therefore the name of the place was called Marah'.

The springs at Ayun Musa discharge from an outcrop of limestone and marls of Neogene age. The extent of these rocks is limited, however, detailed investigations by Israeli hydrologists have placed the source of the ground water far to the southeast and the recharge in central Sinai. The ground water that discharges at Ayun Musa discharges upward under artesian pressure along a fault zone. The water from these springs is rich in sulphate, 500 ppm, mainly calcium sulphate, but also contains magnesium sulphate which gives it the bitter taste. Perhaps the Biblical reference that emphasises that the water was bitter, was therefore, one of the first reported 'hydrochemical' observations in human history. Ayun Musa was used by the Bedouins travelling along the eastern side of the Gulf of Suez long before Moses' time. Therefore the use of this limestone spring would predate 1500 BC The water that discharges from these springs would have to be termed 'fossil' or palaeo-water since it came into the ground tens of thousands of years ago. An age determination by the ^{14}C carbonate content dissolved in the water provides an age of about 20,000 years.

Another spring mentioned frequently in the Bible is Kadesh Barnea. The large springs at Kadesh Barnea issue from Eocene Limestones. The spring is fed by rains and floods flowing over the rocks of the Negev and Sinai during the late winter and spring of the year. Surface water infiltrates into the solution channels in the limestone of Middle Eocene age until it reaches downward to the impermeable chalk layers of Lower Eocene age. There are several sources for the water of the Jordan River. The two main springs, the Dan (Tel Dan – ancient Laish – is situated near this spring) and the Banias (named after the god Pan), emerge from Mount Hermon which is built of limestones of Jurassic age. The limestones are highly permeable and the water from the rain falling on the mountain and from the melting snow which covers the higher stretches of the mountain each winter quickly infiltrate the subsur-

Photo 2.5. The oasis of Al Hasa, Eastern Province, Saudi Arabia, is fed by big karst springs, which have been used by the peoples since Neolithic time (photo: Hötzl 1976).

Photo 2.6. The capture of a karst spring in the oasis of Al Qatif, Eastern Province, Saudi Arabia (photo: Hötzl 1976).

face to enrich the aquifer which gives rise to these springs (as well as to many other springs such as those irrigating the oasis of Damascus). The average annual total amount of precipitation may reach 1200 mm. The Dan and Banias springs, together with some smaller springs flowing from the flanks of the Hermon, contribute most of the water to the upper Jordan River. Due to high permeability and the high rate of precipitation, the discharge of these two major springs is fairly regular.

Another series of references from the Bible are related to the Israelites returning to the promised land and facing problems of removing the Canaanites from their foothold in the area. Interestingly, most of the Canaanite towns (Hazor, Gezer, Megiddo, Gibeon, and Jerusalem) were built on a hill with a spring, or springs, below the city. The Canaanites, and subsequently the Israelites, in each instance were able to carry out engineering practices that would enable them to keep water available within the town. For example, Gibeon had a spring issuing from the limestone rocks of Middle Cretaceous age. The development of the limestone springs beneath the city of Jerusalem has recently been described in great detail by Gill (1994). In essence, these early people were able to construct big shafts, spiral staircases, tunnels, and connecting channels through the limestone rock, taking advantage of existing solution cavity development and thus tapped the water issuing at the spring. These systems were so constructed that when an enemy approached the city or village, the outlet of the spring was blocked and the people would have access to the water via the tunnel. Water from these springs, in part, are still being used for irrigation. Some archaeologists maintain that the tunnels of Megiddo, Jerusalem, and Gibeon were built by the Israelites. Others believe that they were excavated at a much earlier date, possibly during the late Bronze age, between 1425 and 1200 BC.

2.2.5 *Ancient Rome* (Paolo Bono)

In ancient Rome, water was considered a deity to be worshipped and most of all utilised in health and art. The availability of an abundant water supply was certainly considered a symbol of opulence and therefore an expression of power (Photos 2.7 and 2.8). At the beginning of the fourth century AD Rome had the most powerful hydraulic equipment that any city in the world had ever had – a maximum of 11 large baths, 865 public baths, 15 ninpheous, 2 naumachiae (naval battle basin) and 1352 fountains and ornamental pools.

At the end of the first century AD there were nine great aqueducts that crossed the countryside conveying to the city from the catchment areas (most of them located in the eastern quadrant of the region) a total discharge per day of 24,360 *quinariae*, equivalent to 1,010,258 $m^3 \cdot day^{-1}$ (11.7 $m^3 \cdot sec^{-1}$), (Photo 2.9).

The population of Rome at the end of the first century AD was about 500,000 people, consequently a mean supply of 2,020 $l \cdot day^{-1}$ per capita was available. Of course the first step in constructing an aqueduct was the search for a rich spring. *Vitruvius*, the author of *De Architectura* provides us with some empirical methods based on direct observations of the vegetation cover, the lithology and the morphology of the terrain and even the air's humidity. Concerning the identification tests on the resource quality, the author states that 'if you put legumes and water in a pot on the fire and they cook properly, this indicates that the water is wholesome and good'. Since the Romans were dealing with a number of large springs located either near the city or up to 60 km from it, the problem did not consist of discovering massive quantities of water but resources of good standard.

Until the fourth century BC Rome had an estimated population of 150 thousand people who obtained their supply of water directly from the Tiber River and from several small springs including large diameter wells located inside the defensive Servian walls of the city. Incidentally, most of the Tiber base flow discharge in Rome is controlled by a number of large karstic springs (some located at more than 100 km from the city) which sum up to a mean of 80-85 m^3·sec^{-1}. The several but minor springs located among the Servian walls refer instead to shallow aquifers most of them related to volcanic deposits. Although the discharge of such 'urban' springs was small compared to the Tiber River flow rate, it seems they had a strategic importance since the city developed almost entirely above loose and water bearing volcanic deposits.

Such favourable hydrogeological conditions therefore ensured the inhabitants of the necessary supply of good quality water either during the siege of the city or during the floods of the Tiber. Since the nearest major karstic springs are located about 30 km away from the city, the planning of an external water supply that could integrate both the demand of an increasing population and the development of public welfare, had to consider the following basic elements: the altitude of the springs, the available discharge, the distance from the catchment area to the city and finally water quality. So topographic, hydrologic, and sanitary conditions were taken into account to make a plan for a new supply scheme, considering also that all catchments ought be located in a defensible area and therefore concentrated along the most promising valley rich with water, closest to Rome. The strategic choice of the Roman water authority engineers was the Aniene River valley where the river itself represents a 'linear spring' due to seepage from an extensive karst aquifer and where major localised springs also occur along the banks of the stream. Today we know that the minimum base flow discharge of Aniene River at the outlet of the karst basin is above 10 m^3·sec^{-1} while the mean annual discharge and the average of the minimum daily discharge of the driest season are of 27 and 16 m^3·sec^{-1} respectively (Boni et al. 1993). Four aqueducts were built in that area conveying water from Aniene karstic springs to the city between 272 BC (*Anio Vetus*: 2.11 m^3 sec^{-1}) and 50 AD (*Anio Novus*: 2.27 m^3 sec^{-1}), *Aqua Marcia* (2.25 m^3·sec^{-1}) and *Aqua Claudia* (2.21 m^3·sec^{-1}). This yield supplied to the city from karstic springs adds up to about 9 m^3·sec^{-1}, representing 77% of the total water supply of Rome.

S.I. Frontinus the water magistrate in Rome during the period 97 AD to 103 AD in *De Aqueductu Urbis Romae* states that a total of 5505 *quinariae* equivalent to 228,300 m^3 day^{-1} (2.6 m^3 sec^{-1}) were missing at the aqueduct terminals in Rome (Pace 1986). The meticulous and pragmatic civil servant complained that the water losses from the spring catchments were due to leakage and most of all to unauthorised connections along the aqueducts which were difficult to control.

Also today aqueducts bring mostly karstic water resources to Rome, with a yield of 20 m^3·sec^{-1}, about 87% of the total supply. At present, Rome receives 1,987,200 m^3·day^{-1} (23 m^3·sec^{-1}) of groundwater, equivalent to a per capita availability of 500 l·day^{-1} including industrial consumption.

Photo 2.7. The polje of Stymphalia (600 m a.s.l.), NE Peloponnesus, Greece. The karst springs on the northern margin (right), the reason for settlements since ancient Greek time, are feeding the permanent lake (central part), known from mythology as home of the fatal birds of Stimfalia, killed by Herakles (Röckel & Hötzl 1986). The lake discharging to sinkholes was tapped by the Romans (Hadrian conduit) via a gallery through the ridge in the foreground (photo: Hötzl 1982).

Photo 2.8. The Hadrian conduit beyond the tunnel of Stimfalia. It was built for the water supply of ancient Korinth (60 km distant) about 125 AC and restored for irrigation demands in the coastal plain near Korinth during 1881-1885 (photo: Reichert 1982).

Photo 2.9. Aqueducts supplying the ancient Rome with water from karst springs (after a painting from Zeno Diener, reproduction Deutsches Museum, Munich).

Photo 2.10. Pont du Gard, Roman aqueduct crossing the river Gardon, Southern France. The aqueduct with a total length of 275 m and an elevation of 49 m a.r.l. was part of the water supply of Nimes (Colonia Augusta Nemausus) which carried the water of the karst spring Eure over a distance of 50 km to Nimes (photo: Hötzl 1995).

CHAPTER 3

Introduction and overview of human impacts

DAVID DREW
Department of Geography, Trinity College, Dublin, Ireland

3.1 OVERVIEW OF THE IMPACTS AND THEIR INTERACTION

Part 2 of the book examines the impacts on karst waters of particular types of human activity. The grouping of activities used in Part 2, reflect those economic activities whose impacts are most widespread and/or intensive in their effects upon the karst system.

Chapter 4 deals with agriculturally induced impacts, the most areally extensive environmental modifications due to human activity with the possible exception of anthropogenic climatic changes. A distinction is made between two types of agriculturally induced impact. First, changes in land use (vegetation changes or the creation of artificial drainage networks for example), associated with particular agricultural systems and practices. Second, changes in the chemistry or biology of soils and water due to the use of pesticides, herbicides, fertilisers or due to the disposal of agricultural wastes.

The universality of agriculturally caused alterations to the karstic groundwater system is apparent from the global distribution of case studies presented in Chapter 4. Examples from China, the USA, Europe, the Caribbean and Australia demonstrate the sensitivity of karst waters to agricultural activities irrespective of the character of the agricultural or economic system practised.

Because agriculture was the first major change made to their environment by human societies, and because many of the earliest developed societies arose in karstic regions, karst landscapes and karst waters have been exposed to low intensity modification due to agricultural activities for many millennia. More modern farming methods have allowed the possibility of high intensity changes in land and in karst groundwater systems.

The topics dealt with in Chapter 5 (urban and industrial impacts) and Chapter 6 (impacts due to mining/extractive activities) are similar in that generally they involve smaller areas of land but very high levels of imposed stress. One impact addressed in Chapter 6, specific to karst, is that of quarrying for limestone – a valuable resource in carbonate rock areas the exploitation of which often affects the qualities of the other major resource of karsts – groundwater.

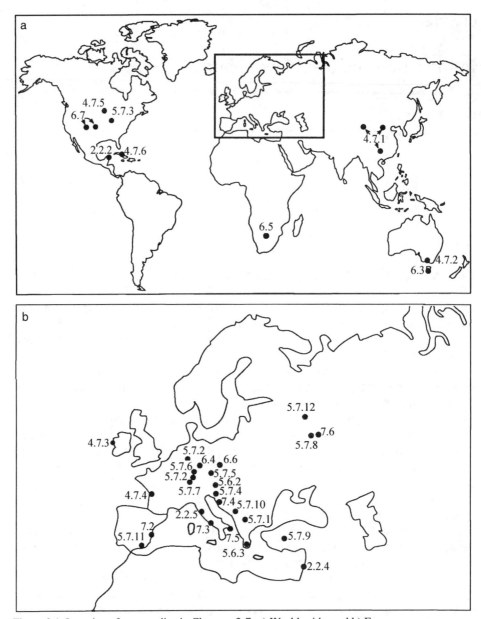

Figure 3.1. Location of case studies in Chapters 2-7: a) World-wide, and b) Europe.

The most direct of all human impacts on karst groundwater is its exploitation for water supply. Demands for water are usually greatest in urban, industrial, touristic centres and in agricultural areas where irrigation is practised. This is true irrespective of the source of the water. However, exploitation of water from carbonate rocks gives the response peculiar to the karst system. The main effects of exploitation such as the lowering of groundwater levels and the potential salinisation of groundwater are common to all aquifers but the response of the karst groundwater is commonly distinctive, immediate and may be less easily ameliorated than in other hydro-geological environments. Chapter 7 deals with the effects on karst aquifers of exploitation of water.

The locations of all the case studies presented in Chapter 2 and in Chapters 4 to 7 are shown in Figure 3.1a (the world) and Figure 3.1b (Europe). If the distribution of these case studies is compared with the world and European distributions of carbonate rocks shown in Figures 1.1 and 1.2 (Chapter 1) respectively it may be seen that whilst Europe is well represented the other major karst areas of the world, for example southeast Asia, are less adequately documented.

The co-ordinators for Chapters 4 to 7 are based in the USA, Germany, Ireland and Spain and to some degree the overview of each topic presented will be coloured by the particularities of the karsts and the cultures of the authors' native region or the areas with which they are most familiar. The differences in emphasis that result reflect real differences in karst types and in degrees of karstification around the world.

3.2 COMPOSITE IMPACTS

Part 2 of the book is oriented towards causes rather than towards effects. Similar consequences may result from different causes. For example, subsidence dolines (collapses) may develop as a result of dewatering of limestones for extractive industry – an indirect impact (Chapter 6), or they may result from over-abstraction of water for consumptive use (Chapter 7). Comparable problems of biological contamination of karst water may result from agricultural practices in an area or equally from touristic pressures or from leaking or inadequate sewerage systems. The topic is therefore treated in both chapters 4 and 5 from rural and from urban perspectives respectively.

It may also be argued that tourism is simply a composite impact similar in direct causes to urban impacts, whilst urban-industrial extractive impacts are themselves often similar in effect, and indeed case studies of touristic impact are included in Chapter 5. However, tourism is a rapidly growing global industry bringing increasingly severe problems to affected areas. Tourism is not an impact in itself but rather it intensifies or exacerbates some of the other types of impact described in Part 2. From the perspective of impact it comprises a localised demand for touristic needs, often in a highly scenic and hitherto undeveloped locale. The tourist attraction may be located in a karst region even though the tourist lure is not karstic, for example the Mayan cities of the Yucatan Peninsula in Mexico (Chapter 2), or the karst feature (cave, gorge etc.) may be the direct feature of interest. The impacts are similar in either instance.

It is ironic that many of the most ancient areas of human impacts on karst waters mentioned in Chapter 2 (meso-America and China for example) are now amongst the world's most visited tourist attractions.

The problems raised by a composite impact such as that of tourism on karst waters are often more complex and more difficult to resolve than those associated with more specific economic activities. There is an awakening interest in the problems generated by karst tourism: for example the studies by Gillieson (1996), by Celli (1995) in regard to the Italian Alps and Chardon (1995) in respect of the Vercors in France.

This book is intended to be accessible and useful to workers without expertise in karst hydrogeology but with a direct or indirect involvement with karst waters perhaps in a regulatory, legislative or risk-assessment capacity and a topic based approach (e.g. mining, agriculture) allows for an overview of possible impacts on karst water resulting from any of these particular economic activities. Thus there are inevitably areas of overlap of content in the separate chapters that constitute Part 2 but it is considered that such an overlap is preferable to an effect-centred structure for the book.

Part 2: The nature of human impacts on karst water

CHAPTER 4

Agriculturally induced impacts

CATHERINE COXON (co-ordinator and main author)
Environmental Sciences Unit, Trinity College, Dublin, Ireland

4.1 INTRODUCTION TO THE PROBLEMS

The impact of agriculture on groundwater has formed the subject of numerous reviews and conference proceedings (e.g. Vrba & Svoma 1982, Vrba & Romijn 1986, Fairchild 1987). Many of the problems found in karst aquifers (e.g. nitrate pollution) are found in all aquifer types, but it is the intention of this chapter to focus on those problems which are unique to karst aquifers, or which occur with particular severity in karst aquifers.

The agricultural activities which may give rise to an impact on karst groundwater are discussed under three main headings: land use change, irrigation and drainage, and agricultural pollutant sources. The latter will obviously produce an impact on groundwater quality, while the former two may produce an impact on both quantity and quality. Alterations to quantity may include changes in timing or amount of aquifer recharge or discharge, and the impacts may vary from minor changes to aquifer water budgets to catastrophic flooding problems. In general, however, the changes to groundwater quality resulting from agricultural activities are of greater significance, and the contamination spans a wide range including suspended sediment, inorganic ions, synthetic organic compounds and micro-organisms.

The sources of contamination can be viewed in terms of their spatial extent, i.e. non-point or diffuse versus point or concentrated. The entry method to the karst aquifer can also differ in terms of spatial extent, i.e. diffuse versus point recharge. This does not necessarily coincide with the spatial extent of the contaminant source, for example a point source (e.g. a silage clamp) might have seepage from beneath it into the diffuse groundwater body, or the silage effluent might run off into a sinking stream and enter a cave conduit. In other words, in any aquifer type the pollutant sources can be diffuse or concentrated but the possibility of concentrated recharge is one of the factors which makes karst aquifers particularly vulnerable.

Because there is no clear distinction between groundwater and surface water in many karst areas, contaminants more frequently associated with surface waters, which would not normally enter by diffuse recharge (e.g. phosphate, or pesticides adsorbed to suspended sediment) may enter karst conduits from sinking streams.

37

Equally, the rapid throughput and lack of attenuation in the karst aquifer mean that groundwater contaminants may emerge at springs and contaminate surface waters to a greater degree than in other aquifer types. Therefore, in addition to identifying the different threats posed by a range of agricultural activities, the chapter includes a discussion of the mobility of agricultural pollutants in karst aquifers, which focuses on how it differs from that in conventional aquifers.

The review includes a brief discussion on the implications for groundwater protection planning, although the reader is also referred to Chapter 8 for further information on karst water resource management and legislation.

Following the general overview under the above headings, the chapter includes short contributions by different authors, which provide more detailed information on particular agricultural impacts in karst aquifers around the world. Lin Xueyu and Liao Zisheng (Section 4.7.1) review the impact of agriculture on Chinese karst water resources, including the hydrological impact of surface and underground reservoir construction and of irrigation water abstraction, and they detail instances of agriculture related groundwater pollution. Ellaway et al. (Section 4.7.2) provide an Australian case study of the impact of land clearance on karst groundwater, examining the evidence for water quality changes following clearance of native Eucalyptus forest for grazing, while Drew (Section 4.7.3) describes land reclamation in the Burren Karst of western Ireland, where scrub and boulder clearance, topsoil spreading and fertiliser application has been accompanied by increased silage making, with the attendant threat to karst groundwater from silage effluent. Audra (Section 4.7.4) discusses land clearance associated with intensive vine cultivation in the Entre-deux-Mers region of France. Libra and Hallberg (Section 4.7.5) discuss agricultural impacts on water quality in a karst catchment in Iowa, USA, and they examine the land-use and hydrometeorological factors controlling nitrate and atrazine concentrations in the groundwater, while Molerio and Gutierrez (Section 4.7.6) discuss nitrate contamination of Cuban karstic aquifers from both point and diffuse agricultural sources. The impact of nitrate and agrochemicals is also discussed by Hardwick and Gunn (Section 4.7.7), who focus on the opportunity provided by speleology to carry out undisturbed sampling of percolation waters.

4.2   DEFORESTATION, LAND CLEARANCE AND AGRICULTURAL INTENSIFICATION

4.2.1   *Soil erosion*

The change in land use from a natural vegetation cover to agriculture is likely to have a hydrological impact in any geological situation, but karstic areas are particularly prone to such interference (Photos 4.1 and 4.2). Soil erosion is easily triggered, both because limestone soils are typically shallow (given that the rock from which they are derived yields little insoluble residue) and because the open joint systems facilitate the washing underground of soil material (Williams 1993). The historical occurrence of soil erosion due to deforestation is reported from many karstic areas. For example, agricultural development by the Maya people in Yucatan from the first millennium BC onwards gave rise to soil erosion in karstic areas (Deevey et al.

1979). Gillieson et al. (1986), studying sediment accumulation in rock shelters in Papua New Guinea, suggest that soil erosion was triggered following disturbance of the primary forest cover 6000 years ago and that the rate increased significantly in the last 300-400 years following horticultural intensification, while Gams (1993) notes the occurrence of soil erosion resulting from deforestation in the Classical Karst area in Slovenia from 500 BC onwards. The problem is not confined to tropical and Mediterranean climates, with even humid-temperate areas being subject to major erosion (e.g. soil loss in Bronze Age times from the Burren Plateau in Ireland, as outlined by Drew (1983). Such changes continue at the present day, with removal of forest cover in some tropical karst areas such as the Hummingbird karst of Central Belize (Day & Rosen 1989), in a large area of sub-tropical China (Yuan 1993) and also in some temperate areas e.g. Vancouver Island, Canada (Harding & Ford 1993) and Tasmania (Kiernan 1989).

Soil erosion may also be provoked by an increase in grazing pressure, particularly in arid and semi-arid areas, e.g. the Nullarbor Plain in Australia (Gillieson et al. 1994a, b), or by the conversion of pasture land to tillage. Audra (Section 4.7.4) describes the erosion of superficial sediments in a karst area in the Entre-deux-Mers region, France, following bulldozing of fields for intensive viticulture, with the formation of collapse dolines and the removal of clays into the underlying aquifer. In some circumstances, however, agricultural intensification can give rise to an increase rather than a decrease in soil cover: in the Burren karst plateau, Ireland, areas of scrub and of limestone pavement have been cleared and a uniform thin layer of topsoil spread and used for silage making. However, this newly established soil cover is itself susceptible to erosion (Drew & Magee (1994), chapter contribution by Drew (Section 4.7.3)). In other areas where agriculture is becoming less intensive, former tillage land may revert to pasture or scrub (e.g. Dougherty 1981), and these changes may have an effect on the karst groundwater.

Several studies concerned with the impact of land use change in karst areas have been concerned with the impact at the ground surface, in terms of loss of soil cover, effect on surface solution features etc. (e.g. Harding & Ford 1993, Kiernan 1987a). The discussion here focuses however on the impacts on the karst water. These can include quantitative changes, i.e. changes in amount and timing of recharge, and qualitative changes, particularly to the suspended sediment load.

4.2.2 *Changes in amount and timing of recharge and discharge*

Changes in land use such as clearance of forest for agriculture, changes from grassland to tillage, or reafforestation of cleared lands, will all have implications for groundwater recharge due to the change in evapotranspiration rates. Lerner et al. (1990) document several case studies of carbonate aquifers where groundwater recharge is related to vegetation cover. In the Carmel Hills in Israel, underlain by karstic dolomite, limestone and chalk of Cretaceous age, areas under pine forest or deep-rooted chaparral scrub have negligible recharge as all rainfall is consumed by evapotranspiration, whereas areas under pasture have an average annual recharge of 420 mm, i.e. 60% of annual precipitation (Rosenzweig 1972). The recharge rate is also dependent on the geology, with higher evapotranspiration rates in areas of softer limestone because of greater root penetration (Rosenzweig 1973). Similarly, in a dolomitic aquifer in a semi-arid region of Zambia, according to Houston (1982), recharge is significantly less in open forest areas than in cleared areas used for agriculture (with an estimated mean annual recharge of 80 mm, versus 281 mm for cleared areas). The plantation of *Pinus radiata* forests on the Gambier Plain, South Australia, has decreased recharge to the underlying Miocene limestone aquifer, with tritium values suggesting that recharge under the forest cover is less than a fifth of that taking place under pasture (Allison & Hughes 1972).

Alterations to groundwater recharge due to vegetation changes and resultant changes in evapotranspiration are not unique to karst areas. However, recharge mechanisms specific to karst may accentuate the change. E.g., whereas increased aquifer recharge due to decreased evapotranspiration following deforestation may be counteracted in some geological situations by an increasing dominance of overland flow over infiltration and percolation, in a karstic situation this overland flow is then likely to enter the aquifer via swallow holes.

Even more important than the change to average annual recharge rates, however, is the alteration in timing of recharge: intense, short duration recharge events are likely to take the place of more sustained inputs, particularly where significant soil erosion has followed the deforestation, and this may increase the severity of droughts.

An example of such changes is documented by Huntoon (1992), who observes that in the stone forest karst aquifers of south China massive deforestation since 1958 has resulted in a major impact on the magnitude and duration of the seasonal recharge pulse. Water that was formerly retained in the forested uplands and gradually released to recharge the aquifers on the lowland now passes rapidly through the system, so that the decline in water level in the aquifers during the dry season has been accelerated, resulting in drying up of wells and springs.

Another area where declining water levels are attributed to deforestation is in the karst area of Batuan on the island of Bohol (Philippines). Here, spring discharges are reported to have declined by 40% in the last 20 years, and Urich (1991, 1993) suggests that deforestation associated with increasingly intensive slash-and-burn agriculture is the primary cause.

Another aspect to the changes in recharge patterns following deforestation is the problem of flooding caused by increased runoff and exacerbated by blockage of the karst drainage system due to soil erosion. Dougherty (1981, 1983) describes a karstified Mississippian limestone region in Kentucky where the change from oak-maple forest cover to tillage of tobacco and corn in the 1930's resulted in disastrous valley floods known locally as 'valley tides'. Areas of bare soil where poor agricultural techniques were practised generated increased runoff, and the increased sediment load of the runoff blocked the already insufficient underground drainage channels, resulting in backing up of runoff and flooding of the valleys upstream of river sinks. The increased hydrostatic pressure in the karst aquifer also resulted in water gushing from boreholes, and the occurrence of ephemeral karst springs. Similarly, Molina & McDonald (1987) report that poor agricultural land management in karstic areas of Jamaica has led to the blocking of swallow holes by plant debris and silt, resulting in flooding problems following heavy rainfall.

Photo 4.1. Karst pasture in the region of Bari, Southern Italy; which is now changed to agricultural land, compare Photo 4.2 (photo: Hötzl 1997).

Photo 4.2. Reclamation of land by changing karst pasture (cp. Photo 4.1) into new bulldozed fields. Frequently the result of a misdirected agricultural policy leading to quantitative and qualitative changes of karst water. Agricultural program in the region of Bari, Southern Italy (photo: Hötzl 1997).

4.2.3 *Changes in water quality*

Land use change in a karstic catchment associated with agricultural development will also have an effect on the water quality. Aside from the introduction of new sources of contamination such as inorganic fertilisers (discussed later in this chapter), the land use change *per se* may give rise to water quality problems. The most significant of these is the change in sediment load. The flooding problems discussed above arise frequently from increased sediment load in conjunction with increased surface run-off.

Sediment loss from karst areas subject to soil erosion can be very considerable: Yuan (1993) records that silt transport rates for rivers in karst basins in southern China subject to deforestation and rocky desertification are 208-1980 tonnes/km^2/annum. However, measurements of suspended sediment load in karst groundwater are relatively rare, and in many instances the increased load associated with soil erosion must be presumed from evidence of soil loss, or of soil accumulation in cave passages. For example, Tucker (1982) notes the accretion of several metres of sediment in Kentucky caves, which she attributes to soil erosion associated with tillage agriculture in areas with erodible soils and point recharge through sinkholes and swallets, and Lewis (1981) records that following storm events, large amounts of mud and fine organic debris were washed into Coldwater Cave, Iowa, from the overlying farmland. Clearfelling in the Florentine Valley, Tasmania, triggered sheet erosion on very low slope angles and gave rise to the deposition of one metre of sediment blocking a cave within a year (Kiernan 1981). The Kentucky karst study by Dougherty (1981) referred to above also describes how a change from forest cover to tillage gave rise to increased sediment load in the karstic flow system, but again the evidence is from examination of cave sediments.

The quantity of suspended sediment in karst groundwaters from agricultural sources shows a large degree of temporal variation. For example, in the Big Spring basin, a karstic catchment in Iowa, USA with wholly agricultural land-use (50-60% of area planted with corn, rotated with pasture/hay), suspended sediment loads vary greatly with spring flow, increasing during flood events from negligible values to concentrations over 4,000·mg l^{-1}, corresponding to a load of over 87,000·kg h^{-1} (Hallberg et al. 1985). (Temporal variations in groundwater quality are discussed further in Section 4.5.3).

Hardwick & Gunn (1990) discuss the mechanisms of soil erosion in karst areas. The greatest sediment inputs are associated with sinking streams, particularly where these originate on non-carbonate rocks. These authors also discuss the extent to which soils may be eroded into karst aquifers by autogenic recharge, noting the importance of throughflow rather than overland flow in temperate doline karsts and drawing attention to the existence of macropores (c.f. Simpson & Cunningham 1982). They consider that sediment transport through the subcutaneous zone (epikarst) is also highly likely, but suggest that vertical movement from the subcutaneous zone through solutionally widened fissures is likely to be very slow, and that soil currently entering the British Carboniferous Limestone with autogenic recharge may take thousands of years to reach cave passages. Nonetheless, this sediment entry route should not be neglected: they cite studies in the Dinaric Karst (Kogovsek &

Habic 1980, Kogovsek 1982) which show suspended load in cave trickles to be of the same order of magnitude as dissolved load.

In addition to changes in suspended sediment load, land use change may give rise to other chemical changes. The chapter contribution by Ellaway et al. (Section 4.7.2) documents the chemical effects of land use change in a karstified Devonian limestone aquifer in Buchan, Victoria, Australia: springs and cave streams from uncleared native forest catchments differ in chemistry from those with catchments of cleared agricultural land. However, the authors note that such changes may be subtle when compared with grosser contamination arising from sources such as waste disposal sites. Kastrinos & White (1986), examining eight springs in the central Pennsylvania carbonate aquifer, found a strong correlation between the spring-to-spring variation in nitrate level and the rural land use in the spring catchment area, with nitrate levels increasing as the proportion of cropland to woodland increased, although this study does not distinguish the effects of the land use change *per se* from the effect of additional nitrogen inputs to cropland. Lichon (1993), studying human impacts on Tasmanian karsts, notes that the release and leaching of forest nutrients following deforestation results in elevated nutrient loading of karst streams and contamination of water resources.

4.3 IRRIGATION AND LAND DRAINAGE

4.3.1 *Changes of soil water content*

In many parts of the world, the soil water content has been altered to facilitate agriculture, with moisture levels being either increased by irrigation or decreased by land drainage operations. Such interference with this part of the hydrologic cycle is likely to have an impact on groundwater resources, and in karstic areas where surface waters and groundwaters are so closely interlinked, the impact is particularly marked. The alteration may be to either quantity or quality; examples of both are discussed below.

The extent of impact will obviously depend on the degree of interference with the natural system. Urich (1993) suggests that wet rice cultivation in karst areas of Bohol in the Philippines was originally a sustainable land-use, with the integration of karst features into the irrigation systems (e.g. the use of caves as irrigation canals and for temporary flood water diversion, and the plugging and unplugging of small swallow holes within the paddies according to desired water levels). However, he contends that changes in land management since the 1970's, including the introduction of inappropriate modern agricultural technologies, have upset a delicate balance and resulted in groundwater quantity and quality problems.

4.3.2 *Changes in amount and timing of recharge and discharge*

Irrigation may cause a hydrological impact at the source of the irrigation water or at the irrigation site. If irrigation water is abstracted from boreholes in karstic aquifers, then obviously this will cause a fall in the water table, as in any aquifer, although as with any groundwater abstraction in karstic aquifers the lowering of the water table may be very uneven and difficult to predict. Alternatively the irrigation water may be provided by the construction of a reservoir, and in karstic areas this may involve subterranean engineering works. Water level regulation in poljes is discussed below, while the contribution by Lin Xueyu and Liao Zisheng (Section 4.7.1) describes the impact of both surface and underground reservoirs in karstic areas of China. The irrigated land may then provide a new source of recharge to the karst aquifer; the water quality implications of this are discussed below.

Many Dinaric Karst poljes have been subjected to drainage schemes because of their importance as flat, fertile lands within a barren karst landscape. The flooding regime of the poljes, which can involve either ponding up of surface water or inflow of groundwater (Mijatovic 1984b), has been altered by various means, with varying degrees of success. An attempt to prevent flooding in Fatnicko Polje by constructing a concrete plug in the ponor led to the water rising at new locations, causing localised seismic shocks (Milanovic 1984). Schemes involving tunnel construction have been more successful, although Bonacci (1985) warns that the decrease in water level and shorter duration of flooding in a polje drained by this method can lead to a worsening of the water regime in downstream and upstream poljes. The multipurpose nature of polje water regulation schemes is stressed by Habic (1987), with schemes being designed for summer irrigation and winter flood relief together with non-agricultural purposes (hydro-electric power generation, water supply and recreation). In some instances, occasionally flooding poljes are transformed into permanent storage basins, which alter the water table in the surrounding karst aquifer, resulting in reactivation of springs or creation of new springs in higher level poljes (Bonacci 1985).

The effects of land drainage on groundwater resources in karstic areas of Ireland are reviewed by Drew & Coxon (1988). In many lowland areas of western Ireland underlain by Carboniferous limestone, channelisation to relieve flooding of agricultural land has resulted in a decrease in aquifer recharge and lowering of water tables. For example, in the Kilcolgan-Lavally catchment in east Galway (area 500 km^2), recharge to the karst aquifer has been diminished by $93 \cdot 10^6 \, m^3 \cdot a^{-1}$ and summer water tables have been lowered by 2-3 m, causing previously perennial springs to become ephemeral. Drainage operations in the western Irish limestone lowlands have also resulted in the draining of turloughs (seasonal lakes having a karstic function similar to poljes), and the cessation of seasonal flooding has serious ecological implications (Drew & Coxon 1988). Drew (1984) documents the hydrological changes to a karst aquifer brought about by land drainage and channelisation in the Clarinbridge catchment in east Galway. In this catchment, drainage operations from the mid-nineteenth century onwards have increased the drainage density from 0.2 km·km^{-2} to 4 km·km^{-2}, and in the lower catchment a wholly artificial river has been excavated into bedrock, linking and draining a series of turloughs (Fig. 4.1a). As a result, surface runoff from the basin has been increased from zero to 40% of effective precipitation and aquifer recharge has been reduced accordingly. The reduction in over-season storage within the aquifer has meant that during a dry summer, the water table falls rapidly and many wells dry up. In addition, the spatial patterns of recharge and discharge have been altered, with the artificial stretches of channel providing line and point recharge and discharge at different times of year.

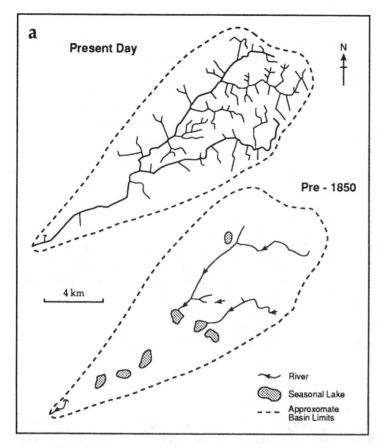

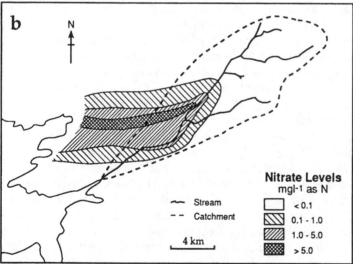

Figure 4.1. Impact of land drainage on the Clarinbridge catchment, Ireland (after Drew 1984):
a) Increase in drainage density, and b) Pollutant plume from line recharge along artificial channel.

4.3.3 *Changes in water quality and ease of entry of pollutants*

Irrigation or drainage of land is generally accompanied by agricultural intensification and increased use of fertilisers and pesticides, the impact of which is discussed in a later section. Irrigation may also provoke groundwater quality changes if it brings about increased leaching of natural or artificial chemical constituents from the soil. salinization of karst groundwaters due to increased leaching of soluble salts has been documented from several areas. In the Flat Crimean karst area, Ukraine, irrigation of agricultural land caused groundwater recharge to increase by up to 200 mm a^{-1}, and increased leaching of chloride and sulphate salts resulted in groundwater mineralisation increasing from 500-1500 mg l^{-1} to 5700-8700 mg l^{-1} (Lushichik 1982). In rural northern Puerto Rico, Day (1993) notes that karst groundwaters contain chloride concentrations of up to 1200 mg l^{-1} as a result of irrigation practices and sea water intrusion, which has been exacerbated by agricultural water withdrawal.

Drainage operations along river channels overlying karst aquifers may also provide entry routes for contaminants. For example, Drew (1984) notes that in the Clarinbridge catchment referred to above, channel excavation into bedrock has enabled line and point recharge with polluted surface waters, resulting in a nitrate plume in the karst aquifer (Fig. 4.1b).

4.4 AGRICULTURAL POLLUTANT SOURCES

4.4.1 *Diffuse agricultural sources*

4.4.1.1 *Inorganic fertilisers*
As in other aquifer types, the main contamination problem in karst aquifers arising from inorganic fertiliser usage is with the nitrogen component in the fertiliser, giving rise to nitrate in the groundwater. This problem is clearly widespread, and this is reflected in its mention in several of the contributions to this chapter: Libra and Hallberg (Section 4.7.5) document an increase in nitrate levels in the Big Spring Basin, NE Iowa, USA, following a three-fold increase in nitrogen fertiliser applications, while Molerio and Gutierrez (Section 4.7.6) discuss problems of nitrate contamination from both point and diffuse agricultural sources in Cuban karstic aquifers. Lin Xueyu and Liao Zisheng (Section 4.7.1) also note an increase in nitrate levels in Chinese karst aquifers following increased fertiliser usage.

The nitrate problem is well documented in Europe, where the European Union drinking water standard of 50 mg l^{-1} NO$_3$ (11.3 mg l^{-1} as N) is frequently exceeded. In Britain, nitrate levels in the Jurassic limestone of eastern England give rise to concern (Smith-Carington et al. 1983). The contribution by Hardwick and Gunn (Section 4.7.7) notes the occurrence of elevated nitrate in cave waters in a Carboniferous limestone aquifer in Derbyshire, the source in this instance being land application of both sewage sludge and chemical fertiliser. Ireland is one of the few European countries where nitrate in karst aquifers remains well below the EU limit, except where there is localised point source contamination; this is due to a combination of relatively low fertiliser usage, a low proportion of arable land and a high effective rainfall providing considerable dilution (Aldwell et al. 1988).

Photo 4.3. Intensive agricultural land use in dolines and other karst depressions in the classical Slovenian karst near Trieste (photo: Hötzl 1994).

In the USA, Berryhill (1989) notes that levels of nitrate above the US EPA drinking water standard of 45 mg l^{-1} (10 mg l^{-1} as N) have been documented in karst groundwater in Minnesota, Pennsylvania and Iowa. Nitrate from diffuse sources is particularly well documented for Iowa karst groundwater (Hallberg et al. 1985, Libra et al. 1986, 1987; Libra & Hallberg, Section 4.7.5).

4.4.1.2 *Land spreading of organic wastes*

Organic wastes applied to agricultural land include agricultural wastes (livestock manure or slurry, silage effluent, farmyard runoff), human wastes (sewage effluent, sewage sludge) and agro-industrial wastes (dairy effluent, blood, offal). In any aquifer type, land spreading of organic waste to excess can give rise to chemical pollution of groundwater. In the case of karstic aquifers, such problems are particularly likely to occur due to the high aquifer permeability, often combined with thin soil cover. Furthermore, in a karstic situation, contamination may not be restricted to the more mobile constituents such as nitrate and chloride; less mobile constituents more usually associated with point source contamination may gain entry. These include cations such as potassium and ammonium, and a particular concern is the risk of contamination by faecal micro-organisms. The latter threat may be removed by pre-treatment of sewage effluents prior to landspreading, e.g. by chlorination, ozonation or ultra-violet irradiation (see Metcalf & Eddy 1991) for a review of wastewater disinfection methods).

The contribution by Lin Xueyu and Liao Zisheng (Section 4.7.1) notes the pollution of karst groundwater by nitrogen compounds originating from wastewater irrigation in Jianshu Province, China. Land application of sewage effluent has also been carried out in a number of karst areas in the USA including Missouri's Ozarks (Settergren 1977, Tennyson & Settergren 1977) and the Floridan limestone at Tallahassee (Slack 1977). In the Floridan study, although the secondary-treated effluent contained 400-80,000 faecal coliforms per 100 ml, none were detected in the groundwater, so these were presumably removed or died off during passage through the sand and clay overlying the aquifer. However, more mobile constituents (nitrate, chloride) did enter the karst aquifer, and nitrate concentrations beneath and downgradient of the most heavily sprayed areas exceeded the drinking water limit.

Land disposal of sewage sludge is likely to become an increasingly popular option within Europe following the phasing out of marine sludge dumping under the Council Directive on urban wastewater treatment (91/271/EEC). European legislation to control the level of heavy metals in sewage sludge is already in place (Directive 86/278/EEC), but further controls are needed, particularly over the situations where spreading is permitted, and it is important that the special vulnerability of karst aquifers is taken into account

Land spreading of waste from agricultural industries can give rise to problems if the nutrient load is excessive, as is likely to be the case since the primary purpose of spreading is disposal of an effluent rather than fertilisation. Irrigation of cheese factory effluent onto pasture land overlying a karst aquifer in Mount Gambier, South Australia, while giving rise to fewer environmental impacts than the original borehole disposal method documented below (Section 4.4.3.2), nevertheless resulted in a substantial increase in nitrate concentration in the groundwater (Schrale et al. 1984). Whereas potassium and phosphorus surplus to plant uptake was largely retained in the soil profile, nitrogen was converted to nitrate over the summer and leached in the subsequent winter, passing through the permeable sandy soils into the limestone aquifer.

4.4.1.3   *Pesticides*

Studies of pesticide contamination in karst groundwaters are less common than nitrate studies because of the expensive and time consuming analytical procedures required. Nonetheless, there is much evidence that such contamination does occur in karst aquifers throughout the world, including Germany (Milde et al. 1988), the Ukraine (Andrajchouk & Klimchouk 1993), China (see contribution by Lin Xueyu & Liao Zisheng – Section 4.7.1) and the USA (see the contribution by Libra & Hallberg – Section 4.7.5, and other examples outlined below).

Pesticides are a matter of particular concern because of their health effects at very low concentrations. This means that only a very small quantity needs to be leached to give rise to a problem in water supplies. For example, although the amount of atrazine herbicide in groundwater in a karst catchment in Iowa was less than 0.1% of the amount applied (Hallberg et al. 1985), this resulted in a weighted mean concentration of 0.18-0.45 $\mu g l^{-1}$ and peak concentrations of up to 10 $\mu g l^{-1}$. The US drinking water limit for atrazine is 3 $\mu g l^{-1}$ (US EPA 1987), while the European limit (under Directive 80/778/EEC) for any individual pesticide is 0.1 $\mu g l^{-1}$.

Atrazine is one of the most commonly reported pesticides. Milde et al. (1988) record the presence of atrazine at concentrations up to 0.5 $\mu g l^{-1}$ in a Jurassic karst aquifer in Germany, where it is applied to areas of maize cultivation. In Big Spring basin, Iowa, it is the most widely used herbicide in the basin and it is the dominant pesticide in the groundwater (Hallberg et al. 1985). Wheeler et al. (1989) also report the occurrence of atrazine in a cave stream in the Coldriver Cave groundwater basin in northern Iowa, southern Minnesota, with concentrations ranging from 0.16 to 0.98 µg l^{-1}; typical atrazine application rates in the catchment are 2.2-3.3 kg ha^{-1} (Huppert et al. 1989). A study of two karstic Mississippian limestone catchments in West Virginia (Pasquarell & Boyer 1996) showed that atrazine and its metabolite, desethylatrazine, were detected in more than 50% of samples. Mean atrazine concentrations for the three springs sampled were below 0.1 µg l^{-1} and the maximum detected concentration was 1.2 µg l^{-1}. While these levels are below the US drinking water limit of 3 µg l^{-1}, they are noteworthy because the catchments concerned are primarily cattle grazing land, and the proportion of tillage land (mainly corn and maize), to which herbicides are applied, is only 9% and 16% in the two catchments. The authors attribute the detection of atrazine and other herbicides, despite the small percentage of total land area to which they are applied, to the more efficient access to the subsurface that contaminants are afforded in a karst area compared to other geological environments.

Other pesticides reported from the Iowa study include the herbicides alachlor and cyanazine, and the insecticide fonofos (Hallberg et al. 1985). The West Virginia study (Pasquarell & Boyer 1996) analysed for nine triazine herbicides in addition to atrazine and its metabolites, five of which were detected: metolachlor and simazine were found in more than 10% of samples at all spring sites, while prometon, cyanazine and prometryn were less common. Miller & Eingold (1987) record the occurrence of the nematocide ethylene dibromide (EDB) in a karst aquifer in northwest Florida, and they note that the sandy soils and karstification render the aquifer particularly vulnerable to EDB leaching. Devilbiss (1988) investigates the risk posed by application of the insecticide methoxychlor to crops in a mature karst in Kentucky, monitoring its presence in the seeps and conduit water below a karst swallet.

The range of pesticides used in agriculture is vast, and increasing continually, so a detailed review of each is not feasible here. However, some general comments on how pesticide mobility in karstic situations may differ from that in conventional hydrogeological situations are made in Section 4.5.2.4.

4.4.2 *Point agricultural sources*

Localised agricultural point sources of pollution can range from groupings of animals at feeding troughs or sheltered or shady locations, providing a localised input of faeces and urine, through to major threats posed by badly stored farm wastes such as slurry and silage effluent. Such point sources can occur above any aquifer, where recharge and flow are diffuse, but in the case of karstic aquifers there is the possibility of concentrated recharge and associated conduit flow, and it is where agricultural point sources coincide with karstic concentrated recharge points such as sinking streams and dolines that the threat is greatest. (Photos 4.4-4.7).

Berryhill (1989) gives one example of the problem of animal groupings, at Laurel Creek Cave in West Virginia, where cattle congregated around the cool, damp cave entrance, resulting in pollution of the aquifer recharged by the cave entrance during storm events. Dead animals disposed of in sinkholes are another aspect of this problem. This is ubiquitous at the scale of individual animals dumped of by farmers, reported for example by Aldwell et al. (1983) and White (1988). It has sometimes happened on a larger scale, for example Emmett & Telfer (1994) record that a sinkhole in the Gambier Limestone (South Australia) was used to dispose of thousands of sheep carcasses following a major bush fire in 1959.

Aldwell et al. (1983), reviewing the impact of agriculture on groundwater in Ireland, note that whereas diffuse source pollution is not generally a problem there, groundwater pollution from farmyard runoff, feedlots and piggeries and silage effluent is quite common. Contamination of Irish karst aquifers by these agricultural point sources is documented by Aldwell et al. (1988): the chief problem is with the presence of faecal bacteria, with two water quality surveys in karstified Carboniferous limestone areas of the west of Ireland both showing more than 50% of groundwater sources to contain *Escherichia coli*.

The contribution by Drew (Section 4.7.3) describes the impact of silage effluent on karst groundwater in the Burren Plateau, western Ireland, where silage production is a new phenomenon associated with land reclamation. Another Irish example of agricultural point source pollution comes from Naughton (1983), who documents the pollution of Teesan Springs in a karst aquifer in county Sligo by the disposal of farmyard slurry and silage effluent through a soakpit 500 m from the springs (the underground residence time established by a tracer test with NaCl being only 12 hours). This incident, and other similar incidents in karst areas of Ireland, resulted in elevated manganese levels in the groundwater, presumably arising by release of manganese from the soil or rock by the acid, reducing conditions in the silage effluent plume.

Danchev et al. (1982) note that groundwater sources with elevated nitrate located in Jurassic limestone in north-east Bulgaria are often connected with settlements and stock breeding farms and they suggest that the nitrate here originates from the disposal of great quantities of waste materials. They note that the nitrate in these sources is associated with traces of nitrite ions, sometimes up to several mg l^{-1}, and also generally with increased chemical oxygen demand. The presence of nitrite is taken as diagnostic of an organic point source, and they state that almost no nitrite has been detected in areas outside settlements and far from stock breeding farms.

4.4.3   *Related point sources*

4.4.3.1   *Septic tank effluent*
In many rural areas, mains sewerage systems are lacking and sewage and household effluent is disposed of to land, generally via septic tank systems. While the impact of septic tank effluent is not strictly agricultural, it is nevertheless important in a rural context, so is discussed briefly here.

Septic tanks are primarily settlement chambers which produce a crude primary effluent, and further treatment depends on the adequate functioning of the second component of the system, the percolation area. Groundwater pollution resulting from septic tank effluent is widely documented, with problems arising from the use of soakpits rather than proper percolation areas (e.g. Daly et al. 1993) or from too high a density of septic tank systems (e.g. Yates 1985). However, in karstic areas, even properly constructed percolation areas may not be capable of preventing pollution because of the lack of attenuation associated with solutionally widened fissures. The

Photo 4.4. Spreading animal slurry on a soil only a few cm thick which overlies highly karstified limestones, Burren plateau, western Ireland (photo: Drew 1994).

Photo 4.5. Silage clamp located on almost bare limestone, Burren plateau, western Ireland (photo: Drew 1994).

Photo 4.6. Winter feeding of cattle on the Burren plateau Ireland, causing point concentrations of organic waste (photo: Drew 1994).

Photo 4.7. Cattle rearing in a karst area of the Swiss Alpine region with a swallow hole in the foreground (photo: Hötzl 1980).

main mechanism of attenuation available in a karst aquifer is simple dilution by clean recharge water. As a result, contamination by faecal micro-organisms is the greatest problem, as illustrated by a simple numerical example: a hundred-fold dilution of septic tank effluent by clean rainwater recharge would bring the nitrogen level down to about $0.5 \, mg \, l^{-1}$, well below the drinking water limit for nitrate, but would still leave hundreds of thousands of faecal coliforms per 100 ml. The presence of conduit flow within the karst aquifer means that viable bacteria may travel for hundreds of metres or even several kilometres from the point of entry. The bacterial contamination of groundwater in rural karst areas of Ireland referred to above is due to a combination of septic tank effluent and agricultural effluents (Aldwell et al. 1988).

4.4.3.2 *Agricultural industries*
Although industrial impacts on karst groundwaters are detailed elsewhere (Chapter 5), reference is made here to agricultural industries as these are often an integral part of the agricultural economy of a region so their impact is closely intertwined with that of the agriculture *per se*. For example, in karst areas of Italy, monocultures of grapes or of olives are often found, and Gams et al. (1993) note the occurrence of pollution induced by olive treating plants, which discharge a range of organic compounds including acids and solvents. The sugar cane industry in Cuba has had a severe impact on quality in karst aquifers used for water supply to Havana and hundreds of other towns and small villages. In addition to the impacts associated with crop production (arising from deforestation, soil erosion, fertilisation and pesticide application), highly aggressive waste waters from the sugar factories have contributed to the deterioration in groundwater quality (Rodriguez-Rubio, pers. comm.).

Another agro-industry which can give rise to problems is the dairy industry. For example, for more than 40 years up to 1976 a cheese factory in the Mount Gambier area of south-east South Australia disposed of whey and wastewater in a well in a karst aquifer (Smith & Schrale 1982, Emmett & Telfer 1994). The pollutant plume in this young (Oligocene-Miocene) limestone appears to have followed a diffuse flow path (velocity approx. 40 m/year) rather than following existing solution features, although the acidic whey appears to have caused selective solution activity (Emmett & Telfer 1994). The plume contains sufficient dissolved solids to be identifiable by surface resistivity surveying (Smith & Schrale 1982), and has a high nutrient loading, with concentrations up to $260 \, mg \, l^{-1}$ total N (mainly in the form of ammonia) in 1976, and total load in the plume in 1980 being 14 tonnes of nitrogen and 14 tonnes of potassium.

The forestry industry can also give rise to groundwater contamination: timber mills can result in serious contamination by trace organics. Emmett & Telfer (1994) document point source pollution of the karst aquifer feeding Blue Lake (the water supply for the city of Mount Gambier, South Australia, population 20,000) by various point sources including timber mills: the paper documents one spillage incident involving a leaking fuel oil tank and another involving creosote and copper chrome arsenate wood preservative.

4.5   MOBILITY OF AGRICULTURAL POLLUTANTS IN KARST AQUIFERS

4.5.1   *Introduction*

The distinctive nature of karst aquifers has important implications for the movement of agricultural contaminants into and through the aquifer. Karst aquifers are particularly vulnerable to chemical contamination due to the occurrence of point recharge via sinking streams and dolines, the presence of an epikarstic zone, and the existence of both conduit and diffuse flow within the aquifer itself (Field 1989, Smith 1993).

The method of recharge to karst aquifers makes it particularly easy for agricultural contaminants to gain access. Swallow holes provide direct access points to the aquifer, allowing little or no attenuation. In addition, many karst areas are characterised by bare rock surfaces or thin rendzina soils, and while rock outcrop and thin soil are not characteristics unique to karst areas, the superimposition of these features on bedrock with solutionally widened fissures creates extreme vulnerability to pollution from diffuse agricultural sources. The epikarstic or sub-cutaneous zone may allow rapid lateral movement of diffuse contamination to vadose shafts. It can also provide significant temporary storage for contaminants, which is unrelated to water table measurements and aquifer testing in the phreatic zone, and the contaminants may then be released from this zone by flood pulses (Field 1989).

The concentration of flow in conduits allows rapid transfer through the aquifer, with minimal opportunity for attenuation by adsorption, ion exchange, chemical breakdown or microbial die-off. The short underground residence time also means that very little time is available for remedial action to avoid contamination of drinking water supplies. Within karst aquifers, however, there is a range of behaviour, often reflecting the age of the limestone. In the unsaturated zone, contaminants may move vertically extremely rapidly through solutionally widened fissures or in dual porosity aquifers there may also be a very slow component in the primary pore space. In the saturated zone pollutants may move rapidly along localised flow paths where conduit flow predominates,or they may form a more conventional plume where there is a significant proportion of diffuse fissure flow. Examples of this range of behaviour are given below in relation to particular parameters.

Speleology provides a unique opportunity to study contaminant movement along different flow routes within the aquifer, with the possibility of sampling from cave streams, vadose trickles etc. Work of this nature is outlined in Hardwick & Gunn (1993) and in the chapter contribution by Hardwick and Gunn (Section 4.7.7).

4.5.2   *Behaviour of particular contaminants*

4.5.2.1   *Suspended sediment*

Sediment derived from soil erosion may enter karst aquifers to a greater extent than other aquifers because of the availability of solutionally enlarged openings allowing input and because of the presence of sinking streams bearing a suspended sediment load. This sediment is important *per se*, in that it may cause problems of turbidity in drinking water supplies, but it is also important in that it can provide a method of entry for strongly sorbed micro-pollutants, and possibly sorbed viruses, as discussed below.

4.5.2.2   *Nitrate*

This ion is highly soluble, and will be mobile in many geological situations so nitrate pollution is not specifically related to karst, but the shallow, patchy rendzina soil cover overlying some karst aquifers increases the risk of leaching. The method of transport within the aquifer will depend on its characteristics – i.e. whether the lime-stone has dual porosity with the possibility of retention in the unsaturated zone for considerable time periods (as per the Cretaceous Chalk – see Foster et al. 1986) or whether there is secondary porosity and permeability only (as per the Carboniferous limestone). Thus the time lag between nitrate release from the soil and arrival in the saturated zone could be anything from hours to decades, and the mechanisms must be properly understood before the impact of any agricultural management changes can be assessed. Movement through any unconsolidated material overlying the aqui-fer must also be taken into account: Krothe (1990) notes that fertiliser-derived nitrate from a tillage area is rapidly flushed into an underlying mantled karst aquifer despite the presence of ten metres of clayey unconsolidated sediments, due to the presence of macropores.

4.5.2.3   *Phosphate*

This is less likely to be leached in carbonate areas than in other rock types as it can react with calcium carbonate to form a precipitate of hydroxyapatite (Fetter 1993). The thin rendzina soils found in some karst areas may allow for the possibility of leaching, but the main reason for concern in karst areas is because of the possibility of point recharge and therefore ready entrance to the groundwater system, followed by rapid discharge to springs which may feed surface water bodies susceptible to eutrophication.

The extent to which phosphate is retained within the aquifer would be expected to vary according to the relative importance of diffuse and conduit flow. The Mount Gambier example documented above provides an example of phosphate retention: in a young karstic aquifer with predominantly diffuse flow: whereas the annual nutrient load of the cheese factory effluent is 50 tonnes of N, 35 tonnes of K and 12 tonnes of P, the nutrient load in the pollutant plume in the aquifer was estimated as 14 tonnes of N, 14 tonnes of K but only 0.1 tonnes of P, and it is noted that 'as anticipated, phosphorus is removed from solution in the alkaline environment' (Smith & Schrale 1982).

Even in a conduit flow situation, phosphate may be retained within the aquifer. For example, Wiersma et al. (1986) note that in the Door Peninsula shallow ground-water system in Wisconsin, with travel times from stream sinks of only a few hours, a seasonal pulse in total phosphorus in the recharge waters did not produce a corre-sponding pulse in the spring discharge. Kastrinos & White (1986) carried out bi-weekly monitoring of both diffuse and conduit flow springs in the carbonate aquifer of central Pennsylvania over one year, but found phosphate to be below the limit of detection (0.02 mg·l^{-1}) in all but two samples.

Thus despite the possibility of phosphate entry via swallow hole waters, there is little evidence of phosphate reaching karst springs in significant quantities. As Hardwick and Gunn (Section 4.7.7) note, even sampling within the karstic flow sys-tem does not always reveal a clear relationship of phosphate levels to agricultural applications.

4.5.2.4   Pesticides

These vary greatly in their properties, so it is not possible to generalise about their mobility in karst situations. Where relatively mobile pesticides are involved, the thin soil and overburden associated with many karst aquifers together with rapid movement through solutionally widened fissures will make contamination more likely than in other hydrogeological situations. However the chief mobility issue specific to karst areas is the possibility of entry adsorbed to suspended particles. Some pesticides are immobilised by adsorption to soil mineral or organic particles and therefore have a low risk of entering groundwaters, e.g. paraquat (Lawrence & Foster 1987), but they may enter karst aquifers adsorbed to such particles, via sinking streams or solutionally widened fissures.

One study which attempts to quantify the behaviour of a pesticide in a karstic flow system is the work of Simmleit & Herrmann (1987a, b), which examines the passage of organic micro-pollutants including lindane (γ – BHC) and its isomer (α – BHC) from their input as airborne contaminants in precipitation, through a karstified Jurassic limestone and dolomite aquifer in Franconia, Germany.

Transient rises in lindane concentration and loading were found to coincide with increases in suspended solids and in dissolved humic material (Fig. 4.2), and it was concluded that the transport adsorbed on particulate matter is very important.

Although measurable pulses of the pesticide were found at the springs, it is worth noting that the vast majority was retained within this karst system (Table 4.1a) Re-

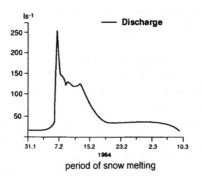

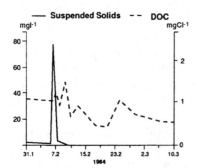
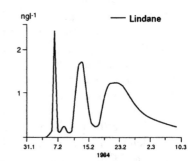

Figure 4.2. Temporal variation of groundwater constituents including lindane at Aufsess spring, Franconia, Germany, during snowmelt (after Simmleit & Herrmann 1987a).

Table 4.1. Retention of BHC pesticides within the Jurassic karst aquifers of Upper Franconia, Germany (Simmleit & Hermann 1987a, b).

a) % retention of BHC within the karst system

Spring	Land use	α – BHC	γ – BHC
During snowmelt in February 1984:			
Aufsess	Agricultural	95.2	87.5
Leipoldstal	Woodland	98.2	96.8
Period 2 February - 6 August 1984:			
Aufsess	Agricultural	99.2	99.4
Leipoldstal	Woodland	99.7	99.7

b) Typical concentrations of BHC in different parts of the karst systems (n.d. = not detected)

	γ – BHC ($\mu g\,l^{-1}$)	γ – BHC / α – BHC
Bulk precipitation	40.0	4.8
Trickling water (2 m depth)	0.2	4.4
Trickling water (7 m depth)	0.1	–
Trickling water (15-20 m depth)	n.d.	–
Spring water (vadose zone)	0.8	3.6
Spring water (phreatic zone)	0.4	2.2

tention was greater in the forested area than in the agricultural area, and was also greater during periods of mainly diffuse flow rather than during rapid throughput following snowmelt. Analysis of percolation waters (Table 4.1b) showed that most of the BHC was retained within the first two metres and the thin rendzina soil layer was thought to be the major pollutant sink (Simmleit & Hermann 1987b).

The work on atrazine by Pasquarell & Boyer (1996) referred to in Section 4.4.1.3 provides an interesting approach to determining the residence time in the unsaturated zone of a karst aquifer, by an examination of the relative concentrations of atrazine (ATR) and its metabolite desethylatrazine (DES), following on work by Adams & Thurman (1991). The ratio of the molar concentration of DES to the molar concentration of DES plus ATR enables identification of periods when atrazine is leached following prolonged storage in the soil and of atrazine by-pass periods, when atrazine transport through dolines and solutionally developed conduits bypasses the process of de-ethylation in the soil.

4.5.2.5 *Micro-organisms*

Faecal micro-organisms are a particular problem of karst aquifers because the lack of filtration within the aquifer and the short underground residence times mean that if organisms manage to pass through or bypass the unconsolidated material overlying the aquifer, they are almost certain to appear in water supplies. As noted in the discussion of septic tank effluent above, the large numbers involved mean that the problem is not solved by dilution. Another reason for concern is that viruses, which are immobilised in many soils by adsorption to clay and iron oxide particles (Gerba

& Bitton 1984), may gain entry to karst aquifers while adsorbed to such particles.

An important aspect of bacterial contamination in karst aquifers is its ephemeral nature (Thorn & Coxon 1992). Temporal variation in contaminant levels is discussed further below; in the case of micro-organisms, where even brief exposure to a pathogenic micro-organism in a water supply could have serious consequences, the need to predict temporal variations is particularly acute.

4.5.3 *Temporal variation*

One important characteristic of agriculturally derived contamination in karst aquifers is the considerable temporal variation in contaminant levels. The greater the degree of karstification and the higher the proportion of conduit flow to diffuse flow, the greater the degree of temporal variation. This is illustrated by Figure 4.3, which shows data from Kastrinos & White (1986) comparing fluctuations in nitrate concentration in four conduit flow springs and four diffuse flow springs from central Pennsylvania. A study of temporal variation in water chemistry in a range of Irish carbonate aquifers found that the greatest variability in nitrate and potassium concentrations occurs where the mean concentrations are greatest (reflecting the degree of agricultural intensification and pollutant loading). However, there also appears to be a hydrogeological control on variation, in that sites with a low mean concentration but a short flow-through time due to a high degree of karstification may show a large temporal variation in these parameters (Coxon & Thorn 1989).

Major recharge events may flush contaminants from the soil and unsaturated zone into the saturated zone. Also, the proportion of spring water originating from point recharge may vary over time. Generally, the higher proportion of point recharge or swallet water in spring flow during flood events will tend to bring poorer water quality, because it gains direct entry to the aquifer, bypassing any protective soil or subsoil layers, although in some instances (discussed below), the swallet water may be of superior quality to diffuse recharge.

Tucker (1982), in her study of agricultural pollution in a karstic catchment in Kentucky, found that pollutant concentrations could be thousands of times greater during high discharge events than during baseflow conditions. In particular, faecal coliform and faecal streptococci counts rose to levels as high as 40,000 colonies per 100 ml during flood events. Thorn & Coxon (1992) note that pulses of bacteria associated with flood events have frequently been reported from Irish karst aquifers. Figure 4.4 provides an example of variation in bacterial levels at a karst spring during a flood pulse; it can be seen that within the space of two hours total coliform counts can vary from zero to over 300 per 100 ml.

The importance of soil moisture levels in controlling temporal variations in faecal coliform levels is stressed by Pasquarell & Boyer (1995). Their work in karstic catchments in West Virginia where cattle are absent in winter (due to the seasonal nature of Appalachian livestock management) found that soil moisture effects had a greater influence on faecal coliform density than the presence or absence of cattle. Dry soil conditions in the late summer limited the movement of bacteria to groundwater, while significant groundwater recharge while the cattle were absent appeared to release bacteria stored in the soil zone.

In some instances, however, a major input of surface water may serve to dilute

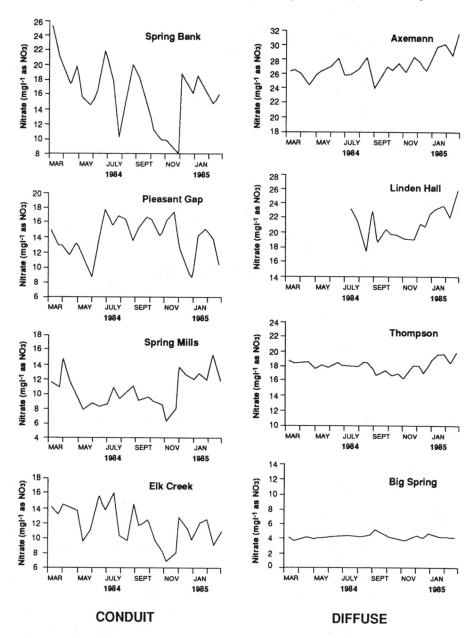

Figure 4.3. Temporal variation in nitrate concentration at conduit and diffuse flow springs in central Pennsylvania (after Kastrinos & White 1986).

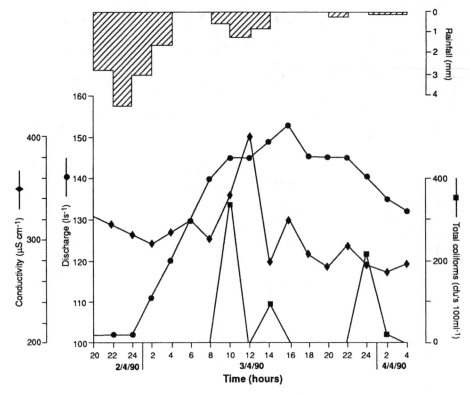

Figure 4.4. Variation in water quality during a flood pulse at a karst spring in County Sligo, Ireland (after Thorn & Coxon 1992).

groundwater contaminants. Wiersma et al. (1986) note that in a spring complex in karstified Silurian dolomite in Wisconsin, nitrate levels are highly dependent on precipitation events: during periods of high flow, the concentrations are low, and coincide with levels in swallow hole waters, while during periods of zero surface water recharge through the sinks, the levels increase 5 to 6 times, and approach to the drinking water limit.

The two sources of recharge water have also been studied in the Big Spring basin in Iowa, where runoff recharge to sinkholes and the conduit system provides constituents typical of surface runoff (e.g. sediment and chemicals of low mobility) while infiltration recharge provides constituents more typical of groundwater (e.g. mobile ions such as nitrate) (Hallberg et al. 1985, Libra et al. 1986). Figure 4.5 illustrates the chemical changes associated with summer discharge events at Big Spring; the pesticide peak corresponds to the suspended sediment peak (as in the work on lindane by Simmleit & Hermann 1987a, b discussed above), whereas the nitrate shows a different pattern, reaching maximum levels during discharge recession when infiltration recharge dominates.

The marked fluctuation in concentration of contaminants during flood events mean that monitoring protocols in karst aquifers must differ from those in conven-

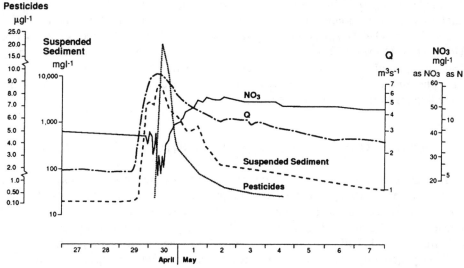

Figure 4.5. Temporal variation in water quality at Big Spring, Iowa (after Libra et al. 1986).

tional aquifers – Quinlan & Alexander (1987) and Quinlan (1988) illustrate this point using data on pesticides and other contaminants from Moth Spring in southern Minnesota and from the Big Spring basin study.

4.5.4 *Prediction of degree of attenuation using tracers*

The use of tracers to predict movement and dispersal of contaminants is by no means unique to karst aquifers. Nevertheless the rapid flow rates in karst aquifers make them particularly suitable to this approach (e.g. Behrens et al. 1992, Hötzl & Werner 1992, Kranjc 1997). Traditionally, tracing in karst aquifers has been concerned with determining underground connections and water flow velocities, so the aim has been to find a conservative tracer which approximates to the water velocity. However, for contaminant studies, the use of non-conservative tracers can enable prediction of the degree of attenuation of different contaminants. As Hacker (1988) comments, sorption, ion exchange and sedimentation, which were previously regarded as negative properties in the search for an ideal tracer, can now be converted into positive factors.

Traditional tracers used in karst flow systems (Rhodamine WT, fluorescein) have been given a new application to predict contaminant mobility in the works of Sabatini & Austin (1991). These fluorescent dyes were used as adsorbing tracers to mimic the behaviour of the pesticides atrazine and alachlor, as fluorescein appeared before and Rhodamine WT after the pesticides. This particular study was in alluvial sands, although such techniques are clearly also applicable in a karstic situation. However some caution is required, as Sabatini & Austin (1991) and Reichert (1991) record evidence that the adsorptive mechanism for the dyes differs from that of the contaminants investigated, and that the relative degrees of adsorption differ significantly in different aquifer materials.

Another group of non-conservative tracers which may be used to mimic contaminant movement are micro-organisms. The use of microbial tracers is reviewed by Keswick et al. (1982). An example of the application of this technique is the work of Henry et al. (1991), who used a range of tracers including two types of bacterium (*Bacillus globigii* and antibiotic-resistant *Escherichia coli*) to monitor the movement of septic tank effluent to groundwater in three different hydrogeological situations including karstified Carboniferous limestone.

4.6   IMPLICATIONS FOR GROUNDWATER PROTECTION PLANNING

A general point in relation to agriculture and groundwater protection (which is not specific to karstic areas) is that agricultural practices are often exempt from planning control. However, this is changing in many countries, and in the European Union, a range of agricultural activities require environmental impact assessment under Council Directive 85/337/EEC, while nitrate from diffuse agricultural sources is now controlled by Council Directive 91/676/EEC. Where activities are not subject to legal control, they may be dealt with by voluntary codes of practice, or grant aid may be dependent on meeting certain standards. Whether they are implemented on a statutory or a voluntary basis, the changes in management practices involved are along the following lines:

– Where deforestation is being carried out in karst areas, whether for agriculture or for timber extraction purposes, it is important that the impact be minimised by good management practices. Kiernan (1987b) outlines the planning requirements for timber harvesting in karst areas, as developed for the Tasmanian forest industry, which involve an inventory of karst features, and the designation of karst reserves within the overall karst area where felling is prohibited. The proposed operational measures (Kiernan 1987c) include no logging on steep limestone slopes or in the vicinity of sinkholes and cave entrances, and the provision of silt traps in situations where there is a risk of karst stream siltation.

– Flooding problems associated with soil erosion may be solved in the long term by improved agricultural management strategies, while in the short term, structures to prevent blockage of swallow holes may be of help. In the Jamaican example cited in Section 4.2.2 (Molina & McDonald 1987), the proposed flood prevention and mitigation measures included and watershed management schemes to minimise soil erosion and the use of debris traps to keep sinks open. The use of gabions (wire baskets filled with rocks) to control surface runoff into sinkholes and swallow holes is advocated by Moore & Amari (1987).

The areas most in need of improved agricultural management can be identified by vulnerability mapping. For example, Tucker (1982) produced a map for the Lost River groundwater basin in Kentucky ranking land according to its soil erosion potential, based on US Soil Conservation Service data, and combined this with a map of groundwater vulnerability based on locations of sinkholes, swallets, lineaments and fracture traces, to identify the areas where the threat to karst waters from non-point agricultural sources was greatest.

Berryhill (1989) reviews US experience on best management practices for controlling non-point source pollution, such as reports by the North Carolina Agricultural Extension Service, and comments on their relevance to karst areas. For tillage areas these practices include conservation tillage with residue cover, crop rotation, reduced input agriculture and integrated pest management, while in grazing areas measures include Planned Grazing (i.e. rotation of pastures with rest periods to allow regeneration) and Livestock Exclusion (i.e. fencing animals away from water bodies and areas subject to erosion).

Management to reduce diffuse source pollution may require reduction or elimination of contaminant inputs. For example, Milde et al. (1988) suggest that substances such as atrazine should not be used in highly vulnerable karst catchments and, for example, its use is now completely prohibited in Germany. The input of nitrate to European karst groundwater is now controlled where such areas are designated as vulnerable zones under the 'Nitrate Directive' (91/676/EEC). However, the time lag between a reduction in fertiliser inputs and a reduction in groundwater nitrate levels can vary greatly, as already mentioned in the discussion of nitrate mobility; the contribution by Libra and Hallberg (Section 4.7.4) also notes that it may take considerable time to observe an improvement in water quality resulting from more efficient nutrient management.

Particular care needs to be taken where land-spreading of manure is being carried out in karst situations. This has recently been acknowledged in Ireland, where current guidelines state that animal slurries should not be applied within 15 m of exposed karst limestone, swallow holes or collapse features. Where karst bedrock is overlain by free draining material (sands, gravels or gravely tills) this must be at least 1 m and preferably 2 m thick where animal slurries are to be applied (Moore 1995).

Agricultural point sources should be somewhat easier to deal with than diffuse sources, given that in many instances they can be eliminated by simple pollution control measures such as constructing adequate storage tanks, rather than requiring fundamental changes to the agricultural system. Berryhill (1989) notes that in the USA large animal production operations are considered point sources which must comply with discharge restrictions, while small facilities should be subject to best management practices including paved animal living spaces and daily removal of manure, together with leak-proof feed and manure storage facilities, runoff control structures and vegetative filter strips for erosion and nutrient control – he cites the example of the Garvin Brook, Minnesota Rural Clean Water Project, where such measures were used to keep feedlot runoff from entering sinkholes (Smolen et al. 1989).

Such controls on particular agricultural activities should form part of an overall protection policy for the karst aquifer. Approaches to groundwater protection in karst aquifers are discussed in greater detail in Chapter 8 and in EC-Cost 65 (1995).

4.7 CASE STUDIES

4.7.1 *Agricultural impact on karst water resources in China*
(Lin Xueyu & Liao Zisheng)

The building of large scale water conservancy projects and the application of fertilisers and pesticides is currently exerting a great impact on karst water resources in

China. Some examples of such impacts are described below (see Fig. 4.6 for locations).

Recharge and discharge conditions are changed by the building of water conservation projects on the surface. The interchange between surface water and groundwater is one of the notable characteristics of karst areas. For example, in Shanxi province – one of the main karst areas in China – there is a very close hydraulic relationship between surface water and karst water (65% of surface water is replenished by karst water). The recharge, discharge and distribution conditions of karst water have been influenced tremendously by the construction of surface reservoirs and engineering works associated with bringing water into the area. Because of building the reservoir on the upper reaches of Zi river in the karst area of Shandong province, the seepage discharge from the karst river course between Zhuya and Huang Zhuang (46 km long) has been reduced by 47%. This has caused the groundwater level to fall by more than 10 m on the lower reaches and 2000 shallow wells have been abandoned.

Of course, karst water resources can be increased due to reservoirs built on karst river courses, where these result in a decreased discharge of seasonal flood waters and increased seepage from reservoirs, rivers and ditches. For example, before building the Haizi reservoir on Ju river of Pingu county, Beijing city, karst ground-

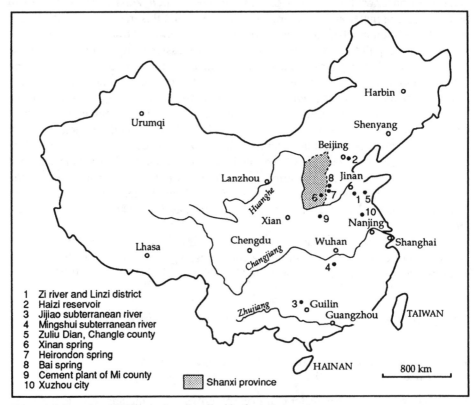

1   Zi river and Linzi district
2   Haizi reservoir
3   Jijiao subterranean river
4   Mingshui subterranean river
5   Zuliu Dian, Changle county
6   Xinan spring
7   Heirondon spring
8   Bai spring
9   Cement plant of Mi county
10 Xuzhou city

Figure 4.6. Location of Chinese sites described in the text.

water discharged into Ju river. Since the reservoir was built, $2.5 \cdot 10^7 \, m^3$ of water in the reservoir has leaked to replenish the karst aquifer.

2. The construction of underground check dams and water impoundment engineering in subterranean rivers changes the run-off and distribution condition of karst water and expands the water storage volume. For example the building of a 9 m high dam in Jijiao subterranean river of Duan county, Guanxi province, for irrigation, resulted in groundwater levels being raised by 34 m, and converted the subterranean river from vadose to phreatic flow (Fig. 4.7).

Moreover, there is a check dam 33.3 m high and 54 m long in Mingshui subterranean river in the south-east of Hubei province, which forms an underground reservoir $2.8 \cdot 10^8 \, m^3$ in volume. At the outlet of the subterranean river, a 5000 kilowatt power station was built. The reservoir provided 1700 ha of irrigation land and provides drinking water for 4000 persons.

3. In many karst areas in north China, a major programme of irrigation well drilling and large-scale water pumping has caused extensive drawdown in regional groundwater levels. This has induced many environmental problems. In the last twenty years, the groundwater level in the main karst water development area in north China has been drawn down by 10-20 m. This depletion of the aquifer has resulted in some springs dramatically decreasing in flow or disappearing in famous karst scenic spots. Shallow wells have been abandoned and engineering works to import spring water have been necessary. Land subsidence and karst collapse have also occurred.

For instance, in Zuliu Dian district of Changle county, Shandong province, the ground water level has been drawn down 110 m in the last twenty years, and water can only be pumped from 3 of 30 wells, making water supply more difficult for domestic and industrial use for a population of several tens of thousands.

Groundwater has become polluted in karst areas due to intensification of vegetable and fruit production. A consequence of this is increased water usage for irrigation

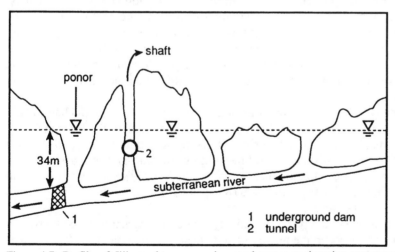

Figure 4.7. Profile of Jijiao subterranean river and water engineering structures, Duan county, Guanxi province, China.

and the application of fertilisers and pesticides. Karst water pollution is also caused by bad well casing or bad well cementation.

Statistical data for the Linzi district of Zibo city, Shandong province, show that 535.5 kg of fertiliser has been applied per hectare of farmland per annum, with nitrogen accounting for 70% of the total. The nitrate in karst water has been increasing yearly as a result of leaching down from the fertiliser remnants. At the beginning of the 1960's, the average nitrate content in karst water of this district was 3.98 $mg \cdot l^{-1}$ NO_3, but by 1989 it had risen to 26 $mg \cdot l^{-1}$, and the national standard of drinking water quality (45 $mg \cdot l^{-1}$) has been exceeded in some areas of Linzi district.

Toxic substances have been entering karst water because of a large amount of pesticide application for fruit growth. The pesticides BHC and DDT have been detected in various karst waters including Xinan spring (Shanxi province), Heirondon spring (Hebei province) and Bai spring (Xingtai city). In the cement plant of Mi county (Henan province), 0.0235 $mg \cdot l^{-1}$ of BHC and 0.005 $mg \cdot l^{-1}$ of DDT have been detected in karst water supply wells.

Water quality problems have also been caused by irrigation with waste water, for example in the buried karst area at Xuzhou city of Jianshu province, where the overburden is thin and waste water irrigation has given rise to pollution of an extensive karst water body by ammonia, nitrate and nitrite.

4.7.2 The impact of land clearance on karst groundwater: A case study from Buchan, Victoria, Australia (Mark Ellaway, Brian Finlayson & John Webb)

The hydrological and water quality effects of land clearing and conversion to rural land uses such as broadacre agriculture are often quite subtle and develop only slowly over time. Usually hydrological and water quality data are lacking for the pre-clearing condition. Because the changes are not dramatic, it can be difficult to separate the impact of land clearance from natural variations and from other impacts such as waste dumping and tip leachates. Groundwater contamination only becomes an issue when the quality of the water has deteriorated to such an extent that there is an adverse effect on users and it is only at this point that investigations begin. In Australia, studies of the effects of agriculture on groundwater have been concentrated almost exclusively on salinization which can develop following the removal of the native tree cover (Smith & Finlayson 1988) with little attention having been paid specifically to limestone terrains.

This case study is based on a small karst area (about 100 km^2) at Buchan in East Gippsland, Victoria, Australia. The Early Devonian limestones of the Buchan Group consist of two major sequences of fairly pure, competent limestone separated by a formation of lime-rich mudstones. The limestones were deposited in a basin formed within an extensive series of acid volcanic rocks which today underlie and surround the limestone. Climate is warm temperate (Köppen Cfb) with rainfall varying across the area from 800-1500 mm.

The main limestone outcrop was selectively cleared for agriculture and converted to freehold title very early in the history of settlement (approximately 150 years ago) because the soils developed on it are superior to those on the surrounding acid volcanics. Some smaller outliers of the limestone have not been cleared and it is therefore possible to carry out a comparative study to gauge the effects of clearing. The surrounding volcanics remain in public ownership as state forests. This area has a sparse rural population engaged mainly in cattle grazing with some sheep and minor cropping and no industrial activity other than sawmilling. Groundwater use in the area is quite limited; the only township in the area, Buchan (population about 200), has a relatively reliable surface source of water for domestic use, so the quality of the groundwater is, for the majority of people in the area, of little concern, hence the lack of historical data. On the limestone most of the important cave sites have been reserved as Crown land and are managed by a government agency. There is only one privately owned tourist cave. This history of government reservation of cave sites has provided valuable protection when compared with caves on private land which have often been used for rubbish tipping or deliberately filled in to avoid the risk of having livestock fall in to them.

Water quality was monitored in springs and cave streams on the cleared land and on the uncleared limestone outliers over a period of six years (Ellaway & Finlayson 1984, Finlayson & Ellaway 1987, Ellaway 1991) using standard collection and analysis techniques. It was subsequently found that the sites could be sorted into three groups on the basis of water quality. These are summarised in Table 4.2.

Sites in group 1 (Table 4.2) are located on limestone outcrops still retaining a dense cover of native Eucalyptus forest with little or no human impact. They are here considered to be representative of an undisturbed catchment and are taken to indicate background conditions.

The group 2 sites are karst springs which drain catchments which have been cleared of native forest and are used mainly for grazing. Median Ca^{2+} and log (PCO_2) are higher than in the group 1 sites and probably reflect higher CO_2 concentrations in the soil atmosphere generated by increased vegetation productivity. The pastures have been improved with high yielding exotic species which are regularly fertilised with superphosphate. Higher temperatures in the cleared catchments arise from higher soil temperatures as a result of increased solar radiation receipts at the ground surface. The higher chloride concentrations in the cleared catchments are believed to be caused by leaching of stored chloride in the soil profile (Smith & Finlayson 1988) leading to a surplus of molar Cl^- over molar Na^+ (lower ratio).

The third group of sites drain a cleared area like those of the second group but this catchment has the Buchan district refuse tip located in it. The tip is sited topographically above the limestone in unconsolidated Tertiary fluvial gravels and leachate from the tip enters the conduits in the limestone aquifer. The connection has been proved by dye tracing. The tip exerts a dramatic influence (Table 4.2). The observed PCO_2 levels are unlikely to have arisen from natural soil processes and result from the high levels of organic material decomposing in the tip. The higher CO_2 levels have led to more carbonate dissolution and higher Ca^{2+} values. Contamination from the tip leachates has also caused the higher chloride and lowered Na : Cl molar ratios.

The Buchan karst groundwater system has a surprisingly high level of water quality variability within a quite small geographic area and most of this variability can be explained in terms of anthropogenic influences. The extent of the influence is significant given the low population density, relative remoteness and low level of human activity. The effects which can be attributed to land clearance and agriculture

Table 4.2. Median values of selected water quality parameters for sites under: 1) Uncleared native forest, 2) Cleared agricultural land, and 3) Cleared agricultural land where the groundwater receives leachate from a refuse tip.

Catchment condition	Ca^{++} (mg l^{-1})	Cl^{-} (mg l^{-1})	Na : Cl (molar ratio)	log PCO$_2$	Temperature (°C)
1. Uncleared (3 sites)	26	19	1.00	−2.40	13.5
2. Cleared (3 sites)	98	39	0.75	−2.18	15.5
3. Cleared + Garbage Tip (5 sites)	167	254	0.54	−1.51	16.5

are relatively minor when compared to the impact of the dumping of domestic refuse from one small rural town.

4.7.3 Impacts of agricultural land use and farm management changes on groundwater in the Burren Karst, western Ireland (David Drew)

The Burren Plateau of County Clare is a classic example of a plateau karst (limestone) characterised by patchy, thin soils, a lack of surface drainage and in the instance of the Burren a rich floristic, archaeological and landscape heritage. The plateau, which abuts the Atlantic Ocean, extends over an area of 370 km^2 and has an altitude of 250 m. The climate is temperate maritime with a mean annual temperature of 10°C and a mean annual rainfall of 1550 mm, of which approximately 65% recharges groundwater.

Until recently land use and agricultural practice in the Burren conformed to that in other less favoured areas of Europe with small land holdings, traditional practices and relatively low standards of living being the norm. However, since Ireland's accession to the European Union and in particular as a result of Common Agricultural Policy initiatives, attempts have been made to raise farm incomes and to modernise agriculture in areas such as the Burren. Beef cattle rearing is the dominant agricultural activity in the Burren and the shift towards more intensive agricultural production has meant more fodder is required, usually as silage, and in turn this requires manageable fields which can be efficiently fertilised and the fodder crop harvested. Thus, radical reclamation of land involving the creation of 'new' bulldozed fields has taken place with the aid of grants.

Intensive reclamation on the Burren has involved uprooting and removing scrub and bulldozing loose boulders, bedrock outcrops and stone walls. The ground is levelled and scrub is heaped for burning or to form a new field boundary. Finally topsoil is spread over the new field, often only to a depth of a few centimetres. The soil used is commonly derived from pre-existing patches or soils lodged in the rock fissures (Photo 4.8). Less commonly soil is imported from surrounding fields or from longer distances. The soil is re-seeded with c. 20 kg perennial ryegrass seed, 5 kg white clover and 5 kg Timothy per hectare. Annual Italian ryegrass is sometimes sown with the perennial grass to cause rapid growth. Fertiliser is spread with the grass seeds.

Photo 4.8. On the Burren plateau the soil is lodged mainly in the rock fissures. In connection with the intensive reclamation loose boulders are removed and new bulldozed fields are created. The soil from the fissures are used to produce a few centimetre thick soil cover (photo: Drew 1994).

Photo 4.9. Effluents from a poorly managed silage clamp escape and infiltrate into the karstic underground polluting the groundwater (photo: Drew 1994).

The new grass swards are maintained by topping and fertiliser application to deter wild grasses from becoming established. In excess of 7% of the area of the plateau was reclaimed in this way between 1981 and 1995 (Drew & Magee 1994).

In part as a consequence of the development of these new pastures and in part due to the availability of grants for silage clamp construction, the practice of silage making has increased markedly over the same period. The use of silage rather than hay as a winter feedstuff in the Burren is desirable given the dampness and unpredictability of summer weather. However, the character of traditional Burren pastures, field size etc., effectively confined silage making to the lowland, dominantly dairy farms of the valleys until recently. If silage clamps are not well managed, the effluent produced in the weeks following clamping may escape and enter water courses; silage effluent is an extremely potent and concentrated pollutant of water (Photo 4.9). The Burren, as with any comparable karst region, is extremely vulnerable to water pollution from silage effluent because any effluent can seep underground via the fissured limestones within a few metres and thereby quickly reach the underground water systems. The thin and patchy soil has minimal ability to purify effluent whilst, when underground, the water flows to springs via fissures or caves which also lack filtration capacity.

Much of the Burren is wholly dependent on this groundwater for water supply and should a supply become contaminated (usually a lengthy occurrence) no alternative supplies are available. The main risk period (summer) is also the maximum water demand period especially for tourism (Drew 1996).

The imposed environmental changes and their likely and possible consequences on ground water are presented in the flow chart below (Fig. 4.8). All of the changes listed with the exception of the 'subsequent induced environmental changes?' have already been observed. For example, some reclaimed areas, formerly free draining, now have standing water during wet periods, whilst contamination of groundwater with what is presumed to be silage effluent, has become a common summer occurrence. In one instance this has rendered the waters of a lake fed by karst springs and a major source of water supply in the south-east Burren, unusable during the height of the tourist season. The extreme sensitivity of this karstic environment to stress and the high degree of vulnerability of the limestone aquifer have allowed agricultural 'improvements' to lead directly and indirectly to a deterioration in ground water quality.

4.7.4 Soil erosion and water pollution in an intensive vine cultivation area: The Entre-deux-Mers example (Gironde, France) (Philippe Audra)

4.7.4.1 Introduction

The Entre-deux-Mers is a wine-growing locality forming a triangle between the Garonne and the Dordogne, to the east of Bordeaux (Fig. 4.9). It is a low plateau, whose altitude varies between 50 and 100 m and into which are cut many steep-sided little valleys. Thus the landscape exhibits large rounded hills, with a flat top. The karstifiable rocks reach a thickness of some metres to several tens of metres.

These consist of little-compacted horizontal, Stampian limestones which are affected by some local fractures (Boyries 1987). The whole plateau is covered by alluvial clay-gravel formations, which can reach a thickness of about 10 m.

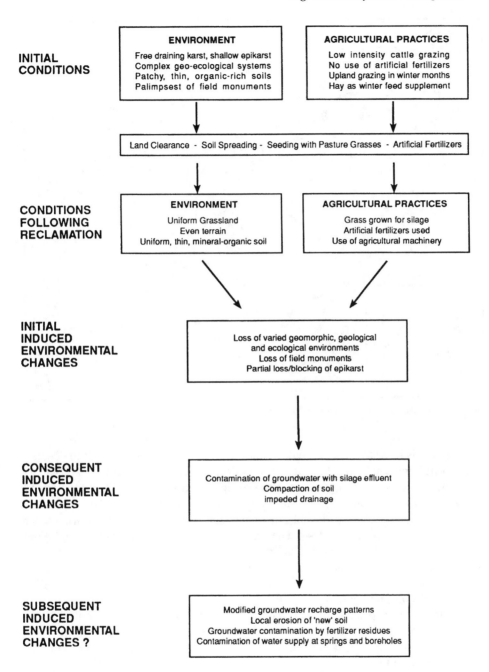

Figure 4.8. Summary of environmental changes in the Burren and their likely consequences.

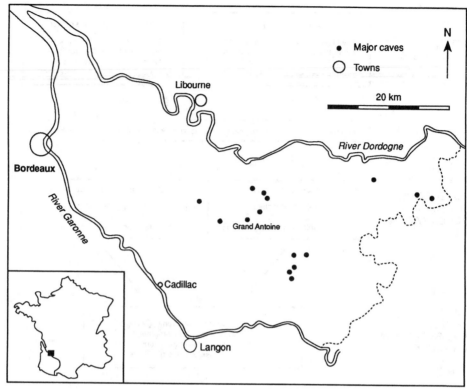

Figure 4.9. Location of the Entre-deux-Mers, with the main cave systems.

This area is intensively karstified: dolines are sometimes so dense that the coalescence of these funnels, reaching ten metres deep, resembles a 'mini cockpit-karst'. Rivulets, formed on the clay cover, often disappear in sinkholes. They feed underground streams developing only a few metres under the vineyards. The Réseau du Grand-Antoine, which includes 11 km of passages, was the largest known French cave in the 1950's (Audra 1994). The karst is young, mainly Quaternary (Boyries 1987, Klingebiel et al. 1993). Its natural rapidity of evolution is however substantially increased by human activities.

4.7.4.2   Human intervention
The karst evolves especially during flood periods. Fed by rivulets, nearly all the cave systems are put under hydraulic pressure. The floods accentuate destabilisation and undermining of the clay covers. Catastrophic openings of sinkholes or collapse-dolines are frequent.

As in the other localities of the Bordeaux area, most of the land is devoted to vine cultivation, except for sloping valleys and some scarce relict woods, which cannot be cultivated because of the doline density. As the vine is a sturdy plant, it can thrive on the clay-gravelly cover, even without soil. So, to intensify and facilitate mechanisation, fields are sometimes levelled with bulldozers over dozens of hectares, thus clearing the superficial soil horizons. Such practices intensify soil erosion, particularly because this area is subjected to especially violent and regular storms. The soil composition favours superficial rill wash and gully erosion. Fine particles are carried away, even on moderate slopes. Eroded patches sometimes reach a thickness of 20 cm. In this case, a field of some hectares can lose tens of cubic metres of sediment during a single storm. These clays fill up the drainage channels, cover fields or roads located downslope and block the karst depressions. Suffosion dolines quickly appear, showing an acceleration of the process of clay removal into the karst.

4.7.4.3 *A destabilised environment*
This soil erosion phenomenon, linked to the particular nature of the local geology and climate, has occurred for a long time. However, it is presently seriously intensified by these excessive agricultural practices. Infiltration of water into the clays decreases to the advantage of superficial rill wash, so overland flow and flooding become more violent and frequent. Parts of cultivated fields are pockmarked by collapse dolines. All of the locality is degraded by soil clearing. If the economic situation imposed an agricultural conversion, the possibility of using crops which are more demanding of soil quality could be strongly compromised.

As for the karst, it is subjected to increased sediment fluxes. However, blockages remain localised, the deposits being continuously reworked. It seems that the karst possesses a peculiar regulation capability in this regard. However, the karst waters are especially affected by pollution: in this rural area, despite being highly populated, treatment of domestic effluent is practically non-existent. To this is added an intense use of fertilisers and pesticides, whose containers are generally thrown in the closest rivulet, or directly in the karst. Indeed, bulky waste (rubble, scrap metal, chemical drums...) is not collected and contributes to the blocking of a number of dolines. This situation is unfortunately not currently improving, local people having a very conservative mind, with few being concerned by environmental questions.

4.7.5 *Impacts of agriculture on water quality in the Big Spring basin, NE Iowa, USA* (Robert D. Libra & George R. Hallberg)

Water quality and discharge, along with agricultural practices, have been monitored since 1981 in the 267 km² Big Spring groundwater basin, located just west of the Mississippi River. Precipitation which recharges the Ordovician Galena carbonate aquifer within the basin discharges back to the surface via Big Spring. This aquifer is overlain and confined by shales in the western part of the basin, but elsewhere is unconfined and mantled by a thin, weathered cover of Pre-Illinoian glacial deposits and/or Wisconsinan loess. Where the aquifer is unconfined sinkholes occur, and surface water in about 10% of the basin drains to these karst features. Infiltration recharge is more important than sinkhole-related recharge. The Galena aquifer exhibits moderate karst development, with both 'conduit' and 'diffuse' flow systems present. Total relief in the basin is about 130 m. Precipitation averages about 82 cm a^{-1}. Typical base-flow discharge rates at Big Spring are 0.7-1.0 m³·s^{-1}, with peak flows of over 6 m³·s^{-1} occurring 1-2 days after large rainfalls or snow melt.

Land use within the basin is essentially all agricultural, and includes about 300 farms. About half of the area is cropped to corn, and about one-third to alfalfa. An additional 10% is pasture, small dairy and hog operations are common. Fertilisers (both purchased and manure) and herbicides are applied to corn primarily during the spring planting period. Nitrate-nitrogen (N), along with the most commonly used corn herbicides – particularly atrazine – are the major contaminants resulting from agriculture. Water samples from Big Spring and up to 40 other surface- and ground-water sites have been analysed for nitrate-N, herbicides, and other parameters on a weekly-to-monthly basis. Discharge is monitored continuously at Big Spring, two surface-water sites, and four tile-drainage outlets.

Agricultural records from the basin indicate a roughly 3-fold increase in nitrogen fertiliser applications between 1960 and 1980. Water quality data from Big Spring, which integrates the groundwater discharge from the basin, showed a similar increase in nitrate-N concentrations, from about 3 mg·l^{-1} in the 1950's-1960's to just below the US EPA 10 mg·l^{-1} drinking water standard by the early 1980's. The initial investigations in the basin also indicated that the nitrate-N discharged by the hydrologic system typically is equivalent to one third of that applied as fertiliser, and exceeds one-half in wetter years. In addition, significant nitrate-N loss occurs in-stream before surface waters exit the basin, suggesting actual losses from agricultural lands are greater than the percentages given. Atrazine is present (in concentrations above 0.1 µg·l^{-1}) in surface- and groundwater year-round, except for extended dry periods. Detectable concentrations of other corn herbicides are more closely linked to the spring application period, but also occur year-round following significant recharge events. These findings, particularly those relating to nitrogen, resulted in an intensive education and demonstration effort, beginning 1986, aimed at reducing fertiliser and pesticide inputs while maintaining crop yields, and therefore improving both economic and environmental performance of basin farmers. This effort has achieved considerable success, with average nitrogen application rates on corn falling by $^1/_3$ over the decade, with no decline in yields.

While significant reductions in nitrogen applications have been documented, relating these reductions to improvements in water quality remains problematic. The effects of nitrogen reductions, occurring incrementally over a decade, are at present largely lost in the year-to-year variations that result from climatic variability – particularly the variability that controls recharge. Figure 4.10 shows annual precipitation, groundwater discharge from Big Spring (a reflection of recharge in this responsive groundwater system), and annual flow-weighted mean nitrate and atrazine concentrations (very similar trends in discharge and contaminant concentrations occur at other monitored ground- and surface watersheds, which vary in size from 1 to 1500 km^2). On an annual basis, nitrate concentrations rise and fall with the volume of water moving through the soil and into the groundwater system. Concentrations at Big Spring generally declined from 1982 through 1989, but so did the overall groundwater flux, which fell from about 50 million m^3 in 1982-1983 to 20 million m^3 in 1989, the second year of extreme drought. While some of the decline in nitrate concentrations may reflect the decrease in nitrogen applications, the magnitude of this effect cannot be separated from the concentration decline caused by a decrease in water-flux. Considerably wetter than average conditions occurred following the drought, including 1993, when major flooding occurred across the upper Midwestern states and discharge from Big Spring was 70 million m^3. Nitrate concentrations increased dramatically during this period. This response resulted from both the increased water volume passing through the soil and groundwater system, as well as

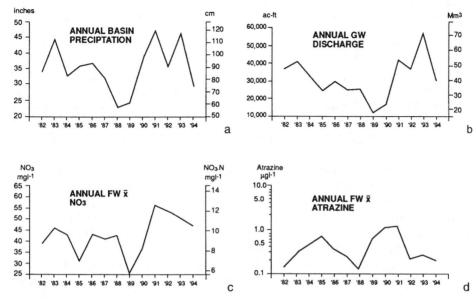

Figure 4.10. Summary of annual: a) Precipitation, b) Groundwater discharge, c) Nitrate-N concentrations, and d) Atrazine concentrations for the Big Spring basin, Iowa.

from the leaching of unutilised nitrogen left over from the drought. Any improvement in water quality that may have resulted from decreased nitrogen applications was again lost in the effects caused by the extreme climatic variations.

During 1983, a government acreage-reduction program provided the opportunity to evaluate the results of a one-year reduction in nitrogen applications of about 40%, a larger reduction than has accrued year-by-year over the last decade. Statistical analysis of the relationship between discharge and nitrate concentrations at Big Spring suggests the significant decline in concentrations during 1985 was related to the major reduction in nitrogen inputs during 1983. Figure 4.10 also shows annual flow-weighted mean atrazine concentrations. Unlike nitrate, annual mean atrazine concentrations do not rise and fall with the overall water-flux in the system, although concentrations do increase, within individual years, following rainfall events. Significant increases in atrazine concentrations occurred following the drought of 1988-1989, and this may reflect the mobilisation of a significant amount of atrazine not degrading during the previous dry years.

The long-term record available at Big Spring illustrates the complexity of documenting the improvements in water quality that result from more efficient input management. Declining nitrate concentrations will ultimately result from more efficient nitrogen management. However, decreases in nitrogen inputs of one third in the Big Spring basin have been overshadowed by several-fold variations in annual recharge.

4.7.6 *Agricultural impacts on Cuban karstic aquifers*
(L.F. Molerio León & J. Gutiérrez Díaz)

4.7.6.1   *Introduction*
Cuban groundwaters occur, predominantly, in karstic terranes. Up to 65% of Cuban territory is underlain by highly karstified carbonate rocks (Fig. 4.11). On the other hand, economic development of the country is basically sustained by an efficient agriculture. Irrigation, drainage and the use of fertilisers and pesticides are the main practices allowing for sustainable success in this field. However important water quality problems are often related to the karstic nature of the main aquifers, those resulting from agricultural land use being amongst the most significant.

Karstic landscapes suitable for intensive agricultural development are mainly gentle coastal flatlands where groundwater lies at shallow depths, commonly not deeper than 30 m, in unconfined, thin soil covered aquifers. A very thin and occasionally, negligible unsaturated zone is the only geological barrier, (almost ineffective) to pollution. Intensive karst development in this unsaturated zone allows for two main infiltration paths to groundwaters.

1. Fast concentrated recharge takes place along vertical groundwater-connected shafts or following horizontal or sub-horizontal caves. Surface flow to those karstic features occurs in two main ways: a diffuse overland flow and a concentrated or channelled flow with streams entering caves.

2. Slow, diffuse recharge takes place following pores, joints and fissures that are less karstified.

Nitrate contamination of groundwaters is associated with both point and diffuse sources. Point sources are likely to be effluents from latrines, animal wastes and from other direct discharge of high nitrogen waste waters, including the incorrect storage and manipulation of nitrogen fertilisers. Diffuse sources are mainly regional fertiliser use associated with irrigation practices. Because of the already known implications for public health, drinking water standards accounts for nitrate concentrations. The Cuban drinking water standard for nitrate is 45 mg l^{-1}, the same as that recommended by WHO's Guidelines for Drinking Water.

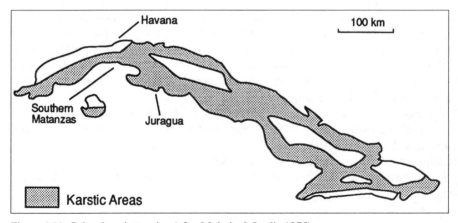

Figure 4.11. Cuban karstic terrains (after Molerio & Leslie 1975).

4.7.6.2 *Nitrogen sources*

The nitrate concentration in Cuban groundwaters is due to the following sources (Gutierrez et al. 1982):
− Contributions from nitrogen-rich leachates from intensive agricultural areas;
− Point-sources of septic tanks and latrines and from husbandry concentration areas;
− Infiltration of poorly treated human and animal wastes.

Cuba is mainly an agricultural country. Some 48% of the country is dedicated to agricultural purposes and approximately one million hectare are under irrigation. Since 1960, significant changes in the political and social sphere have taken place. Therefore, there has been a qualitative change in the use of chemical fertilisers and in particular, those of nitrogenous type, thereby increasing nitrogen concentration in the environment.

With some variations, between 1965 and 1990, nitrogen application rates nationally averaged 120-150 kg ha^{-1}·a^{-1}, comparable to application rates in relatively developed countries. Table 4.3 shows the values of each of the sources of nitrogen affecting groundwater quality in two karstic aquifers of Western Cuba (Fig. 4.11). In both of them, intensive agricultural activities started in the 1970's. Nitrate peak concentrations occurred after ten years of agricultural practices, showing a 4-6 times increases with respect to the original values. In other words, this represents variations ranging between 15-28 mg·l^{-1} of nitrates.

Table 4.4 shows the nitrate contribution from irrigation waters in the two selected aquifers accounted for by for the original nitrate concentration in those waters and the irrigation volumes. This fact is one of the most important causes of increased nitrate concentration in groundwaters because of the recirculation of nitrified groundwater.

Another matter of concern is the potential impact of groundwater drainage on surface waters, since the groundwater inflow is known to be a considerable source of nitrogen in lakes, reservoirs, estuaries and coastal areas The main effect is the change of the trophic status of surface water bodies which in turn causes eutrophication problems.

Table 4.3. Sources of nitrogen contribution to groundwaters in two Cuban karstic aquifers in kg·ha^{-1}·a^{-1} (after Gutiérrez et al. 1982).

Aquifer	Year	Atmospheric fixation	Precipitation (rainfall)	Human and animal wastes	Inorganic fertilisers
Southern Matanzas	1972	20-30	3-5	10-20	20-40
	1979	30-40	3-5	20-30	40-60
Juragua	1970	20-30	3-5	10-20	40-60
	1979	30-40	3-5	20-30	60-100

Table 4.4. Nitrogen load from irrigation waters (after Gutiérrez 1983).

Aquifer	Nitrate in irrigation waters (mg·l^{-1})	Nitrate supplied by the soil (kg·ha^{-1}·a^{-1})
Southern Matanzas	2	12
Juragua	32	160

4.7.6.3   Conclusions

Nitrate groundwater contamination in Cuba is always related to the agricultural economic framework. Evidence of an increase in nitrate concentrations have been observed in monitoring wells in the last decades. The karstic nature of Cuban territory and of its main aquifers means that there is a high vulnerability to nitrate pollution.

A knowledge of the factor involved in leaching process allows an assessment of the magnitude of nitrogen supply to groundwaters and this is estimated at 40-80 kg $N\ ha^{-1}\cdot a^{-1}$. Local variations are related to the type of water used for irrigation, the kind of crops and the nitrate concentration of irrigation waters.

In Cuba, where high level nitrate groundwaters are used for irrigation purposes it becomes evident that the nitrogen loads supplied from irrigation are relevant in computations and must be included in any mathematical model. Values as high as $30\ kg\ N\ ha^{-1}\cdot a^{-1}$ have also to be considered in fertiliser balances.

Once the nitrogen enrichment process is established concentrations will tend to fluctuate considerably about the computed concentration because of the annual vagaries of rainfall and the effective recharge, associated with local intense precipitation or major droughts typical of tropical climates. Recovery processes are slow and under Cuban conditions, would only reduce by between 10-15% nitrate concentrations in about 10 years, figures that are much lower than the estimates for other climates or type of aquifers.

4.7.7 *Agricultural impacts on cave waters* (Paul Hardwick & John Gunn)

4.7.7.1   Introduction

The majority of investigations of the impacts of agriculture on karst groundwaters have considered the karst aquifer as a lumped model or 'black box', monitoring agricultural inputs to a delimited groundwater basin and gauging the aquifer system response at spring or rising output sites. However, a small number of studies have considered the response of the vadose sub-system to such inputs, utilising natural caves as a means of gaining undisturbed access for monitoring and sampling. This section considers the impacts of agriculture on levels of the plant nutrients nitrogen and phosphorus in cave recharge waters and on potential inputs of agrochemicals normally immobilised on soil materials.

4.7.7.2   Plant nutrients

There is considerable evidence from mainland Europe that nitrate-nitrogen levels are enhanced in autogenic recharge to caves beneath chemically fertilised arable land and orchards, when compared to caves beneath woodland (Stefka 1990, Bolner et al. 1989, Bolner & Tardy 1988a, b). This is not surprising given the conservative nature of the nitrate ion. Once the ion enters groundwater it will only undergo further change under anaerobic conditions where bacterial denitrification is significant.

In Britain, the only known detailed investigation of nitrate levels in cave recharge is that by Hardwick (1995) in the P8 Cave, Castleton, Derbyshire. The study revealed that autogenic recharge to a cave beneath land treated with chemical fertilisers and with digested sewage sludges, contained mean and minimum nitrate concentrations of 38.6 $mg\cdot l^{-1}$ and >20 $mg\cdot l^{-1}$ respectively, both of which are an order of magnitude greater than concentrations in recharge to caves beneath unimproved pasture, heather moorland or limestone pavement elsewhere in Britain. Temporal variation was also evident, the pattern following the 'classic flush' of nutrients during late summer and autumn exhibited by surface rivers and streams (e.g. Klein 1980). The only comparable levels of nitrate to P8 were in Swildon's Hole in the Mendip Hills, where Hardwick (1995) recorded a maximum concentration of 114 mg·l^{-1} in a sample from an aven known to have been regularly polluted by slurry from dairy farm waste (Barrington & Stanton 1977).In contrast to nitrate, there have been very few investigations of phosphorus in cave recharge or in karst groundwater. This is surprising as phosphorus is an important plant nutrient occurring in chemical fertilisers and organic sewage sludges (Hall 1988); is known to impact on calcite deposition (Reddy 1977, Morse 1983, Lorah & Herman 1988); and is also considered the important limiting nutrient on algal growth in rivers and streams (Vollenweider 1982). In an investigation of phosphate concentrations in recharge to British caves, Hardwick (1995) found that phosphate concentrations in cave recharge were massively greater than natural inputs from rainfall and generally increased following applications of sewage sludge to the overlying field. However, some sites in the cave discharged water containing levels of the ion seemingly unrelated to agricultural applications. These are probably related to multiple adsorption/desorption mechanisms in the overlying soils; to geochemical changes from insoluble forms of P to soluble forms; and to soil percolation waters flushing soluble P into the cave.

4.7.7.3 *Transmission of agrochemicals to caves*

Increased background levels of plant nutrients or agrochemicals in groundwater can follow inputs of clastic sediment to cave passages, the contaminants slowly leaching into underground watercourses (Tucker 1982). Such inputs may occur not only via sinkholes and other point sources (Tucker 1982) but also areally. Areal inputs occur where overlying soils and drifts are heterogeneous. At Castleton, Derbyshire, agricultural soils have developed on loess parent materials, and are characterised by macropore structures which connect the soil surface with the underlying bedrock thereby forming extensions of karst drainage networks to the land surface (Hardwick 1995). Measurements of ^{137}Cs concentrations in soil materials and in cave sediments inwashed by percolation waters indicate that clastic sediments are transported through the macropore networks and bedrock to cave passage some 17 m beneath the surface over a period of less than 30 years (Hardwick 1995). The migration velocity in the order of 0.5 m·a^{-1} is an order of magnitude greater than reported by other workers on other lithologies (e.g. Denk & Felsmann 1987). It thus appears possible that agrochemicals normally considered immobilised on soil materials could be transported into cave environments thereby potentially increasing background concentrations in underground waters.

4.8 CONCLUSION (Catherine Coxon)

The above review, together with the contributions on specific topics, demonstrates the wide range of impacts on karst groundwater associated with agricultural activities. The initial impact associated with the clearance of forest vegetation for agriculture can be very great, while further intensification of land use involving overgrazing or high intensity tillage, with large inputs of fertilisers and pesticides, can have a detrimental effect, particularly on groundwater quality. The unique characteristics of karst aquifers, including the presence of conduit flow within the aquifer and the occurrence of concentrated recharge, together with the thin soil cover found in many karst areas, render karst groundwater particularly susceptible to such impacts. Whereas it may be possible to prohibit some industrial activities in karst areas, moving them to hydrogeologically less sensitive areas, this is clearly not an option with agriculture. Therefore it is important that farming should be carried out on a sustainable basis which will minimise the adverse impacts on karst groundwater.

CHAPTER 5

Industrial and urban produced impacts

HEINZ HÖTZL (co-ordinator and main author)
Department of Applied Geology, University of Karlsruhe, Germany

5.1 GENERAL PROBLEMS DUE TO HUMAN SETTLEMENTS IN KARST AREAS

5.1.1 *Direct and indirect impacts*

Distortion of the ecological balance of the natural karst systems results from the different activities of man (Williams 1993, Kranjc 1995, Günay & Johnson 1997). Therefore it is not surprising that urbanised areas with dense populations and therefore increased human activities show special risk potential when they are situated in karst areas or in catchment areas draining to karst systems.

The effects of urbanisation and industrialisation on karst in general and specially on karst groundwater include on one hand direct impacts. That is where the relationship between the trigger mechanism and the consequences for the karst water is very clear. The main impacts endangering karst systems are:
– Emission of air pollution;
– Discharge of waste water and nonaqueous organic liquid phase;
– Storage and deposition of solid waste;
– Excavations in connection with foundations and construction work.
Equally widespread are indirect effects resulting from the existence of and the activities in, such densely populated areas. For example the effects on the surrounding area can be very crucial, yet the synergetic effect between urbanisation and such changes in the surrounding landscape is frequently neglected, though it is very important (Williams 1993, Sauro et al. 1991).

Examples are the increased demands for additional land for construction purposes leading perhaps to deforestation and destruction of the natural ecosystems (Day 1993, Furley 1987). Another effect is the demand for construction material increasing the number of quarries. Their spatial extension leads to more areas with destroyed surface, removed cover systems and extremely exposed rocks. From the past we know the devastating effects of the deforestation of the Dinaric Karst (Gams 1987) due to the strong demand for wood as a construction material.

Another effect might be the rapid growth of recreational sporting or touristic areas

81

with different technical facilities in the surrounding of cities, which jeopardise and destroy the natural karst ecosystems (Goldie 1987, Lozek 1990). The huge food supply for urbanised areas intensifies the agricultural efforts in the surrounding areas (Drew 1996). All these activities induced by settlements and industrial areas may have decisive influences on the underlying or adjacent karst water system (Williams 1993). This is partly explained in the other chapters and should be kept in mind, even though it is not discussed in these chapter as it is intended to concentrate on the more direct effects of urbanisation and industries on karst water.

5.1.2   *Special vulnerability*

The high vulnerability of karst systems, which is understood as the intrinsic property of such groundwater systems (EC-COST 1995) was already discussed in previous chapters. For the assessment of possible impacts from urbanised area it seems appropriate to demonstrate which variations of the properties of the system are specially susceptible to risks resulting from urbanised areas.

With regard to the numerous possible hazards emanating from the urban area the most important safety element seems to be a thick covering system protecting the karstifiable rocks (Zötl 1974, James & Choquette 1988). An impermeable rock sequence interrupting direct hydraulic contacts between the land surface of an urbanised area and the karst aquifer are the best protection. Increased thickness reduces the vulnerability to give a reasonable amount of protection (Hölting et al. 1995, DVGW 1995). The thickness of the cover system has to extent at least several metres below the foundation depth.

Vice-versa, the lack of a cover system increases the vulnerability so that potential flow of contaminants from the different risk-sources is directly into the karst system. The thin cover layer like the non-soluble remnants of limestones formed by weathering processes is frequently washed down in small depressions but this offers little protection in urban areas. Due to excavations, e.g. for the foundation of buildings, cuts for traffic systems or sewer systems, thin layers frequently become leaky.

Vulnerability increases with special respect to urban hazards when a high degree of karstification exists and especially with the existence of a well developed epikarst system or vertical shafts (Doerfliger & Zwahlen 1997). Another point is the thickness of the unsaturated zone and the position of the karst water level below the surface. Though self-purification in karst is known to be unimportant, a thick unsaturated zone might cause certain adsorption and retardation which can lead to attenuation of possible pollutants. However in the case of long-term pollution or large quantities of contaminants, this cannot prevent significant breakthroughs into the karst groundwater.

5.1.3   *Shallow and deep karst systems*

Another important aspect endangering karst aquifers are the flow conditions present, depending on the development of a shallow or deep aquifer. According to their position compared with the relevant base level karst aquifers are subdivided in two main types (Wagner 1954, Bögli 1980, Villinger 1977). Shallow karst aquifers have their karst basis above the base level (Fig. 5.1). In this case rather flat aquifers de-

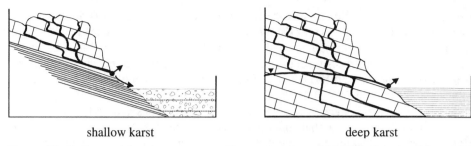

shallow karst   deep karst

Figure. 5.1. Shallow and deep karst systems with regard to the position of the base level (expanded after Bögli 1980).

velop with fast flow mainly along conduits or preferential karstic flow paths on the base of the karst sequence so that the vadose zone is dominant. In the case of an impact from urban areas or due to an accident in an industrial plant this may cause within few hours, a breakthrough of the pollutant at the springs.

The fast transmission of the pollutants is generally connected with relatively high concentrations of the pollutants. That affects strongly the use of karst water. The only advantage of such rapid contamination might be that it lasts generally only for a short period before it becomes attenuated and the pollutants are flushed out.

In deep karst systems, which sometimes extend to great depths below the outlet of the drainage system the phreatic zone is well developed. The recovery time from pollution and the rates of travel depend on the distance of the infiltration site from the possible outflow. For short distances the breakthrough curves may be comparable with those from shallow aquifers but with a more pronounced tailing. With increasing distance the flow paths follow deeper circulation according the theory of potential flow. Thus the beginning of the contaminant outflow from the system becomes more and more lagged as well as attenuated. On the other hand contamination will continue on a relative high level for a long period.

The effect is even intensified, if such an aquifer becomes confined with discharge through the confined part of the system. In such cases the contamination of the outflow of the system may start years after the original impact occurred. The exchange time of the contaminated water may increase exponentially with travel time.

5.2 EFFECTS OF AIR POLLUTION
(Neven Kresic, Radisav Golubovic & Petar Papic)

5.2.1 *Acidic deposition and precipitation*

Air pollution and its effects have been intensively studied in various scientific fields. Although 'air pollution' is a more general term, and it refers to the contamination of the atmosphere with products of anthropogenic origin, it is often in environmental and hydrologic studies used interchangeably with terms 'acidic deposition', 'acidic precipitation', or 'acid rain'. Acidic deposition is referring to the removal of acidic sub-stances from the atmosphere through the combined processes of wet and dry deposition. Acidic precipitation is the removal of acidic substances (primarily dilute sulphuric and nitric acids) from the atmosphere in precipitation as the result of oxidation and hydrolysis of sulphur and nitrogen dioxide. Finally, acid rain is the removal of acidic substances from the atmosphere in rain during atmospheric transport. Rain is naturally acidic (pH 5.0-6.0). Precipitation with an acidity <pH 5.0 is generally considered 'acidic' (NAPAP 1991).

Effects of acid rain have been studied since the late fifties and early sixties, but a real 'breakthrough' in a global (international) research was made after the publication of works by Scandinavian scientists who were greatly concerned with a rapidly deteriorating conditions in surface waters and forests of Norway and Sweden: Oden (1968), Granat (1972), Overrein et al. (1980) and others (after Cowling 1982). The majority of these and works that followed focus on four groups affected by acidic deposition:

1. Terrestrial ecosystems (mainly soils, forests and crops);
2. Surface waters (mainly lakes) and aquatic ecosystems;
3. Materials and cultural resources; and
4. Human health.

Few scientific papers since the pioneer work of Hultberg & Wenblad (1980) address the effects of acid deposition on groundwater. Based on survey of 1,300 wells in western Sweden, Hultberg and Wenblad postulated acid precipitation as the probable cause of groundwater acidification, heavy metal accumulation and plumbing problems associated with these wells.

Similarly to groundwater in general, the impact of air pollution and acidic deposition on karst aquifers has not been extensively studied. One of the possible 'additional' reasons why karst aquifers were not considered vulnerable to air pollution is a common perception that acidic deposition is expected to be fully neutralised by percolation through soil and bedrock rich in calcium carbonate (limestone and dolomite). However, karst groundwater can be affected by acidic precipitation if certain conditions, not that uncommon for karst aquifers, are present. These may include absence of soil and vegetation cover, a rapid flow of the infiltrating water through the porous medium between aquifer recharge and discharge areas, or simply a high load of more conservative pollutants such as nitrates (nitrates have small adsorption power and they constitute soluble calcium nitrate). In addition, ions of heavy metals, either present at the surface due to some other processes (artificial or natural), or being deposited together with the precipitation, may be mobilised by a low pH and introduced into the saturated zone.

5.2.2 *Sources of air pollution*

Anthropogenic and naturally produced sulphur dioxide (SO_2) and oxides of nitrogen (NO_x) are the primary air pollutants that cause acidic deposition (Fig. 5.2). Oxidants and volatile organic compounds (VOCs) are also major air pollutants that play an important role in the formation of acidic precipitation. Oxidants and hydroxyl radicals, formed by complex chemical reactions involving NO_2 and VOCs, are essential for converting SO_2 and NO_x into sulphates and nitrates. In the gaseous phase (clear air), SO_2 and NO_x can be transformed into sulphates and nitrates by hydroxyl radicals. In the aqueous phase (in clouds and precipitation), oxidants such as hydrogen

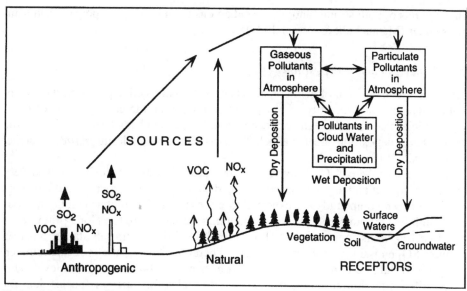

Figure. 5.2. Processes involved in source-receptor relationship (adapted from NAPAP 1991).

peroxide and, to a lesser extent, ozone, convert SO_2 to sulphates and acids. Beyond direct contribution to acidic deposition, NO_x is an important component in the formation of oxidants, which convert SO_2 to sulphates.

The use of fossil fuels, such as coal and residual oil for power generation, is the most significant source of SO_2 emissions in the United States. Electric utilities account for over $2/3$ of total US anthropogenic SO_2 emissions and over $1/3$ of total US anthropogenic NO_x emissions. Other major contributors to SO_2 and NO_x emissions are the industrial sector's use of fossil fuel for power generation and process heat, the smelting of copper ore, the refining of oil, the combustion of fuel for space conditioning, and the use of fuels by the transportation sector. The transportation sector is the most significant source of NO_x and VOC emissions, contributing over 40% of both. VOC emissions also come from solvent use, dry cleaning, paint application, other uses of organic chemicals, and residential wood combustion (NAPAP 1991).

Rainfall is naturally acidic: distilled water in equilibrium with carbon dioxide in the atmosphere has a pH of about 5.6. Based on research in the United States, background rain in humid temperate latitudes has an estimated pH of about 5.0 (Albritton 1987, Malanchulk & Turner 1987). Therefore, it is common to consider rain with pH less than 5.0 as acidic.

Although acidity is commonly expressed as the concentration of hydrogen ions, water containing acidity also contains anions which may be associated with acidity and are therefore termed 'acid anions'. They may be anions of strong acids which include sulphate, nitrate and chloride, the first two of which are the dominant anions in anthropogenic acidic deposition, but which also occur naturally in watersheds. Acid anions may also be anions of weak acids such as bicarbonate or organic anions, which are produced through bio-geochemical processes. Most waters contain differing amounts of both strong- and weak-acid anions from both anthropogenic and natural sources (Malanchulk & Turner 1987).

5.2.3 Effects of acidic deposition

Atmospheric deposition of acidic materials on to a watershed depends on the chemical composition and quantity of wet precipitation and on the concentration of gases and particles that contribute the dry deposition. Important characteristics of the acidic atmospheric deposition are (Malanchulk & Turner 1987, Albritton 1987, Barchet 1987, NAPAP 1991):

– In the atmosphere, NO_x converts to nitric acid (HNO_3) which is removed very rapidly by wet and dry deposition, so that distances between sources and receptors for nitrogen compounds are expected to be less than for sulphur compounds (although fine nitrate particles could remain in the atmosphere for a long time).
– More NO_x emissions are emitted at ground level than from tall stacks, whereas most SO_2 emissions are from tall stacks. Therefore, fewer anthropogenic nitrogen species would be introduced into the stronger winds above the ground level. This contributes to shorter source-receptor length scales for nitrogen species than for sulphur species.
– Deposition loading depends on land cover type. A forest canopy scavenges reactive gases from the atmosphere more efficiently than grassland, and a deciduous canopy is a more efficient scavenger when leaves are present than when they are absent.
– Hardwood forest canopies retain a significant portion of the wet and dry hydrogen and nitrates ion deposited during the growing season, but they retain very little of the sulphur. As a result, the acidity of precipitation which passes through the canopy is less than that of wet precipitation above the canopy. In contrast, coniferous forests typically increase the acidity of water as it passes through their canopies.
– The ratio of dry to wet deposition increases closer to major emissions sources.

Hutchinson & Whitby (1974, after Cowling 1982) established that strongly acid rain near Sudbury, Ontario (Canada) is accompanied by deposition and/or mobilisation of heavy metals (especially nickel, copper, cobalt, iron, aluminium, and manganese). They have found toxicity of these metals sufficient to inhibit germination and establishment of many native and agricultural species and plants. In hydrogeologic studies the main interest is focussed on soils and groundwater as the most important receptors of acidic deposition. These two receptors are discussed separately in Sections 5.2.4 and 5.2.5.

5.2.4 Interaction with soil and rocks

Soils and rocks vary widely with respect to their properties (type of bedrock they are derived from, thickness, biological, chemical and mineralogical content), they support different vegetation communities, are situated in different climatic zones, and are exposed to a broad spectrum of acid loadings. Therefore any generalisation regarding the impact of acidic deposition on soils and rocks should be avoided and theoretical calculations should take into account particular in situ characteristics. Although it is now a commonly accepted fact that acid precipitation can cause soil acidification, but in spite of the many experimental studies, it has been difficult to determine the comparative contributions of anthropogenic versus soil-derived (natural) acids to soil acidification.

The most important reactions that occur between percolating water and soils (or rocks) include reactions affecting: 1. supply of acidic and basic cations through cation exchange and weathering of minerals, 2. the mobility of anions through soils, including dissolution of carbon dioxide from respiration of soil biota, dissolution of organic anions in the soil water, retention of strong and weak-acid anions in the soil (Malanchulk & Turner 1987). All of these reactions plus the thickness of the soil in which they occur can affect the acidity of soil water recharging aquifers or discharging into streams and lakes.

Under conditions of acidic depositions or natural acid production, base cation 'buffering' (holding solution pH roughly constant) will occur until the soil exchange sites are essentially filled with hydrogen ions and the soil is thus acidified. This 'buffering capacity' is large in many soils, and in the soils derived from highly fossiliferous sediments (or metamorphic equivalents), limestones or dolomites this capacity is considered 'infinite' (Glass 1982).

The weathering of most common soil minerals consumes hydrogen ions and releases base cations and aluminium in different proportions, depending on the mineral. This includes direct neutralisation by reaction with carbonates. Weathering thus buffers soil solutions and replaces base cations on the soil exchange sites as vegetation, cation exchange, and solution leaching remove them. (Malanchulk & Turner 1987).

As it can be seen, the acidification of soils and lowering of pH of soil water (or maintaining already low pH levels of the infiltrating precipitation) is not to be expected in carbonate terrains. However, under certain conditions, other negative effects of acid precipitation can cause deterioration of karst groundwater. These include introduction of nitrates into groundwater and demobilisation of heavy metals cations already present in the soil/rock.

5.2.5 *Interaction with groundwater*

The effects of acidic precipitation on waters include some or all of the following (NAPAP 1991):
– Sulphate increases;
– Base cations increase;
– ANC and pH decrease and aluminium increases;
– Dissolved organic carbon (DOC) decreases.
The main negative effects of acidic deposition on groundwater quality in karst terrain include introduction of nitrates into groundwater and demobilisation of heavy metals cations that may be (are) already present in the soil.

Nitrates have a special place among groundwater pollutants because of their specific properties, in particular their negative sorption. Nitrates are almost inert in oxidising conditions that are commonly present in karst aquifers. If the percolating water has a high load of nitrates (e.g. 'acid rain'), they can quickly and easily enter the aquifer, especially if the soil cover is thin or absent, and/or there is no a significant vegetation cover that would withdraw nitrates from the solution.

Ions of heavy metals are an important potential pollutant of karst groundwater since they are frequently found and leached naturally from rocks and soils, and can also be introduced with acidic deposition as mentioned earlier. As with organic matter, the ions of heavy metals can not be decomposed into inert substances and are able to exist for a long period in karst media. Negatively charged ionised groups of organic matter present in soils can bind the cations of heavy metals thus deactivating them. As such, they can be removed from the overlying soil through various processes, the principal one being flushing by percolating water. The flushing is especially intensive if the percolating water has a low pH (acid rain). Acid rain infiltration causes the carbonate and hydroxide solution of heavy metal ions and increases the desorption to ions through substitution with hydrogen ions. This is also a mechanism of remobilization of previously retained heavy metal ions due to sorption, ionic exchange, sedimentation, filtration, bioaccumulation, and especially hydrolysis.

5.3   SOURCES OF LIQUID POLLUTION

5.3.1   *Waste water and sewer systems*

The most frequent reason for qualitative problems with karst groundwater results from the seeping of waste water into the underground (Ford & Williams 1989, Quinlan & Ewers 1985, Yuan 1983, White 1988). There are four main reasons why waste water is the most important risk factor for the groundwater in general:

1. Due to the liquid phase, waste water is a substance of high mobility. It can flow down through small pores and fine fissures and it does not need an additional transport medium.

2. As water with mainly dissolved contaminants and fine grained particles (colloids and micro-organisms) it does not require additional processes of solution or mobilisation. Therefore it is just a question of mixing, which occurs under the conditions of natural groundwater flow and causes a widespread and sometimes rapid distribution in the aquifers.

3. Almost by definition waste water includes a lot of undesirable substances, which are alien to the natural system and frequently cause pollution of the groundwater resources.

4. Waste water is produced by humans in huge quantities and occurs nearly everywhere where human settlements exist. The output reaches especially high rates in urban areas and at industrial sites. Highly toxic substances are produced in connection with special industrial production processes.

In karst areas these general factors impact in a special way. Due to the high vulnerability of karst aquifers by direct and fast connections between the surface and the karst water table (Aldwell et al. 1988, Hötzl 1996, EC-COST 1995) processes in the unsaturated zone, like filtration, adsorption or biodegradation leading in porous soils to a reduction of the contamination (Appelo & Postma 1993, Domenico & Schwartz 1990), are rarely efficient in karst, so that the full pollution loading reaches the karst groundwater. With the high yield of waste water in urban, touristic or industrial areas this creates an extreme risk to the safety of underlying or adjacent karst water resources.

With regard to the pollution degree, waste water from urban areas may be distinguished into different categories. From the less problematic to the stronger polluted components these are:
- Rainwater runoff from roofs and traffic routes;
- Domestic waste water;
- Waste water from commercial enterprises and industries;
- Special toxic waste water from different sources.

In general we consider *rainwater* as a clean natural recharge component for the groundwater. In hydrogeologically unfavourable areas, like many karst regions, it is even collected in cisterns for direct use. However besides its possible contamination by the wash-out of air pollution (cf. Section 5.2) the rain water may react immediately with the available material at the surface. The normal hydrogeochemical interactions with natural soils leading to the characteristic mineralization of ground water is not a matter for discussion here. The problem is the enrichment of the emitted contaminants on the surfaces by dry deposition, from where they can be washed away by rain water and become dissolved. Soot and other industrial dusts with a high content of heavy metals or even organic components may accumulate on the roofs and later be washed away and incorporated by the rain runoff (Table 5.1). Special organic contamination occurs in the rain water discharging from traffic routes and other paved areas (Fig. 5.3). The accumulation of combustion residues from car mo-

Table 5.1. Concentration ranges of organic components ($ng \cdot l^{-1}$) in precipitation water of Central Europe (after Schleyer et al. 1991).

Organic substances	($ng \cdot l^{-1}$)
Volatile chlorinated hydrocarbons:	
Trichloromethane	5-300
1,1,1-Trichloroethane	4-20
Tetrachloromethane	1-10
Trichloroethylene	< 5
Tetrachloroethylene	2-20
Trichloroethane acid	200-6500
Aromatic hydrocarbons:	
Toluene	< 15-200
Xylene	< 15-200
Naphthalene	< 15-20
Medium or non-volatile chlorinated hydrocarbons:	
Chlorobenzene	< 15-30
Dichlorobenzene	< 15-45
Trichlorobenzene	≤ 3
Hexanchlorobenzene	< 0.5-2
α – Hexachlorocyclohexan	< 2-4
γ – Hexachlorocyclohexan	10-90
Hexachloroethan	< 1.2-1.,8
Surfactants:	
Dibutylphthalate	60-1000
Dietylphthalate	< 900-1500

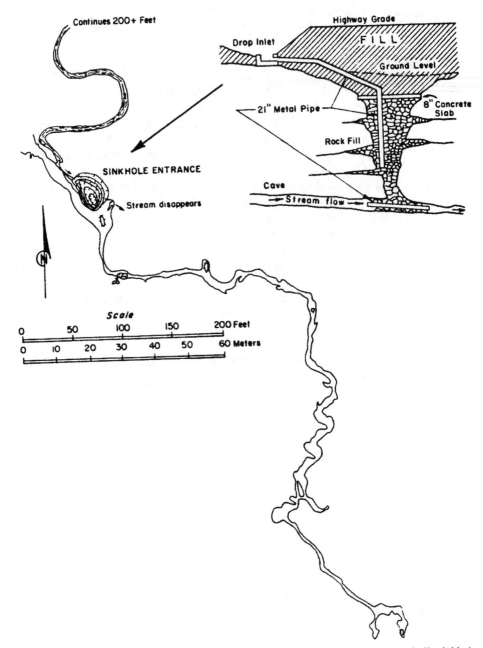

Figure 5.3. Drainage of the runoff from a highway (US 65) into a natural, now rebuilt sinkhole, which is connected with a stream cave, Steury Cave in Springfield, Missouri (Thomson 1997).

tors or the rubber from tyres can cause a high concentration of contaminants. If there is no natural filtration system available as in many of the karst areas, serious pollution problems may result from the untreated rain water runoff. Sinkholes in urban areas (Photo's 5.1 and 5.2) are frequently used for stormwater runoff (Barner 1997, Thomson 1997). Though storm floods may contribute to a certain dilution of the contaminants, the high risk of polluting therewith important groundwater resources is frequently underestimated.

Domestic waste water coming from toilets, bath and showers, dishwashers and washing machines contributes a main source of liquid pollution in urban areas. The amount of waste water depends on the living standard and can be estimated from the water consumption giving in well developed cities an average value of 120 up to 200 litre per capita and diem. In rural areas and less developed countries the daily amount decreases while under special technical and climatic conditions it might be much higher. Domestic waste water is generally characterised by increased content of the inorganic components, chloride, nitrate and ammonium as well as high rates of decomposable organic matter leading to high oxygen demand (Table 5.1). High concentrations of special toxic substances, like heavy metals are not very common. Therefore the most negative effects result from the high content of micro-organisms causing frequent bacteriological pollution of the karst groundwater. In Section 5.2 it was already mentioned that in the past and partly nowadays in the developing countries pollution of the groundwater by pathogenic bacteria were or still are responsible for epidemic diseases.

Industrial waste water composition strongly depends on the source materials used, the production processes and the resulting products and residuals. Inevitably they show great variations in composition and concentration of toxic substances. High risks result from untreated water from mining industry, chemical and pharmacological industries as well as from paper mills.

Special toxic waste water originates from different sources, which can not be included in the three previous categories. They are either obtained from reactions with special substances, e.g. radioactive material, or special processes like leachates from sanitary landfills (Table 5.2).

The classification of waste water is important for the assessment of the possible risks. However it should be recognised that waste water is a by-product of our lifestyle and industrial production processes, which cannot be avoided completely. Therefore the reduction of the possible impacts on groundwater becomes more a question of treatment and disposal, than of avoidance. A wide range of different handling methods for waste water are still in use:

– Untreated and uncontrolled release, which means that waste water is just left to infiltrate or sink into the underground where it occurs. This procedure is still very common in rural areas and in some remote touristic centres, but also in some towns among them some rapid growing mega-cities (LeGrand, 1973, Yuan 1983, 1991, Day 1993).

– For the specific induced discharge of untreated waste water into the underground, karst areas seem to offer a wide range of possibilities for those who are not aware of the pollution problems. Open karst joints, cave and shafts can take large amount of waters which disappear from the surface within a short time and so disappears also out of the responsibilities of the producer. This apparent simple solution

Photo 5.1. A newly constructed stormflood runoff system with open channels and buried pipes leading to the sinkhole, Valley View Sinkhole, Springfield, Missouri (photo: Hötzl 1997).

Photo 5.2. Sinkhole in the southern part of Springfield, Missouri engulfing natural rain discharge and flood runoff from land and commercial areas (photo: Hötzl 1997).

Table 5.2. Domestic waste water and examples of leachates from solid-waste deposits in Germany. Comparison of domestic waste water (DWW) from a small city (Rastatt) and the chemical composition of the leachate from a municipal (A) landfill (Grötzingen, SW-Germany) and from a hazardous (B) waste site (Malsch, SW-Germany).

Parameter	Unit	DWW	A (municipal)	B (hazardous)
Total dissolved solids	$(mg \cdot l^{-1})$	1350	15,500	158,200
Specific electric conductivity	$(\mu s \cdot cm^{-1})$	1371	15,500	150,400
pH		8.36	8.33	6.8
COD	$(mg \cdot l^{-1})$	1100	3840	15,550
BOD	$(mg \cdot l^{-1})$	450	325	4090
Chloride	$(mg \cdot l^{-1})$	85	1650	91,710
Sulphate	$(mg \cdot l^{-1})$	46	114	3445
Ammonia	$(mg \cdot l^{-1})$	55	1470	5363
Iron	$(mg \cdot l^{-1})$	1.5	1.9	157
Manganese	$(mg \cdot l^{-1})$	0.2	0.2	6.2
Zinc	$(mg \cdot l^{-1})$	0.8	0.8	3.8
Nickel	$(mg \cdot l^{-1})$	0.03	0.12	0.25
Chromium	$(mg \cdot l^{-1})$	0.01	0.01	0.32
Arsenic	$(mg \cdot l^{-1})$	0.01	0.01	0.03
DOC	$(mg \cdot l^{-1})$	300	386	3,020
AOX	$(mg \cdot l^{-1})$	0.05	2.1	89
Phenolindex (Σ phenols)	$(mg \cdot l^{-1})$	–	0.06	203
Chlorophenols	$(mg \cdot l^{-1})$	–	0.006	68
Σ VOX	$(mg \cdot l^{-1})$	–	0.007	66
Σ PCA	$(mg \cdot l^{-1})$	–	0.001	8

was applied commonly in the past, but is still in wide use all around the world without considering the negative consequences for the karst water quality (White 1988, Quinlan & Ewers 1985). Photo 5.3 shows the uncontrolled inflow of the undiluted sewage of the city of Tripolis in Southern Greece into a sinkhole dump, whilst Photo 5.4 show a reinforced sinkhole for the input of untreated waste water from a nearby village in the western part of Bavaria, Germany.

– Septic tanks and cesspools are designed to discharge waste water into the subsurface (Fig. 5.4). They are used in many regions (Fetter 1993) where due to the originally dispersed distribution of houses connecting sewerage systems still seems to be expensive. One of the risks in karst areas is that there is frequently no soil cover, which could help to purify somewhat the waste water by natural filtration or adsorption processes (White 1988). The effluent from the septic tanks in karst areas may therefore reach the karst water table more or less unaltered due to the fast flow in the wide karst channels. Groundwater contamination from septic tank effluent is therefore a particularly common feature in karst regions (White 1988, Quinlan 1983).

– Sewer systems have the task to collect the waste water and divert them to a treatment plant or to another safe disposal site. It is still a common practice to divert the untreated sewage into nearby rivers and to expect its fast attenuation by the river discharge. In karst areas or in areas where the river flows on downstream to a karst landscape, such polluted rivers frequently supply the karst aquifer by partial seepage or sinking or flowing completely into swallow holes. As an example a case study is

Photo 5.3. The Katavothre Kanatas (swallow hole) in the Polje of Tripolis. It swallows the untreated waste water of the city of Tripolis, Peloponnesus, Greece (photo: Hötzl 1986).

Photo 5.4. A rebuilt sinkhole with seepage installations for the untreated waste water of a nearby village, Triassic Muschelkalk karst near Rothenburg o. Tauber, Bavaria, Germany (photo: Hötzl 1988).

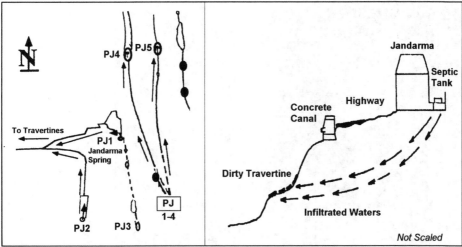

Figure 5.4. A leaking septic tank on the famous Pamukkale travertine hill (right figure), which pollutes the re-infiltrating thermal water and discolours the travertine at the outflow; the underground connections could be proved by tracing experiments (left figure), Pamukkale, Turkey (for details see Section 5.7.10).

described by Kogovsek and Kranjc in Section 5.7.4 from the famous Reka River in the Slovenian karst. Industrial waste water with high content of organic pollutant caused strong contamination of the Skocjanska Timavo karst system, but which clearly improved after the industrial plants went out of operation.

Another common practice is the diversion of untreated sewage into nearby sinkholes or swallow holes, without consideration of existing downstream abstractions for water supplies. Examples are numerous in the literature (Yuan 1991, Day 1993, Quinlan 1983, Hötzl 1973, Zupan & Kolbezen 1976, Kogovsek 1997). A case study from Central Kentucky is presented in Section 5.7.3 by Worthington. The discharge of waste water into the underground of karst areas requires in any case a sufficient treatment which in turn depends on the type of pollution.

Big problems are caused by leaky sewers. Leakage of collector pipes may result from a break in the pipe or impervious sealing or wrong pipe connections (Photo 5.5). Leakages are so widespread in urban areas that it is appropriate to speak of diffuse contamination rather then from point sources.

Disposal wells are used to discharge liquid wastes into the underground. In order to get a recharge capacity from the well the injection has to take place in highly permeable rock sequences (Hickey 1982). Unfortunately karst aquifers are frequently used for waste injection. Some of them are at great depth but generally they are still connected with the hydrologic circulation systems. In South Germany for example the Jurassic Limestone below the Molasse basin at depths of 1200-1500 m was used for the disposal of industrial and municipal sewage. Detailed investigation has shown that even this aquifer system is connected with the active hydraulic base level system Villinger (1977). Though the circulation time might be several thousand years it produces a real risk for future water demands. Therefore several countries in Central Europe have forbidden the injection of untreated waste water into the underground.

Photo 5.5. Municipal sewage system with broken sewage pipe (photo: Kramp 1993).

5.3.2 *Oil and gasoline*

5.3.2.1   *Risk potential of petroleum distillates*

The maintenance of high energy consuming industry and society demands the availability of tremendous energy resources. To a large degree these are supplied by petroleum and its distillates. Beside certain thermal power plants most of the fuels are used for house heating and for keeping vehicles and engines running. The high mobility of traffic systems as well as the supply and storage of the fuels for the individual households create in urban areas a high risk potential for uncontrolled seeping of the fuels into the underground. Leaking tanks and accidental spillages of these non-aqueous phase liquids (NAPL) are amongst the most frequent impacts on ground water resources. Karst aquifers with their nearly unlimited options for liquid infiltration are endangered by these organic contaminants in a special manner.

Petroleum distillates including light and heavy gasoline, kerosene, gas oil and heavy oil consist of a mixture of hydrocarbons with varying molecular weight. The main groups are alkanes, naphtenes, aromatic and polycyclic aromatic hydrocarbons (PAHs) (Voigt 1989, Fetter 1993). Their toxicity varies whereby the last group, the PAH, are carcinogenic. The liquid phases are characterised by a strong smell, which is recognisable even in dilution of 1 ppm and makes water unsuitable for human consumption. The immiscible phase liquids have generally a low water solubility, which amount up to 200 mg·l^{-1} for gasoline and 20 mg·l^{-1} for heavy oils. The specific gravity is less than that of water. They have high chemical stability. Biodegradation is possible under aerobic conditions, but because of the special microbial toxicity of the low molecular hydrocarbons the biodegradational turnover remains generally small (Alexander 1985, Barker et al. 1987, Chiang et al. 1989, Klecka et al. 1990).

Beside their properties it is the widespread use of fuels, which makes them so dangerous for groundwater resources. Starting with the extraction of crude oil (leaky production wells), their distillation in refineries, their transport, their transfer, storage and use as well as the disposal of remnants create so many opportunities that in spite of all regulations for careful handling, seepage of fuels is one of the most frequent environmental damages. Among the diverse pollution sources are refineries, pipelines, transport on roads and railways, freight depots, storage tanks in industry and private households and of course petrol(gas)stations are the most frequent places, where accidents occur with loss of oil and gas into the underground.

5.3.2.2 *Transport behaviour in the karstic underground*
Petroleum distillates consist mainly of substances with relatively low water solubility. While more soluble materials may have a greater potential mobility in the environment, the lower mobility of the fuel substances leads to slower migration and increased residence time especially in the vadose zone of karst systems. In the case of an uncontrolled seepage of the spill the NAPLs percolates downward through the unsaturated zone toward the water level. The NAPL spread through the pores, cracks, fissures, subcutaneous drains, karst conduits and shafts according the hydraulic gradient (Fig. 5.5). The solid rock matrix bordering the pores and other openings are wetted by the natural water. This water is not displaced by the NAPL, which moves forward via the empty pore spaces step by step after each residual saturation is exceeded (Farmer 1983, Schwille 1971, Jury et al. 1987).

In karst areas the spill may flow directly along nearly vertical karst fissures, vadose shafts or other preferential flow paths to depth so that the water table may be reached quickly. The flow can be also collected by very well developed epikarst systems close to the surface, from where, in highly permeable karst channels, the spillage may approach the main aquifer. Often, flow occurs in vadose conduits oriented down the dip of the rock, whereby the spill can be deflected from the vertical to flow in lateral directions for distances that may exceed several kilometres (Field 1989).

On the other hand even in well developed karst areas high percentages of the surface of the carbonate rocks still show no features of high permeability so that a slow percolation into the fine network of small fissures and pores can take place, where in spite of relatively small porosities a large proportion of the spillage may be retarded. From there it migrates slowly or by small dilution processes toward the groundwater. With regard to the low porosity of some carbonate rocks, one has to consider the sometimes very thick unsaturated zones in karst areas. Due to the residual saturation a considerable part of the contaminants will be trapped in the fissures and voids of the karstified rocks. Thus the quantity of mobile fluid decreases with the increasing quantity of immobilised NAPLs, which mean that for complete contamination down to the groundwater level a certain quantity of the NAPL is required depending on the porosity and cavern frequency as well as on the thickness of the vadose zone (Domenico & Schwartz 1990).

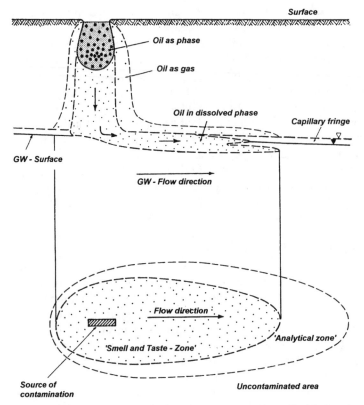

Figure 5.5. Idealised spreading of light nonaqueous phase liquids in a homogeneous unsaturated zone (Schwille 1971).

In karstified rocks the irregular shapes of conduits and caverns, their size and the geometric type of their connections create siphons or pool-like structures. They function like traps, where large quantities of the NAPLs can be stored. From there, removal or further transport is sometimes severely restricted so that contaminant reduction may occur only by microbial degradation. Although in high permeable porous aquifers intensive and fast biodegradation especially of the aromatic compounds can be observed (Fetter 1993), the conditions in karst systems with accumulation of the nonaqueous phase liquids in pockets of karst openings or half closed fissures are unfavourable for microbial activity. The rates of degradation in the vadose karst zone are rather small and take a long time. Besides biodegradation, dissolution and volatilisation occur, their intensity depends upon the relevant properties of the contaminants. Whilst the water solubility for petroleum and most of the distillates is relatively low, some of them show a tendency to become volatile by passing into the gas phase and spreading from there upward or downward according their density and the actual gradient of the vapour concentration. When the NAPLs finally reach the karst water level, the pattern of spreading becomes more complex. The light nonaqueous phase liquids(LNAPL s), like gasoline, kerosene and diesel remain on the top of the water and spread mainly laterally (Fig. 5.5).

Compared with a porous aquifer the situation in karst can be very different. There are three main possibilities:

1. In the first case, which is rather rare and untypical for karst areas, the network of fine fissures can form a more or less similar situation to that in a porous aquifer. The hydraulic situation can be described in terms of the water table and the capillary fringe. When large volumes of LNAPLs reach the capillary fringe in a short time, the fringe below the main percolation cone will be superseded and replaced by a flat disk-like bubble consisting of LNAPLs. The extent of the bubble depends upon the quantity of the LNAPL and its density. In addition the LNAPL spreads as a thin layer in the upper part of the capillary fringe. Depending upon the recharge the spreading in the capillary fringe continues until the total spillage is at residual saturation.

2. In the second case the hydraulic situation resembles more a fissured aquifer. Open fissures, partly enlarged by corrosion or even small karst channels form the water paths. Matrix porosity can be neglected. Below the water table the single fissures or karst conduits are hydraulically connected in the sense of communicating tubes, which means that they reflect water level changes but that parts of the network are not under continuous flow conditions. These parts are in a more or less stagnant or immobile condition floating only according the water level changes. The LNAPLs will be trapped in openings above the water table. More or less isolated pools of contaminants with different thickness will result. Flooding of the karst system may cause arise in water levels and thus an uplift of the contaminants, which spread over the walls, fill up small cavities or may be driven into higher cave levels (Field 1989, 1993).

3. Case three describes the typical karst system with preferential flow along widened fissures, caves and other karst conduits. The seeping of a fuel spill may lead directly into an open vertical karst shaft which can be connected with open channel flow in one of the conduits in the vadose zone or in the transition to the phreatic zone. In this case the immiscible liquids are transported over long distances via convective transport. This may be associated with extended pollution of the aquifer, but can result also in a fast flushing of the contaminants out of the aquifer. The spreading pattern of the NAPLs is completely different if the preferential water paths lead down to conduits, which continue down into the saturated part, where siphon section may completely interrupt the discharge of the LNAPLs similar the situation described under point 2. Once again they would be trapped above the water table and the emissions are restricted to the slow dissolution or degradation processes.

5.3.2.3 *Site Remediation*

Accidents with hydrocarbons or leaking tanks are extremely common. Frequently they will not be discovered or by chance only months or years later. Fortunately most of them involve only small quantities of contaminants which got stuck in the cover system and do not break through to the groundwater. In case of unrecognised leaking storage tanks great amounts of fuels may seeping into the underground by continuous loss.

Accidents with seeping fuels are well-known to hydrogeologists, some of them are reported in the literature. A huge spill with 216,000 gal of gasoline is mentioned by White (1988) from a karst area south of Harrisburg in Pennsylvania. It started from a tank farm and finally formed a 500 m wide and 2000 m long pool with thickness up to 2 m, which was floating on the water table in a solutionally modified carbonate aquifer. Kogovsek (1995) described a oil spill, which caused a narrow oil accumulation of 30 m thickness above the karst water table in the south-eastern karst area of Slovenia. After four years and intermittent pumping activities the fuel layer was still 0.6 thick, showing strong variation according the changes of the water level (Fig. 5.6). Emmett & Telfer (1994) reported on point source pollution by leakage of fuel oil from underground storage tanks contaminating the Gambier limestone aquifer in SE Australia, which is used for water supply. Other impacts in karst areas are described by Crawford 1988, Quinlan & Ray (1991), Recker (1991), Chieruzzi et al. (1995) and others.

For the planning of remediation measures and for selecting an appropriate technique the type of contamination (nonaqueous phase liquid, dissolved phase or gas phase), the exact distribution pattern of the contaminants and the hydrogeologic setting have to be studied carefully. In managing the remediation four phases have to be considered:

1. First, all ongoing sources of contamination have to be stopped. Source control will be necessary to prevent the continuing release of contaminants. Removal of installations on the surface or of soil from the cover system affected by the contaminants may be required.

2. In the second step priority has to be given to remove the accumulations of mobile nonaqueous phase liquids being trapped in karst pockets or floating on top of the karst water table. Extraction wells (Fig. 5.7) are required in hydraulically optimal po-

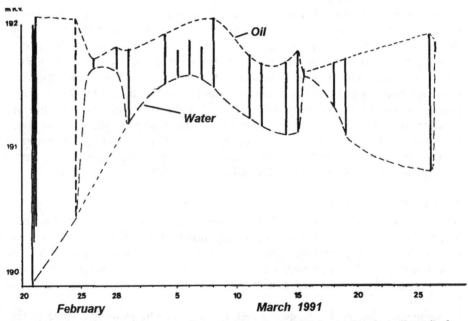

Figure 5.6. Changing oil and water levels in a karst well of SE Slovenia. The fuel oil results from leakage in a tank and reached a thickness of 30 m in the well. After remediation pumping reduction occurred to less than 1 m (Kogovsek 1995).

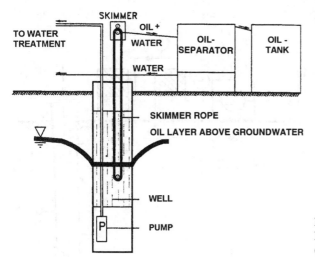

Figure 5.7. Remediation well for LNAPLs with skimmer system (LFU 1995).

sitions. In karst systems finding the right location can become very difficult and may require a great number of wells. The point will be to get access to the main openings (joints and caves) just at the karst water table. Careful study of the joint pattern and karst features on the surface together with geophysical surveying, evaluation of borehole logs and hydraulic tests may be a help (Barker & Nicholson 1993, Quinlan 1989, Quinlan et al. 1992, ASTM 1995, EC-COST 1995). Even a detailed study will not guarantee access to all NAPLs especially, if there is a thick vadose zone with high heterogeneity. The efficiency of extraction wells can be increased by additional injection wells. For the stabilisation of the plume careful planning of each hydraulic measure is necessary. To partly reduce residual saturation long-term flushing is required. Nutrients can be added to promote bioremediation (Hinchee et al. 1993, LFU 1995).

3. In a third part of the remediation procedure, which can run parallel to the first two steps, attention has to be paid to purification of dissolved phase contaminants in the karst aquifer. The most convenient technique is pump and treat. Once again the difficulties in karst aquifers arise from the heterogeneous flow conditions. Conduit flow can cause on one hand strong dilution, on the other hand it may require a significant depletion of the karst water level to draw by the plume from less permeable parts of the aquifer. Both can lead to high pumping rates, which makes the remediation less efficient and increases the costs. Additional air sparging (Böhler et al. 1990) can be used to accelerate the efficiency of the measures, but there are no results available from deep karst aquifers thus far. The same is the case with more advanced techniques like reactive walls (Gilham & O'Hannesin 1994, Starr & Cherry 1994), which seem hardly economic in the hard rocks of the karst aquifers.

4. As an additional step the removal of the contamination in the gas phase has to be considered. Vapour extraction (or soil venting, Fig. 5.8) is an effective and successfully applied technique for the remediation of contaminations with gas phases in the vadose zone (Harress 1990, Pederson & Curtis 1991). In cases where it is difficult to catch the NAPLs itself this technique is even used to remove the liquid phase, which

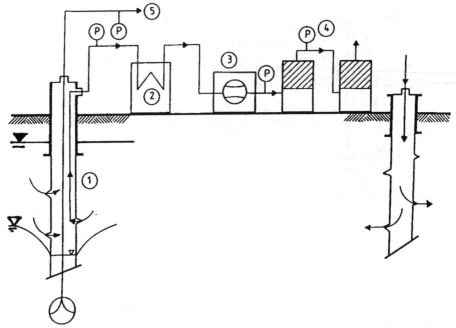

Figure 5.8. Schematic outline of an installation for soil vapour extraction: 1 = Special cased well in karst aquifer, 2 = Water separator 3 = Compressor, 4= Charcoal filter, 5 = Pipe to water treatment plant, P = control sampling (Nahold 1996).

may be necessary in connection with the residual saturation. The application of these techniques demands volatile contaminants.

5.3.3 Organic solvents

5.3.3.1   Groundwater relevant substances

Beside the petroleum distillates, the halogenated hydrocarbons are one of the largest and most important groups of organic contaminants found in groundwater (Barbee 1994, Fetter 1993). They include aliphatic and aromatic substances. Included in the first groups are solvents such as methylen, chloroform, carbon tetrachloride, 1,1,1,-trichlorethane, trichlorethylen as well as industrial chemicals as methyl chloride and methyl iodide and some pesticides. The aromatics include some environmentally dangerous groups like chlorbenzene and polychlorinated biphenyls as well as some pesticides like DDD. Though there are some additional other important organic pollutants like phenols, polynuclear aromatic hydrocarbons or organometallic compounds (Domenico & Schwartz 1990), the following discussion is restricted to urban and industrial impacts which exemplify the problems with some organic solvents.

The frequent occurrence of halogenated hydrocarbons in groundwater results from different pollution sources. Their application as solvent or thinner, their use for degreasing and extraction processes, for dry cleaning, for refrigeration, for disinfecting or for preservation causes a wide dissemination of these substances especially in urbanised and industrialised areas. With regard to their diverse deployment, the production, transport, storage, use and disposal act as multipliers for possible uncontrolled losses of these substances into the underground. According the quantity of the halogenated hydrocarbons used most serious contamination results from industrial production and storage facilities, from degreasing installations, especially in metallic branches, as well as from dry cleaning enterprises.

In the urban area contamination by organic solvents is generally via point sources, where the sources are normally of well defined restricted local extent (Mackay & Cherry 1989, Mercer & Cohen 1990, Barner & Uhlmann 1995). Only if there are secondary spreading processes for instance in leaky sewer systems, the original source may be hard to detect. Beside the point sources, volatile organic solvents can be found in low concentration in the air and precipitation of some urban areas leading to a general low concentrated diffuse source (Bock et al. 1990). From urban areas the plumes can be traced downstream in the groundwater flow up to several kilometres. Frequently, overlapping plumes of different sources and substances may restrict groundwater use in extensive areas.

The high risk potential of the organic solvents is based on two facts. On one hand the varying toxicity of the substances can cause real problems for drinking water as well as for the food chain as some of them are carcinogens. Some are not toxic by themselves but high toxic risk may result from degradation products like vinyl chloride, which may be derived from trichlorethene (Fetter 1993). The special risk potential results on the other hand, from poor biotic or chemical degradation. The most commonly found products of microbial degradation come from reductive dehalogenation, while abiotic degradation tends to involve hydrolysis and oxidation (Whelan et al. 1994). The average half lives vary considerably, but is in general in the range of months and years. Therefore the contamination remains for long periods in the unsaturated or saturated zone.

5.3.3.2 *Transport behaviour in the karstic underground*

The transport behaviour of organic solvents in the underground is mainly influenced by two physical properties, the water solubility and the specific gravity (Schwille 1984, Jury et al. 1987, Mercer & Cohen 1990, Domenico & Schwartz 1990). Like most of the petroleum distillates the organic solvents are immiscible but of higher density than water, and this determines their downward movement (Fig. 5.9).

The movement of such dense nonaqueous phase liquid (DNAPL) through the vadose zone of an karst aquifer is more or less equivalent to the LNAPLs with gravity movement downward leaving behind a cone of residual saturation, which can be accompanied by a zone of dissolved phase and another zone of gas phase (Fig. 5.10). The development of such cones might vary over a wide range due to the inhomogeneities in the overburden or due to the irregularities of joints and karst paths (Schwille 1984, 1988). Under extreme conditions it might be just the wall of a corrosive extended joint or of a karst shaft. Depending on such special karst conditions or due to the high quantity of the spill or constant or repeated spill flow the penetration cone will finally reach the water table.

Unlike LNAPLs spreading mainly within the capillary fringe DNAPLs can penetrate through the whole water column by gravity flow down to the bottom of the aquifer, leaving behind in the descent zone a residual saturation (Schwille 1984, Mercer

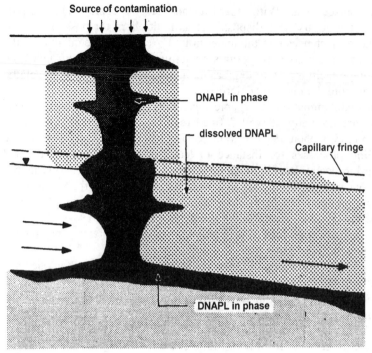

Figure 5.9. Schematic sketch of the percolation of a dense nonaqueous liquid phase through an aquifer (Schwille 1984).

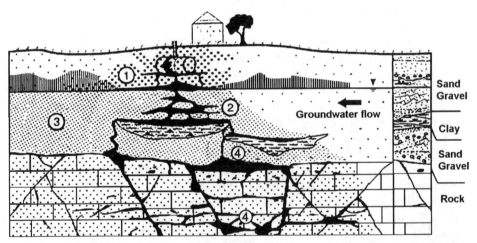

Figure 5.10. Percolation of dense nonaqueous liquid phase into the underground: 1 = The input in the vadose zone occurs in liquid phase with direct emission in gas phase, 2 = The downward spreading occurs mainly in liquid phase, where impermeable layers may cause lateral distribution both in the vadose and phreatic zone, 3 = Dissolved compounds may spread all over the downstream aquifer; and 4 = Nonaqueous liquid phase will accumulate in depressions or trap-like structures on the bottom of the aquifer.

& Cohen 1990). In the saturated zone the residual saturation is clearly higher than in the unsaturated zone, where balance is reached between the water, the air and the phase liquid. When there is enough supply of the DNAPLs they will accumulate on the aquifer bottom where the phase liquid flow follows the gradient of the aquifer bottom towards the deepest point (Figs 5.9 and 5.10). Thereby the flow direction might be different from the groundwater flow.

At the deepest location a pool of contaminants is formed which can become stagnant unless a significant volume of additional phase contaminant exists to maintain a head to induce a more or less horizontal flow. As a result the DNAPLs can spread in unexpected directions following cracks and karst paths. Surprisingly Chieruzzi et al. (1995) reported from a contaminated limestone in south-western Kentucky that a diffuse flow of dense phase liquids took place during pumping in a nearby well. This requires relatively homogeneous permeability conditions. Under dominant conduit flow completely different transport conditions may occur (Hötzl 1989). Especially under turbulent flow with high velocities in the conduits close to the water table parts of the phase liquids can be swept away or flushed out. The more usual case however will be the trapping of the DNAPLs due to their special gravity and the karst features.

Trapping of DNAPLs in the phreatic karst zone may occur under different conditions. Deep cavities, cave sections and shafts with higher overflow channels may act as contaminant pools. The same may occur in siphon sections or other caverns with pools. The trapped contaminant may continue to act as a source of dissolved phase constituents emanating from the pool of hydrocarbons until water levels drop (Ewers et al. 1991). Dissolved phase contaminants will flow as and with karst groundwater, with transport velocities up to several hundred metres per hour (Hötzl & Werner 1992). The volume transported in the karst fractures, channels and conduits established in a mature karst setting will efficiently flush more rapidly due to the shorter residence times characteristic of these karst systems. However the complete recovery of non aqueous phase liquid contaminants is not feasible, because plumes of groundwater contaminants emanating from trapped NAPL sources are potentially infinite in dimension (Barner & Uhlmann 1995).

5.3.3.3 *Site remediation*
The main objective of each remediation plan is to remove the contamination source first, in order to stop further emission. This includes two steps. The first is to remove existing contamination sources like inappropriate industrial installations, a leaking storage tank or unqualified handling of the relevant substances. The other important step is to remove the affected uppermost soil below the sources, where the organic solvents may be still accumulated in phase and serve as a secondary pollution source for the aquifer.

In karst areas the second step is frequently hardly to be realised, due to the special features and the resulting vulnerability of the system. The problems arise when there is no overburden or cover system which is able to keep back the mass of the contaminants by residual saturation or just forming a barrier by extreme low permeability. These demands are realised with an sufficient clayey, silty or sandy cover system, which can keep back the whole amount of the NAPLs, so that only dissolved components are flushed down into the karst aquifer. The normally very thin weathering layer on top of limestone sequences usually does not meet this condition. In contrast the frequent direct outcrop of karstified rocks with open fissures or shafts favours the quick penetration of the contaminants into the deeper part of the karst system.

Because of the difficulties posed by both the nature of contaminant migration and the groundwater flow in karst systems, appropriate remediation will largely depend upon an accurate characterisation of the karst water flow and the distribution of the contaminants (Barner & Uhlmann 1995). Lack of experience of how a karst aquifer system works can lead to irrelevant data or solutions. For appropriate investigation techniques the reader is referred to special karst literature, like Quinlan & Ewers (1985), Quinlan et al. (1992), EC-COST (1995), USEPA (1992), White (1988), and Ford & Williams (1989).

In the case that the spill is still in the vadose zone kept back as residual saturation, vapour extraction techniques are the most appropriate and economical measures to clean the site. The practicability as well as the necessary time depends greatly on the distribution of the contaminants, the permeability of the vadose zone, the local and technical arrangements as well as the volatility of the substances. Examples of successful remediation by vapour extraction and the details of this technique are given in Barbee (1994), Baehr et al. (1989), Gierke (1990), Nyer & Morello (1993), Rathfelder et al. (1995), and Roth & Peterson (1994). Experiments are going on with regard to biodegradation techniques but till now there is no really satisfactory technique for the complex karst aquifers.

Remediation of non aqueous phase organic solvents from the phreatic zone of karst systems presents a special challenge. At the outset it has to be mentioned that there is no technique available up to now, which can almost certainly give a guarantee of a successful remediation. Quite the reverse, as most of the experiments failed to achieve an acceptable reduction of the contaminants within a reasonable time frame (Quinlan & Ray 1991). The DNAPLs can be transported deep within the aquifer and may be distributed over several more or less isolated pools in the karst system. Without the exact position being known it is nearly impossible to catch the pools by a well. Due to the strong heterogeneity of the permeability of the karst system hydraulic measures, like pumping, flushing or air sparging generally achieve only partial success (Fig. 5.11). Even if the main nonaqueous phase liquid can be drawn away, residual saturation remains. Therefore a complete recovery of DNAPLs from the phreatic zone is not feasible because plumes of contaminants are potentially infinite in dimension. They tend to be adsorbed and retained indefinitely in the interstitial spaces of the matrix and the network of the fine fissures (Mackay & Cherry 1989, Barner & Uhlmann 1995).

Residual DNAPLs act as continuous sources of ground water contaminants and if not removed from contact with water, will continue to generate a plume of dissolved chemical constituents (Fig. 5.12). Dissolved phase contaminants of organic solvents (independent of their origin, from the surface or vadose zone or from residuals in the phreatic zone) will generally flow as and with groundwater. Some differences with velocities and path lines may occur due to mechanical dispersion and chemical diffusion, but with this convective transport of the dissolved components a flush from the aquifer will take place. There is certainly an advantage for karst aquifers in that with the flow in channels and conduits the flushing is relatively fast. With regard to the

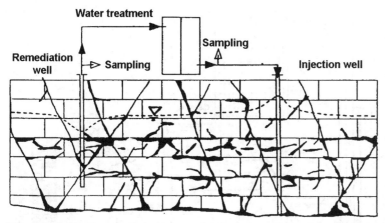

Figure 5.11. Simple remediation concept 'pump and treat' for a karst aquifer polluted by dissolved hydrocarbons.

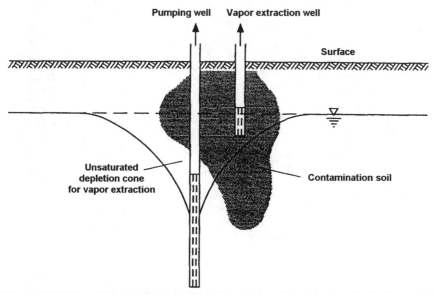

Figure 5.12. Schematic sketch of a combined hydraulic and pneumatic remediation technique for the residual saturation of NAPLs with depletion of groundwater level and vapour extraction in the depletion cone (LFU 1995).

continuous generation of a plume from the residuals the fast conduit flow will cause strong dilution so that the concentration may be comparatively low.

The aims of remediation in karst aquifers have to follow in a certain way this natural flush. The most appropriate remediation techniques for dissolved phase contaminants are pump and treat (Fig. 5.11) or air sparging in combination with vapour extraction and partly together with additional hydraulic measures (Figs 5.12 and 5.13). These techniques were used in a successful remediation in a Triassic karst formation

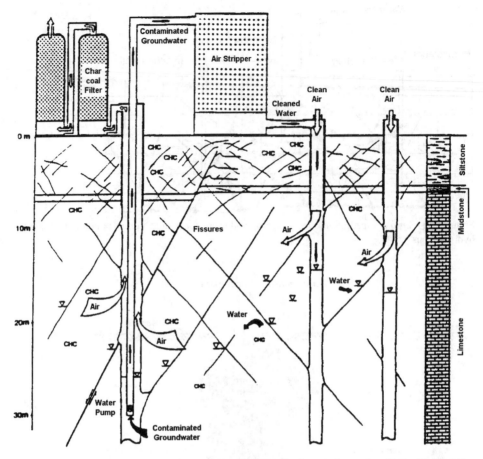

Figure 5.13. Remediation of dissolved and gaseous DNAPLs from a fissured and karstified aquifer: remediation well for air and water pumping, injection wells for water and air, cleaning installations with charcoal filter (air) and air stripper for the contaminated water. The groundwater is depleted and air is extracted from the fissures (Nahold & Hötzl 1993).

in Southwest Germany (Nahold & Hötzl 1993, Nahold 1996), which is described as a case study in more detail in Section 5.7.5.

5.4   SOLID WASTE AND WASTE DISPOSAL

5.4.1   *Sources and disposal of waste*

Human activity produces a huge amount of hazardous and non-hazardous solid waste, which was almost entirely dumped in the past. Although today there are efforts to avoid waste accumulation by reuse and recycling or incineration for energy production, the amount of waste being dumped is still very high. Especially in urban

areas with large populations and dense industrial structures the accumulation of solid waste is particularly large. In terms of the degree of harmfulness it can be classified as non-hazardous, controlled, special, hazardous or toxic (radioactive) waste. With regard to the facilities for disposal, the environmental regulations in many countries, for example England and Germany, recognise:
- Inert waste: that will not pollute under normal circumstances;
- Semi-inert waste: incorporating small amounts of leachable but non-toxic components;
- Household waste: private garbage and trash as well as general waste from commercial and industrial premises with high organic content capable of reactions and biodegradation;
- Hazardous toxic waste: risky even during handling and perhaps requiring special underground storage conditions.

The disposal of solid waste is still unregulated in many countries of the world. Thus disposal occurs without reference to the harmful composition of the material or the appropriateness of the selected disposal location. A widespread practice is still to use municipal or industrial wastes in order to fill up hollows, rehabilitate abandoned pits or even to throw into open waters. In the developed world approaches to the management and disposal of waste have changed drastically since the early seventies (Howard et al. 1996). The introduction of laws for the treatment of waste and improvement of these regulations (for example HMIP 1989, USEPA 1991, TA-Abfall 1991, EU-Council 1997) advances in landfill design technology including the introduction of sophisticated leak controls as well as rigid selection criteria for waste sites, now provide considerably improved environmental protection. Also, these reuse and recycling programs have diverted significant quantities of material from the waste stream. Nevertheless due to the incomplete measures taken both active and inactive landfills and hazardous waste sites remain a serious threat to groundwater and provide little or no protection of the aquifers. In karst areas contaminants can easily enter the aquifer system through solution fissures and collapsed structures. Once in the subsurface flow system contaminants, are distributed rapidly through conduits and enlarged passageways and natural filtering processes are nearly non-existent. Therefore karst aquifers require especially strict regulations and management to avoid karst water pollution. For this reason waste disposal on karstified limestone has always been regarded as problematic by hydrogeologists. But on the other hand sinkholes and karst depressions as well as abandoned quarries, which from the view of landscape restoration and agricultural landuse should be refilled, offer a place to deposit unwanted materials at first glance (Photos 5.6 and 5.7). The problems were not obvious to many rural residents, who have traditionally used sinkholes and caves as convenient dump sites because they are areas that cannot be farmed and were therefore attractive sites for trash disposal. Apart from the disposal of dead, often diseased animals, the use of sinkhole dumps was relatively benign in the past. However, the general use of these dumps for all kind of waste and their use as sanitary landfills for urbanised areas have made their use much more dangerous. Figures 5.14 and 5.15 show the result of two detailed studies from Virginia, USA (Slifer & Erchul 1989) and from southwest Germany (Hötzl 1995) on the distribution of such uncontrolled dump sites and sanitary landfills in karst areas. Similar figures were published by Edwards & Smart (1989). Although nowadays the

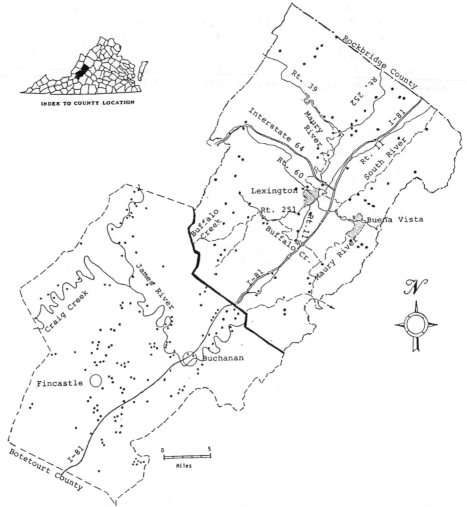

INDEX TO COUNTY LOCATION

Figure 5.14. Distribution of illegal dumps in the karst area in Botetourt and Rockbridge Counties, Virgina, USA (Slifer & Erchul 1989).

disposal of waste in developed countries is much more controlled these two figures, still give an impressive indication of the extent of the misuse of karst features as well as quarries. The drainways from many sinkholes connect directly with underlying conduit systems White (1988).

5.4.3 Waste leachates and karstwater interaction

Modern waste sites are designed to minimise the adverse effects of waste disposal. However, this was not the case in the past when the main objective was to remove it from sight. Hence, every depression and hole was an appropriate dump, sometimes covered with only a thin soil layer. Even now only a few landfills are designed such

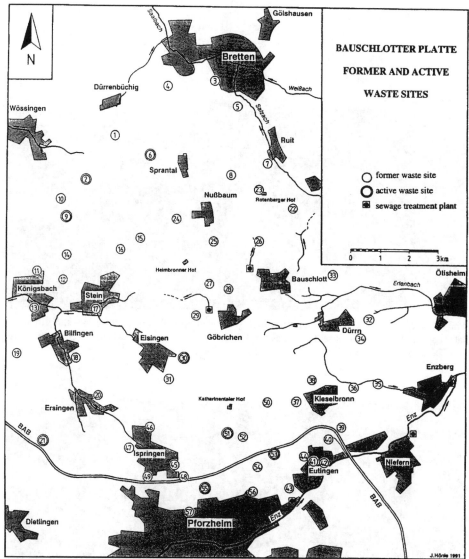

Figure 5.15. Map of the karst area of Bauschlotter Platte (Middle Triassic limestone) with the distribution of inactive and active dumps and sanitary landfills (Hötzl 1995).

that no emission can damage the surrounding environment. The direct impact on karst groundwater by solid waste is less important. Most dangerous are the leachates; in the case of leaking waste sites they can migrate to the aquifers and contaminate an otherwise usable water source. In karst aquifers the discharge may occur very quickly and thereby pollute the aquifer system over large distances in a short time.

Leachate is formed from the liquid in the waste as well as by infiltrating rain or surface water leaching the solid waste (cf. Table 5.2). The organic component of garbage can have a high water content, which becomes liberated by biodegradation

Photo 5.6. A newly developed sinkhole, which was filled up with bulky refuse in a short time and later covered with soil to reuse the area as farmland; Triassic Muschelkalk Karst overlain by gravel near Donaueschingen, SW Germany (photo: Hötzl 1972).

Photo 5.7. Disposal of different types of waste (garbage, unused hay, cadavers) in a sinkhole outside Springfield, Missouri (photo: Hötzl 1997).

and decomposition. The dewatering of such a sanitary land fill can take tens of years even if it is completely protected from water access. In addition, waste sites frequently contain sludges of different origins, soil contaminated by hydrocarbons or even liquid hazardous waste. The resulting leachates are complex liquids with high content of salts, metals and organic compounds. The volume and chemical character of leachate produced by land fills varies considerably. Depending on the origin of the waste the amount of total dissolved solids may exceed 100,000 mg·l^{-1} and it may have a BOD of more than 150,000 mg·l^{-1}. High levels of nitrogen and phosphate compounds, heavy metals, toxic trace elements as well as hydrocarbons with organic solvents and pesticides may be included in the leachates from waste sites (Fetter 1993, Barker et al. 1989, Howard et al. 1996). Frequently strongly acidic and with a high reducing capacity, the leachates can induce additional solution and mobilisation in the underground.

In general many years may elapse before the landfill waste reaches field capacity and leachate is generated. Also, time is required before high concentrations are attained. The decline in contaminant concentration takes place exponentially with time, depending on the solubility of the contaminants.

Karst aquifers are generally affected very directly by leachate contamination. If there is direct access to conduits through the epikarst or via karst shafts and sinkholes the migration of the contamination may be fast. There is a considerable potential for spring pollution due to the unattenuated movement of the leachate through conduits (Edwards & Smart 1989). However, contrary to single pollution events, where conduit flow may lead to a fast flashing out of the contaminants, landfills contribute to a more or less continuous input of contaminants over long periods as was mentioned above. In case of waste sites which do not have direct connections to conduit flow, as in some former quarries, the movement of the leachate is slower thus allowing more attenuation by the time that the groundwater body is reached.

In spite of the numerous dumps in karst areas the chemical impact on karst water is comparatively small. It mainly comprises a general increase in mineralisation with high concentrations of alkali ions and magnesium, as well as of chloride and sulphate, but with most of the waters remaining within the drinking water standard (Hötzl 1995, Edwards & Smart 1989, Davis 1997). The reasons for the relatively insignificant influence of such old dumps are partly due to the type of waste thrown into the dumps (little garbage, old construction material, nearly no hazardous waste. It is also due to the generally oxidising and neutralising conditions found in many karst waters as well as to the attenuation affect at the spring outflow which rapidly dewaters even large catchment areas.

With the increasing use of toxic chemicals during the last thirty years and their partly uncontrolled disposal in dumps, the quality of karst groundwater has decreased. Even disposal in controlled and licensed landfills in karst areas is no safeguard, due to the inadequate design of the waste sites. Hoenstine et al. (1987) reported contamination of the Floridan aquifer system by synthetic organic compounds including trichlorethylene, methylene, pesticides and herbicides) from a landfill in sands and clays covering the aquifer. A clear relationship with precipitation was established, with the rainfall, flushing out the contaminants from the landfill into the karst aquifer. An interesting relationship between fires in landfills and increased flushing of contaminants from waste sites (Fig. 5.16) by fire-fighting water was pub-

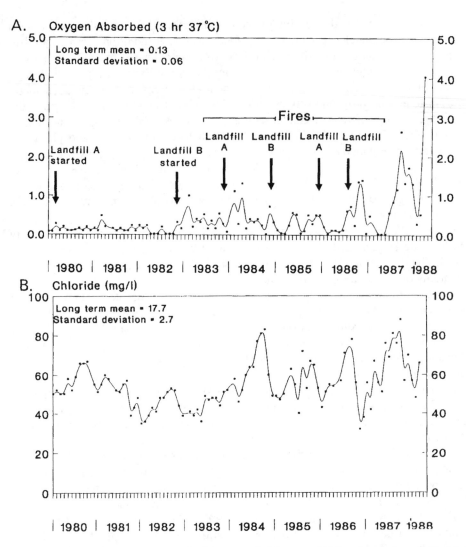

Figure 5.16. Variation in oxygen demand (adsorbed in 3 hours at 37°C) and variation of chloride in Cold Bath Spring Broadfield Down, Mendip Hills, UK., due to the build-up of two landfills about 1,2 km distant as well as due to fire events and the use of fire-fighting water in the landfills (after Edwards & Smart 1989).

lished by Edwards & Smart (1989). A further example of the effects of waste sites and their rehabilitation by remediation work is presented in a more detailed case study in Section 5.6.7.

5.4.3 *Siting landfills in karst areas*

In karst areas the possible failure of even a well designed and monitored landfill is compounded by the possibility of rapid migration of contaminants into the karst water supply through shafts, karst pipes and flow conduits. Landfills in karst areas are frequently situated in doline areas, where subsidence might suddenly be provoked and even lead to a the formation of a collapse sinkhole. The associated loss of structural integrity of the dump and the release of leachate can cause contamination of the karst water resources and may even affect human health by flowing towards the spring or well abstractions. Ultimately it would be easiest to say that no landfills should be constructed in areas of karst (Davis 1997).

In fact some of the new and more restrictive regulations regarding the construction of landfills more or less forbid the development of new waste disposal sites in karst areas. For example, in Germany new regulations were passed in 1991 (TA-Abfall) and 1993 (TA Siedlungsabfall). Both of the orders say clearly that no additional waste sites should be located in karst areas or in areas with strong jointing and preferential water paths along these joints. Due to this regulation special studies are required to be carried out in areas with slope instability as well as in areas with possible sinkhole activity.

However a general ban on waste sites in karst areas is impractical due to the prevalence of karst landscape in many countries of the world. Construction of waste sites in karst areas is a reality. In order to avoid possible impact on karst groundwater, regulations for better safeguards have to be established, which can be used to prevent the loss of structural integrity and the uncontrolled release of the leachate (Glover et al. 1995).

Consideration of possible sites for waste disposal in karst terranes requires careful screening of the selection criteria and demonstration that the selected site is hydrogeologically suitable (Hughes et al. 1994). Table 5.3 includes a comprehensive list of generic categories of information that should be considered during evaluation of potential sites for land disposal. Hydrogeologic conditions discovered during the evaluation process and which would lead to the rejection of a site include:
– Areas that contain well developed karst features;
– Areas with recent karst activity (for example: development of new sinkholes);
– Recharge areas for karst water supplies;
– Specific geologic structures (faults, seismic impact zones).
For selected sites it is necessary that the design of a landfill provides and guarantees safeguards that no emissions into the underground can take place. The necessary requirements vary depending on the special hydrogeologic conditions of each proposed waste site. That is one of the reasons that regulation even in neighbouring states or countries, differ. They are frequently rather general and do not refer to special karst hydrological conditions. Only the regulations of few countries take special care for the protection of karst aquifers.

With regard to a suitable multi-barrier concept, the design of a landfill or hazardous waste site in a karst area should include the following components:
– A low permeable or impermeable cover system emplaced as soon as possible to limit the infiltration of rain water;
– A lateral sealing system, which can result from the connection of the cover system and the basic liner system;
– A collection system for the leachate so that it can be taken to wastewater treatment plant;

Table 5.3. Important characteristics of karst terrain, which have to be considered in the selection of waste sites in karst (after Hughes et al. 1994).

Stratigraphy
(Regional and Local)
  Stratigraphic Column
  Thickness of Each Carbonate
Unit
  Thickness of Non-Carbonatic
  Interbeds
  Type of Bedding
    Thin
    Medium
    Thick
  Purity of Each Carbonatic Unit
    Limestone or Dolomite
      Pure
      Sandy
      Silty
      Clayey
      Siliceous
      Interbeds

Overburden
(Soils and Sub-Soils)
Distribution
Origin
  Transported
    Glacial
    Alluvial
    Colluvial
  Residual
  Other
Characteristics and Variability
  Thickness
  Physical Properties
  Hydrologic Properties

Hydrology
  Surface Water
    Discharge
      Variability
        Seasonal
        Gaining
        Losing
  Ground Water
    Diffusive Flow
    Conduit Flow
    Fissure Flow
    Recharge
    Storage
    Discharge
  Fluctuation of Water Levels
  Relationship of Surface-Water and
  Ground-Water Flow

Geologic Structure
(Regional and Local)
  Nearly Horizontal Bedding
  Tilted Beds
    Homoclines
    Monoclines
  Folded Beds
    Anticlines
    Synclines
    Monoclines
    Domes
    Basins
    Other
  Fractures
    Lineaments
    Locations
    Relationship with
      Geomorphic Features
      Karst Features
      Structural Features
  Joint System
    Joint Sets
      Orientation
      Spacing
      Continuity
      Open
      Closed
      Filled
  Faults
    Orientation
    Frequency
    Continuity
    Type
      Normal
      Reverse
      Thrust
      Other
  Age of Faults
    Holocene
    Pre-Holocene

Activities of Man
  Construction
    Excavation
    Blasting
    Vibration
    Loading
    Fill
    Buildings
  Changes in Drainage
  Dams and Lakes
  Withdrawal of Ground Water
    Wells
    Dewatering
    Irrigation

Geomorphology
(Regional and Local)
  Relief-Slopes
  Density of Drainage Network
  Characteristics of Streams
    Drainage Pattern(s)
      Dentritic
      Trellis
      Rectangular
      Other
    Perennial
    Intermittent
  Terraces
  Springs and/or Seeps
  Lakes and Ponds
  Flood Plains and Wetlands
  Karst Features-Active,
    Historic
    Karst Plains
    Poljes
    Dry Valleys, Blind Valleys
      Sinking Creeks
    Depressions and General
      Subsidence
    Subsidence Cones, in
      Overburden
    Sinkholes
    Roof-collapse
    Uvalas
    Caverns, Caves, and
    Cavities
    Rise Pits
    Swallow Hoes
    Estavelles
    Karren
    Other
  Paleo-Karst

Climate
  Precipitation (Rain and Snow)
    Seasonal
    Annual
    Long-Term
  Temperature
    Daily
    Seasonal
    Annual
    Long-Term
  Evapo-Transpiration
  Vegetation

- A basic liner system which normally comprises either a composite liner or even a double liner;
- A geologic barrier, which means in karst areas at least a sufficient overburden above the karstified carbonate rocks to avoid seepage of the leachate into the underground;
- A control system including regular observation and/or measurement of the relevant parameters.

Photos 5.8 and 5.9 show the preparation of the extension to a hazardous waste site in a karst depression of Triassic limestones in SW Germany. Examples of design and their realisation in practice are discussed by Edwards & Smart (1989), Hughes et al. (1994) or Davis (1997).

5.5 IMPACTS FROM CONSTRUCTION MEASURES

5.5.1 *Urbanisation and construction*

Urban expansion in karst environments has a diverse and lasting influence on the groundwater in the karst systems (Newton 1984, White 1988, Gams 1987, Fabre 1989, Reeder & Crawford 1989, Gams et al. 1993). First, all the effects of the overall development of constructions in an urban area like buildings, roads and carparks have to be considered. All of these surfaces are largely impervious and therefore they reduce infiltration and increase surface runoff. Beside the possible flooding of the cities (Day 1993, Nicod 1991) the channelled runoff that is frequently discharged to rivers or the sea can be responsible for a reduction of groundwater recharge and the lowering of groundwater levels. On the other hand the runoff from the towns is frequently left to its own devices outside of the urban area or even diverted into existing sinkholes. Both possibilities might lead after a certain period to the same effects, because the release of the water outside the cities will favour point infiltration with development of sinkholes and conduits. Urban storm water runoff flowing into caves often exceeds the surface water criteria for public water supplies regarding the faecal coliform, ammonia, oil and grease, chromium, lead, zinc and iron contents (Crawford 1982, Stephenson & Beck 1995). Hand in hand with the change of flow condition different types of contamination due to the use and operation of the buildings can occur. In the previous chapters diversion or leakage of sewage systems, of local septic tanks or leaking of hydrocarbons from petrol stations or from road accidents have been described already.

Attention may be drawn here to construction work and single buildings. This applies to the excavation, construction of foundations and erection of a structure and to any directly connected modification of the land surface. In general, construction work and single buildings do not have an immediate and direct negative influence on karst groundwater, insofar as special care is taken that no pollution can result either from construction material nor leaking vehicles and engines or from any sanitary release from the construction camps and buildings. The major problem is caused by excavation and embankment cuttings during which the mostly thin soil cover on karstified rocks is removed and cavities or shafts are frequently opened, so that surface water has direct access to the deeper karst system (Photos 5.10 and 5.11).

Photo 5.8. The toxic waste site at 'Billigheim' situated in a karst depression in Triassic limestone and covered with loessic loam, north of Heilbronn, SW-Germany. The photo shows reconnaissance drilling for the extension area and toxic waste (left) of the first deposition phase (photo: Hötzl 1986).

Photo 5.9. Extension area of the toxic waste site at 'Billigheim', north of Heilbronn, SW-Germany. Due to the location in a karst depression a multi-layered liner system (clay liners, HDPE liner) with a checkable collector system for waste leachate was built (photo: Hötzl 1987).

Photo 5.10. Karst cavities opened by excavation work on a construction site for a public building, Triassic Muschelkalk, Ispringen, SW Germany (photo: Hötzl 1974).

Photo 5.11. Karst joints exposed by construction measures. In the background is a reinforced concrete beam bridging the karst feature, Triassic Muschelkalk, Ispringen, SW Germany (photo: Hötzl 1974).

Another effect of construction measures is the triggering of sinkhole development by blasting or excavating and so impairing rock stability or by diverting and discharging surface water underground. Similar effects may be caused by the use of completed buildings, if drainage, e.g. roof water or sewage or leaking swimming pools or septic tanks (Moore 1995), is diverted underground. Frequently it may be observed that a few years after the activity has been completed sinkholes occur, where before none or only few widely distributed karst features were known. Less attention was paid to the aspect of endangering groundwater in the past. Quite the reverse; the opened cavities were used for discharging surface water and effluents from septic tanks or even for direct inflow of waste water and manure.

Engineers and hydrogeologists were more concerned that possible sinkholes or cavities might impair the construction or the completed buildings (Reuter & Tolmacev 1990, Beck 1984, Beck & Wilson 1987, Prinz 1982, Mishu et al. 1997). Devilbiss (1995) has focused on the necessity to address adequately the impact of karst on development and as well as the impacts of development on karst. Guidelines for construction practises in carbonate rock areas are required, in order to avoid unnecessary risks for the karst water. These shall include recommendations regarding:
– The uncovering and excavation of karstified rocks;
– The filling and sealing of exposed open karst features;
– Drainage diversion especially of contaminated water.

5.5.2 *Traffic routeways*

Road and railway tracks do not differ substantially from other constructions with regard to the problem of karst groundwater pollution. Their construction can involve deep cuttings into the karstified rocks removing the protective cover and exposing karst joints, shafts and caves. At some places tunnels are necessary and are constructed penetrating the mountain ridges sometimes partly below the watertable, to shorten the distance and to avoid steep gradients.

Beside the exposure of karst features, construction practices can cause the development of additional sinkholes as mentioned in Section 5.5.1 (Newton 1987, Moore 1995). Emplacements of weights above thin-roofed existing cavities; vibration resulting from blasting or heavy traffic or the concentration of water by drainage diversion to and through existing openings in the bedrock may all cause rock failure. Collapse also can result where surface water gains access to uncased or unsealed boreholes. The resulting collapse mechanisms are piping process and saturation. If storm runoff is diverted to sinkholes or other karst openings careful design and construction measures are required (Fig. 5.17).

The main concern for the karst water resources does not derive from the development of collapse structures, which is of course important for the safety of the traffic system (Reuter 1962, Prinz 1980), but from polluted water or toxic liquids, which might flow quickly through the open karst features down to the groundwater. Road discharge is known for heavy metals like lead, zinc or copper, for polyaromatic hydrocarbons (PAH), for benzene as well as for gasoline, diesel, oil and grease (Stephenson & Beck 1995). The impact of salt used for de-icing of the roads in winter may cause strong seasonal effects (Reeder & Day 1993, Bakalowicz et al. 1996, Werner 1983, Saleem 1977). Discharge from railway tracks might be polluted by release of

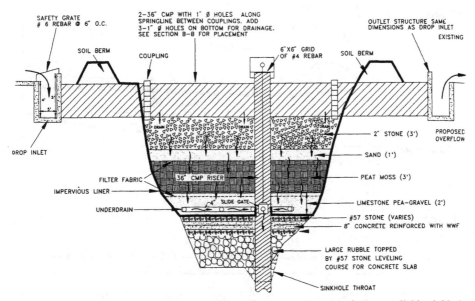

Figure 5.17. Design for a highway stormwater runoff treatment system using an available sinkhole. Proposed for the stormwater runoff at the I-40/I-640 interchange in eastern, Knoxville, Tennessee (after Stephenson et al. 1997).

waste water as well as by the application of herbicides to reduce plant growth along the tracks.

In order to avoid direct impacts from contaminated runoff from the roads, protection measures for karst groundwater are required. The scope of the required safety measures depends on the hydrogeological situation, the degree of karstification and the importance of the karst groundwater resource. In general, large amounts of road runoff should pass through an oil separator. Special treatment systems are required for improving the quality of groundwater. The treatment system will involve a combination of sedimentation, filtration and adsorption especially designed to treat the first flush of stormwater runoff, which is the most heavily contaminated (Beck et al. 1996, Stephenson et al. 1997).

Road and highways need special safety measures if they are planned to cross catchment areas for drinking water abstractions. In Germany for example, the regulations forbid, even in areas of porous aquifers, the construction of new roads within the inner protection zone (zone II) of existing water supplies. In any case special constructions are required to avoid, in case of an accident, the outflow and infiltration or sinking of liquid contaminant or polluted water underground (Fig. 5.18). With regard to the protection of used karst groundwater a certain roadway design should become standard:
– No discharge of untreated road runoff into the underground;
– Impermeable road pavements;
– Impermeable linings for drainage swales and water courses;
– Marginal demarcation walls or dams to avoid an accidental break out by vehicles.
Existing regulations or recommendations still pay more attention to highway safety

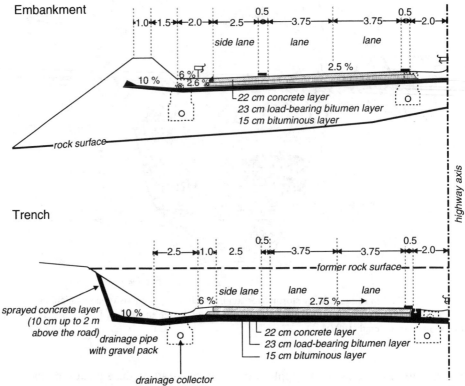

Figure 5.18. Standard profile of the motor highway A7 (Ulm-Würzburg) for sections crossing groundwater protection zones (after Flinsbach & Drescher 1984) (distances in m).

than to environmental aspects (Fischer et al. 1993, Bonaparte & Berg 1987, Moore 1984).

Special attention has to be given to traffic tunnels. In the past they were mainly constructed for the railway but they have become a more frequent solution also for highway planning today. Tunnels in karstified rocks in any case present a high risk potential independent of whether they penetrate the vadose or phreatic zone of the karst system. In the vadose zone tunnels remain frequently without any reinforcement and sealing due to the stability of the host rocks. But in the case of a loss of liquids or a leaking vehicle or container, fast contamination of the karst water will occur without any realistic possibility of reduction. Therefore within the catchment of water supplies tunnels should be designed and constructed as impervious pipes.

In the phreatic zone the excavation of galleries demands either a lowering of the water level or sealing measures ahead of the gallery. If the sealing is not done ahead, far reaching depletion of the water level will occur within the karst system leading to a drying out of springs and rivers. The inflow of water into the gallery can happen surprisingly quickly when karst conduits are opened. Frequently the sudden intake has catastrophic consequences for the whole construction enterprise as well as for water supplies in the surroundings.

A more detailed example is described in Section 5.7.7 by Mundry and Chauve referring to the excavation for the Mont d'Or Railway Tunnel in 1912. The main impact of such a tunnel on karst water is the strong quantitative reduction of the natural resources by the artificial drainage. Many negative examples are known from different countries (Nicod 1995). Photo 5.12 shows the constant outflow of the Karawanken Highway Tunnel (excavation 1987-1991) at the Austrian-Slovenian border with more than 300 $l \cdot s^{-1}$ which is contributed by a karst system opened by the tunnel below the local base level. As a result springs dried up and the discharge of the river upstream of a small electric power plant showed a water loss of the same amount as that which is now flowing out of the tunnel. The blocking of the inflow into the tunnels and the sealing of the gallery can restore the old discharge conditions in the karst system.

5.5.3 *Impoundments and reservoirs*

The storage of water in depressions on the surface usually impounded by a dam, serves different objectives: for hydroelectric power plants, for collection of irrigation or drinking water, for installation of recreation centres or for flood control. The construction of dams and reservoirs in karst areas requires special engineering and hydrogeologic preconditions regarding the tightness of the basin as well as of the dam site.

The principal problem is karstification which is associated with deep solution

Photo 5.12. The north portal of Karawanken highway tunnel, Tauernautobahn, near Villach, Austria-Slowenian border. On the left side is the outflow of karst groundwater (about 300 $l \cdot s^{-1}$) draining from the tunnel (photo: Hötzl 1993).

cavities providing bypasses or drainways along which the water may escape from the reservoir or from smaller impoundments. There are several examples around the world where the original aim to store water completely failed due to the strong water losses into the underground and the further discharge from there. The increased leakage can induce sinkhole development or new solution channels which may threaten even the mechanical integrity of the dam. Though the problems are huge there are many impoundments and large reservoirs in karst areas (Photos 5.13 and 5.14), which function very well and are sufficiently water-tight, frequently due to additional grouting and sealing of the ground (White 1988, Heitfeld 1991, Beriswill et al. 1995).

In general the storage of discharge water and thereby the balancing of seasonal variations in flow is an advantage especially in karst areas with high seasonal variations. However, damage can result from modifications to the natural recharge conditions, due to major water diversions or by the triggering of new sinkhole development and piping, which completely can change the discharge behaviour by reducing retention capacity or by changing the flow direction. As a consequence, far-reaching depletion of karst water levels with drying up of whole regions can occur. On the whole quantitative aspects are predominant, while qualitative aspects of karst water composition are commonly less altered.

The most traditional human intervention in the storage and discharge behaviour of the natural water cycle in karst areas is connected with the discharge of swallow holes (ponors). In periods of high precipitation some of the ponors are not able to engulf the whole amount of inflow so that backing-up can cause flooding of large areas. These effects are particularly well-known in the case of the inundation of poljes. In order to avoid a blockage of the ponors by timber debris and fine sediments and to encourage the fast discharge of the stored water people started, even in historic times, to build gabions or gate-like swallet protections (Gospodaric & Habic 1986). The concept is to improve the runoff flow into subsurface cavities by removing debris and trees from around the throat of a swallet and to prevent the cavity opening from clogging with debris. Photos 5.15 and 5.16 show such simple construction measures in front of the Katavothre of Kapsia in the Polje of Tripolis, Greece, while Photo 5.17 shows a more reinforced 'gate' in front of the ponor of the Elmali polje, west of Antalya, Turkey. Rapid discharge through the main ponor allows the land to drain more quickly but the total retention capacity of the karst aquifer is reduced proportionally.

5.6   IMPACTS FROM TOURISM (Hans Zojer)

5.6.3   *General problems*

Human impacts on karst groundwater due to outdoor leisure and recreation activities fall into two categories:

1. People take part in touristic or sports activities, which are not specific to any geological environment. Thus the settings for skiing and golf courses for example are not related to any geological features.

2. People sightseeing, visit fascinating scenery, perhaps specifically related to

Photo 5.13. Construction work for the Mosul Dam, Tigris, Iraq. In the centre is the excavation in Tertiary limestone and construction of the main control gallery for the dam (photo: Hötzl 1984).

Photo 5.14. Trebisnjica river with a concrete arch dam, East of Dubrovnik in the Dinaric Karst (photo: Hötzl 1985).

Photo 5.15. The Katavothre of Kapsia (swallow hole) at the northwestern margin of the Polje of Tripolis, Peloponnesus, Greece. The swallow hole is protected by a wall with small openings to prevail a blockage by wood and other debris during the flood period (photo: Hötzl 1985).

Photo 5.16. The Katavothre of Kapsia (swallow hole) at the northwestern margin of the Polje of Tripolis, Peloponnesus, Greece showing details of the swallow hole with part of the protection wall built in order to improve swallow hole capacity (photo: Hötzl 1985).

Photo 5.17. The sluice-gate structure for controlling and improving the water intake in the main ponor of the Elmali polje, west of Antalya, Turkey (a) and the opposite view with the ponor and the sluice gate in the foreground and the Elmali polje in the background (b) (photo: Hötzl 1990).

karst phenomena. They are impressed by surface landforms (karren, towers) and also by subterranean phenomena (caves, collapse structures), impressive springs or travertine deposits.

The practice of such recreational activities like sports undoubtedly affects the environment in many ways (Photos 5.18 and 5.19). It is of course not the sport or the tourism itself, which is the problem, but are the accompanying features like increased traffic, construction work, additional settlements, houses and hotels with all their typical features producing more waste and waste water in areas which show special sensitivity and vulnerability and which were not touched before. In a certain way it is similar to the onset of urbanisation with all its negative impacts in a certain area, which may have remained in a natural condition until then. In karstic areas the high permeability of the rocks causes a rapid infiltration of contaminants, which might result from touristic activities, by the subsequent fast flow it may lead to wide contamination of these important water resources.

Touristic activities in karst regions are frequently associated with uncontrolled waste and sewage disposal from single refuges, tourist lodges and hotels. For the disposal of refuse, swallow holes and dolines are used (cp. Section 5.4) Hence rapid transport of dissolved contaminants along solution conduits and fractures takes place. Pollutants can migrate underground also from diffuse superficial sources like alpine pastures, which are becoming more and more important to tourism in combination with individual touristic activities (Photos 5.20 and 5.21).

5.6.2 *Point and diffuse sources*

In mountain regions the movements of people for vacation may be seasonal (summer and winter sports) or all year round, depending on the interest of the landscape and the transportation infrastructure. The consequence of this development is the construction of refuges, hotels and restaurants in high elevated regions. The running of such business generates a certain amount of solid and liquid wastes which are rather rarely disposed of on the basis of legal regulations. Two examples may be given from the Austrian Alps: the Dobratsch karst massif and the Höllengebirge.

The Dobratsch mountain, located in the Southern Alps of Carinthia, is well served by an asphalt road on the one hand and a number of chair-lifts on the other hand, hence readily transporting a large number of people up to elevations of 2000 m a.s.l.. The massif itself consists mainly of Mesozoic limestones and dolomites crossed by major faults, cutting the mountain into several blocks with different altitudes (Stini 1937, Anderle 1950, Holler 1976). The tecto-morphological development of the area causes a general subsurface karst water drainage to the east to a number of springs. The main spring, called Union spring, provides drinking water for the town of Villach and has been studied with regard to its recharge characteristics and especially whether the touristic activities on Dobratsch mountain are affecting any changes in the water quality (Kahler 1983, Zojer 1980, Zojer & Zötl 1993, Poltnig et al. 1994).

Investigations on hydrochemistry, environmental isotopes together with artificial water tracings have been carried out. Oxygen-18 data show that the recharge area of Union spring is located in the medium altitude region of the mountain's flanks. The summit area and the lower foot hills of the karst massif drain towards other sources belonging to different systems of aquifer recharge. The results of stable isotopes

Photo 5.18. Tennengebirge, Northern Limestone Alps, Austria. The vegetationless karst plateau with some large karst depressions is a favoured area for mountaineering and climbing. Due to a general karst water protection programme the area will be kept free from more intensive touristic facilities (photo: Hötzl 1982).

Photo 5.19. The Dachstein Plateau, Northern Limestone Alps, Austria. In the foreground the Hall-stätter Glacier, in the background the peak of the Hoher Dachstein (2995 m a.s.l.). The plateau is highly touristically developed with access by cable railway. The glacier is used for cross country skiing (tracks in the snow) for the whole year. In spite of a sewage pipeline down to the valley negative effects on karst groundwater can be observed (photo: Hötzl 1988).

Photo 5.20. Sinkhole in the Rax Mountain, Northern Limestone Alps, Austria, used as a dump by mountain hikers and alpine shepherds (photo: Stummer 1990).

Photo 5.21. Kaiserbrunn, capture of a big karst spring at the base of the Schneeberg, Northern Limestone, Austria. The spring has been used for the Vienna water supply for more than 100 years. The protected catchment is strictly controlled by the water supply company (photo: Hötzl 1995).

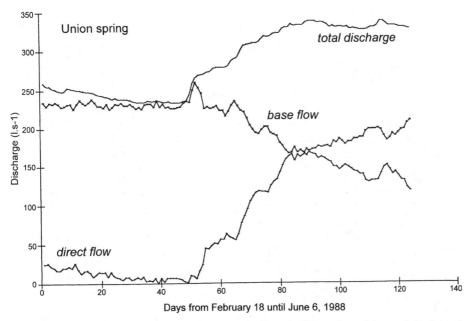

Figure 5.19. The Union spring, Dobratsch Mountain, Austria. Separation of the total discharge by calculation from the magnesium content in direct flow and basis flow component all in $l{\cdot}s^{-1}$. The 140 days of the basis line represent the period from February 18 until June 6, 1988.

were confirmed by a combined tracing experiment to provide clear evidence that the replenishment area of Union spring covers a wide belt of intensive touristic utilisation. The effect of snow melt in spring time on the karst water dynamics was studied using sodium and magnesium (Fig. 5.19) as natural tracers. It shows that at the end of the dry period (late winter time before snow melt) practically the whole discharge of the spring originates from a well mixed karst system. Generally speaking it is evident that even in the time of maximum aquifer recharge the portion of base flow does not fall below 50%. and suggests a high storage capacity for the karst aquifer.

As a consequence of this knowledge of aquifer dynamics and the catchment it is possible to establish appropriate measures for the protection of karst water resources. In particular it highlights the conflict between the impacts of summer and winter tourism on the karst landscape and the utilisation of karst water for drinking purposes. It is not possible to restrict certain activities to those areas which are outside of the spring catchment.

The Höllengebirge is a typical alpine karst massif with a distinct plateau landscape. Water quality problems at the springs located radially around the mountain derive from a number of refuges and shelters on alpine pastures which have become more and more attractive for individual tourism.

The upper part of the Höllengebirge consists of Triassic limestones and is heavily karstified (Benischke 1993, Benischke et al. 1988) in contrast to the lower unit of mainly dolomites (Fig. 5.20). Deep faults gave rise to the development of large cave systems with almost 900 m differences in elevation. These caves are situated on the northern edge of the plateau, and dye tracer tests show clearly a rapid subsurface

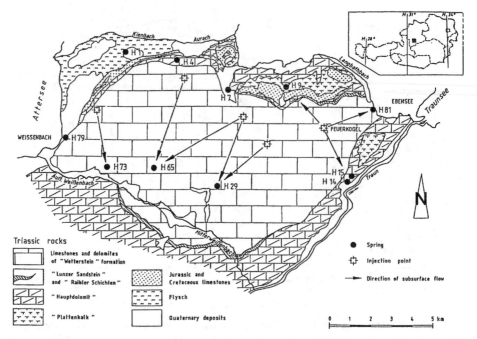

Figure 5.20. Geological sketch map of the Höllengebirge, Austria showing flow directions of the underground karst system derived from tracer experiments.

flow to the springs in the South. This fact indicates that the catchment areas of these springs comprise the whole plateau of the karst massif, where the precipitation infiltrates in situ, following deep karstified faults.

Environmental isotope measurements and chemical analyses were carried out from selected springs and demonstrated that the storage capacity of the aquifer is rather limited due to the advanced karst development. Thus the turnover time of karst water does not exceed half a year and the maximum water volume of the major springs ranges between 30 and 40 $m^3 \cdot s^{-1}$. These hydrogeological features make it very difficult to reconcile the demand for high water quality at the springs with a sustainable utilisation of the landscape. The treatment of solid and liquid wastes even in remote areas of the karst plateau in any case is essential.

The development of skiing tourism in alpine regions during the last decades has created a number of ecological problems and locally even of disasters, some of which are specific to karst areas (Hoblea 1995):

– The construction of skiing courses in difficult terrains requires the removal of soil cover. Then the exposed rocks must be revegetated in order to stabilise the hydrology and the eco-system. In extreme high altitude regions such attempts are difficult because special resistant seeds are not available. Unfortunately this lack is in some cases over-compensated for by a too intensive fertilisation of the artificially prepared soil thus effecting a diminution of karst water quality by rapid infiltration of contaminants.

– The snow-preparation of skiing courses with chemicals like calcium chloride or nitrogen fertilisers also endangers the quality of karst water resources since surface runoff is minimal and practically all melted snow infiltrates underground.

– The production of artificial snow raises ecological problems as in many cases contaminated surface water is used for snow-freezing and this migrates to the subsurface during the snow melt. Especially in karstic areas, filtration processes are minimised if the soil cover is diminished or removed and pollutants can reach the karst groundwater almost without impediment.

5.6.3 *Historic and sightseeing sites*

Because of the gathering of people at historical places, considered as attractions for the tourist industry, such sites are often the location for groundwater pollution. With reference to karst regions, such locations are usually located preferentially at springs in the lowlands or in the foothills of mountains. Some of them, especially castles for defence, have been constructed in higher elevations or even on the top of mountains. Since they are situated in the immediate recharge area of karst springs and groundwater, they represent a hazardous source for contamination. These problems can be recognised in many places associated with Greek and Roman history but also at tourists attractions in other societies all over the world. The Yucatan Peninsula with its Mayan antiquities, described in Chapter 2, is an excellent example of such pressure on karst water resources.

Caves are most fascinating as underground scenery which makes them attractive not only for cavers but also for ordinary tourists (Photos 5.22 and 5.23). Many of the caves are speleologically well investigated and prepared for visitors with specific lighting effects and exotic designations of calcite forms:

– The Postojna cave is one of the most famous underground karst locations in the world that is accessible for visitors. Here environmental hazards generally originate not from the controlled tours in the cave system itself but on the surface above (Knez et al. 1995). Pollution comes from camping sites, small villages without sewage treatment and an abandoned missile base. Investigations in the cave from percolation as well as from trickle water even show a clear reflection of the daily pattern of the use of the camp showers.

– The Hölloch in the Swiss Alps is not developed for mass tourism but is much favoured by cavers and speleologists (Bögli 1980, Bögli & Harum 1981, Jeannin et al. 1995). From the environmental point of view the well known cave system is rather out of the way of the main traffic routes and, which might be even more important, there exists no intensive landuse in the surroundings and within the watershed contours. Therefore pollution is very limited and is additionally suppressed by strict environmental laws.

– Yucatan cenotes are in some cases the result of collapsed dolines, located on tectonic lines. Many of them can be visited by tourists, if the walls are not vertical and steps have been carved along sloping sides down to the karst water table. These features are also windows into the main karst water systems are thus highly vulnerable to pollution.

Photo 5.22. Sinter (speleothem) formations in caves can attract mass-tourism with the consequences of a disturbed environment and polluted karst water (photo: Hötzl 1991).

Photo 5.23. The spring of Düden basi, Travertine Plateau, Antalya, Turkey. The impressive collapse structure and spring cave form a special tourist attraction. The great number of visitors cause environmental problems with water pollution downstream. (photo: Reichert 1995).

5.6.4 *Impacts in the surroundings of karst springs*

As mentioned above the surroundings of karst springs during all historical periods have been used preferentially for settlements. Many of the big springs are captured for water supply and partly for irrigation. Adverse effects on the quality of spring water can take place not only in the recharge area but also at the outlet itself, when the water is not used properly. The Planiteros karst springs are an example of such a misuse. They emerge from Triassic limestones in the high northern part of Peloponnesus peninsula, Greece (Morfis & Zojer 1986). The location of the individual outlets, is shown in Figure 5.21. The total discharge ranges between 1 and 6 $m^3 \cdot s^{-1}$. Water temperature and conductivity of the single springs differ to a degree due to the different recharge patterns. Within recent years the use of spring water has changed considerably. In the early 1980's only a small amount was taken for water supply (especially at the northern margin of the spring district). In addition one fishery was supplied. Since then the spring area has been commercialised to the disadvantage of water quality. Additional fisheries and restaurants were established, retail shops were erected offering typical touristic products, a dyer's enterprise for clothes was set up and local people enjoy themselves at a large variety of official places for picnics. All these activities decrease drastically the quality of the water at the spring sources and

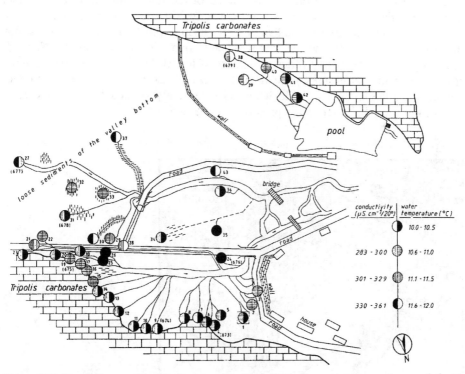

Figure 5.21. Detailed map of the Planiteros spring district including water temperatures and electric conductivities (April 27, 1983) Peloponnesus, Greece. Springs in the eastern part and in the valley floor are endangered by increasing numbers of settlements and touristic activities.

downstream (Reiss et al. 1986). Appropriate measures to regulate these touristic impacts are urgently required.

Thermal springs from karstic aquifers indicate a deep circulation of the infiltrated water and usually a reasonable storage capacity of the aquifer. Pollution of the thermal water occurs in practice only after its usage for recreation purposes and this can be avoided by a careful utilisation of the fluid. The precipitation of calcium carbonate as a consequence of over saturation with respect to carbonate minerals creates fascinating and different coloured landforms in combination with overflowing water close to the karst springs. Therefore it is obvious that such regions become an attraction for tourism. On the other hand, free access to travertine sites greatly increases the possibilities for environmental pollution. Three examples, the famous Pamukkale sinter terraces, Turkey and the famous Lakes of Plitvice, Croatia as well as the Nerja cave in southern Spain are described in the subsequent Sections 5.7.9 to 5.7.11.

5.7   CASE STUDIES

5.7.1   *Air pollution in the Lelic karst of Serbia*
(Neven Kresic, Radisav Golubvic & Petar Pavic)

The Lelic Karst belongs to the Ophiolite Belt of the Inner Dinarides in the former Yugoslavia (Fig. 5.22). It occupies approximately 250 km^2 of the intensively karstified outcropping limestones. The karst aquifer itself is much bigger and extends further to the northeast below the low-permeable ophiolitic rocks and the Neogene sediments (Fig. 5.23). The average thickness of the Middle and Late Triassic limestones, the main aquifer units, exceeds 600 m. Lelic Karst owes its name to a famous Serbian karstologist Jovan Cvijic who had extensively studied it during the early 20th century, and had used it as a prototype of merokarst.

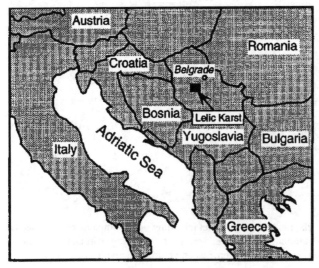

Figure 5.22. Location of the Lelic Karst.

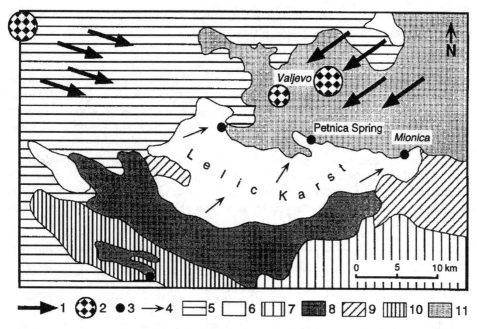

Figure 5.23. Simplified geologic map of the Lelic Karst with major wind directions and sources of air pollution. 1) Direction of prevailing winds, 2) major source of air pollution, 3) large karstic spring, 4) general direction of karst groundwater flow,; 5) Paleozoic shales: impermeable aquifer base, 6) karstified Triassic limestone: karst aquifer, 7) ultramafics: side-barrier and/or low-permeable cover, 8) Diabase-Chert Formation: side-barrier and/or low-permeable cover, 9) Cretaceous flysch: low-permeable cover, 10) Cretaceous limestone, and 11) Neogene sediments: low-permeable cover.

The recharge zone of the Lelic Karst aquifer (its limestone outcrop) is a thinly populated low-mountainous rural area without any industrial facilities or intensive agricultural/fertilising practices. Groundwater discharges from the Lelic karst through several large springs along its northern edge. The annual discharge is almost 90 million m^3, only a small part of which is being used for the public water supply of the City of Valjevo and other urban areas. However, this karst groundwater usage is planned to be more intensified in the near future. The quality and quantity of the Petnica karstic spring discharge and the precipitation in its drainage area have been studied continuously since 1990, while occasional measurements exist for a longer period (Papic et al. 1991, Kresic 1991, Kresic et al. 1989).

Sources of air pollution
The Petnica spring is located 7 km south of the city of Valjevo, one of the major industrial centres in Yugoslavia. Near the city is also one of the major coal-burning power plants in the country (Fig. 5.23). In the path of the prevailing winds from north-east are several other industrial centres including Belgrade as the largest one, as well as other large coal-burning power plants, all within 100-km range. Another direction of prevailing winds from west brings air pollution from the North-Central

Bosnian industrial basin which lies at a distance of 30-60 km, and is 'crowded' with ore smelters and coal-burning power plants.

Precipitation quality
Chemical composition of the precipitation at a representative gauging station in the Petnica spring drainage area has been monitored on a daily basis since 1990. Figure 5.24 shows scattered diagrams of several interrelationships between the amount of precipitation, total dissolved solids, pH, and nitrates which are, together with sulphates, the main air pollutant in the area.

As it can be seen, almost 30% of all registered amounts of daily precipitation belong to 'acid rains' (data are for 1991). Even though it is not statistically obvious (correlation coefficients are statistically insignificant), there is a certain 'visual' shift on the graphs pH versus the amount precipitation, and pH versus nitrates. This shift indi-

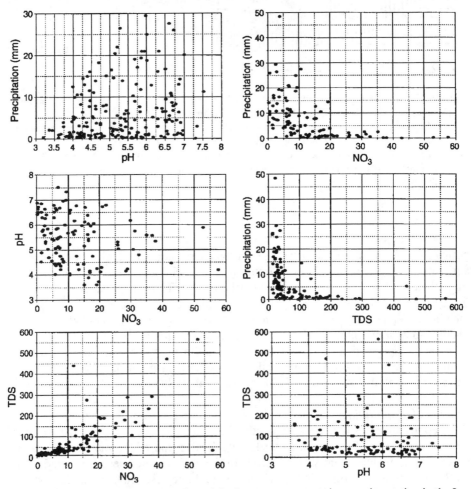

Figure 5.24. Characteristic elements of precipitation at a respresentative gauging station in the Lelic Karst. TDS = total dissolved solids (mg·l^{-1}), NO$_3$ = concentration of nitrates (mg·l^{-1}).

cates two facts that are to be expected: pH is generally lower with the lower precipitation (dilution effect is less pronounced), and lower with an increased contents of nitrates. Much more obvious are non-linear interrelationships between total dissolved solids (TDS), nitrates and the amount of precipitation, both visually and statistically. Nitrates and TDS increase with the decreasing amount of precipitation which, again, is a consequence of lesser dilution.

Response of the karst aquifer to precipitation
The influence of precipitation on the Petnica karst aquifer is easily determined through cross-correlation and cross-regression analyses of the relationships between various elements of precipitation and spring discharge. Figures 5.25 and 5.26 show input-output relationships between the amount of precipitation and nitrates as inputs, and the spring discharge and nitrates as outputs. Nitrates are chosen because of their very low reactivity in the carbonate environment compared to sulphates, and because some historic data on their contents in the Petnica spring water were available.

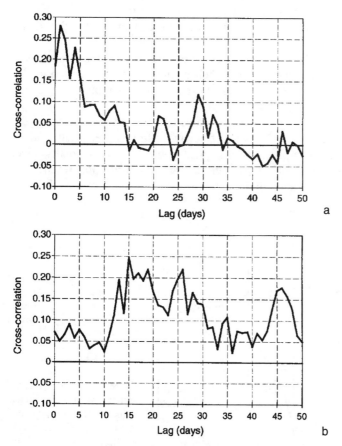

Figure 5.25. Cross-correlograms between the Petnica spring discharge and the amount of precipitation (a), and between the content of nitrates in the precipitation and in the spring water (b).

The cross-regressive model of order q used to examine the input-output relationships is:

$$Q_t = a + c_1 P_t^{-1} + c_2 P_t^{-2} + ... + c_q P_t^{-q} + e_t$$

where Q_t is the predicted output (spring discharge or the contents of nitrates); P_t^{-1}, ..., P_t^{-q} is the input (amount of precipitation or the contents of nitrates) for 1, ..., q preceding days; and a, c_1, ..., c_p are model parameters; e is the model error.

The cross-regressive model (Fig. 5.26) uses only gross precipitation and the time distribution of effective infiltration remains unknown. This introduces a lot of zero values into the multiple cross-regression and significantly decreases its coefficient. However, although the model, in its simple form, would not be a good predictive tool, it does provide a valuable information on the time dimension of the input-output relationships.

A comparative analysis of cross-correlograms in Figure 5.25 and cross-regression in Figure 5.26 shows the following:

– The response of spring discharge to precipitation is almost immediate, with the lag of just one day. The influence of precipitation on discharge decreases quickly, and judging from both graphs it is statistically insignificant 5 to 6 days after the precipitation events. In other words, the aquifer has a 'short memory' for this input-output relationship.

– For about 13 days after the main precipitation event, the contents of nitrates in the spring water is completely independent of their contents in the precipitation. After 13 days, however, this dependence is apparent and increases rapidly, remaining statistically significant for the next 2 weeks or so.

This mechanism can, at least in part, be explained by an initial response under pressure to heavy rains, and a gradual inflow of newly infiltrated precipitation afterwards. In addition, common precipitation subsequent to heavy rains is usually lower and with a higher concentration of nitrates. Its decelerated infiltration reaches the spring after 2 weeks on average and continues to directly contribute to the discharge for next 2 weeks.

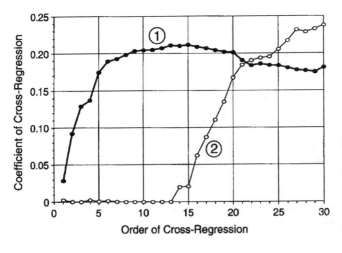

Figure 5.26. Cross regression between the Petnica spring discharge and the amount of precipitation (1), and between the content of nitrates in the precipitation and in the spring water (2).

Conclusion

The results of precipitation quality monitoring in the Lelic karst area indicate a frequent occurrence of acid rains with pH lower than 5.0. An increased content of sulphates and nitrates in precipitation (up to 60 mg·l^{-1}) is attributed to urban and industrial sources in the nearby city of Valjevo, a major power plant in the vicinity, as well as to several large industrial complexes north-east and west of the area. Inputs of nitrates into the karst aquifer cause a steady and statistically significant increase of the nitrate contents in the Petnica karstic spring waters. This increase, on an annual basis, has a general lag of about 2 weeks after the precipitation events. The highest recorded concentration of nitrates in the spring water is 27 mg·l^{-1} (in 1991) which is considerably more than 6 mg·l^{-1} recorded in the 1976-1978 period, the last one with data available. This dramatic change clearly is a consequence of a rapid industrialisation of this part of the former Yugoslavia. It also shows that non-industrialised and unpopulated drainage basins in karst do not necessarily provide good quality groundwater if they are exposed to acid precipitation. The negative influence of air pollution on karst aquifers may be apparent, as is the case with the Lelic karst, or completely attenuated, with all possible interstages. The final outcome depends on numerous factors, including presence, nature and thickness of the soil cover (or overlying rocks), vegetation, and the degree of karstification which governs groundwater flow velocities and residence times.

5.7.2 *Historical report on typhoid epidemics* (Heinz Hötzl & Werner Käss)

5.7.2.1 *Bacteriological problems in karst*

The infiltration and flow conditions in a karst system favour the transport of microorganism. Due to the wide openings of the water paths the efficiency of mechanical filtration as well as of adsorption is very low in karst aquifers. Additionally the high flow velocity leads to a fast transport through the whole system, which can cause breakthrough of the micro-organism within their survival time. Breakthrough and output of entero bacterias, bacteriophages and viruses normaly do not occur after longer underground water passages especially in porous aquifers but are frequently observed in karst water outflows according the above mentioned aquifer properties.

In the case of the use of karst water for drinking purposes, generally a careful hygienic control is carried out. The relevant routine examination refers mainly on the total number of micro-organisms and includes additional observations on entero bacterias like Escheri coli or coliform bacterias. They are an indicator of the possible occurrence of other pathogenic micro-organism. Karst waters which have shown such hygienic problems in the past are sterilised continuously by treatment with chlorine or ozone. In the case of temporary contamination, which is observed frequently with high water a continuous monitoring of the turbidity might be used to know when additional chlorinating is needed. In connection with the treatment of karst water, epidemics have disappeared nearly completely, especially in countries with high hygienic standards. Therefore the awareness of risk caused by bacteriological pollution is beyond the memory of people, especially in urban areas.

But the danger still exists and pathogenic micro-organism, e.g. of cholera or typhoid, are still very widely distributed in some subtropical and tropical regions. Two examples from Central Europe from the past show on one hand the health problem of contaminated water on the other hand that contamination of karst water is not only an impact of our high industrial society of today but occurred also in the past.

5.7.2.2   *Typhoid epidemic of Lausen 1872*

Lausen (Fig. 5.27) is a small village near Basle in Switzerland. It consisted of about 90 houses with about 800 inhabitants in the seventies of the nineteenth century. From August up to October 1872 144 persons fell ill with typhoid, many of them died. The hygienist Hägler (1873) from the nearby city of Basle was investigating the origin of this epidemic. He found out that only those people had become sick who took their water from the few public fountains (Photo 5.24) being supplied from a nearby karst spring. Peoples from houses with their own hand dug water well were not involved. Though at that time nothing was known about the bacterial sources of typhoid, Hägler concluded that the 'toxic substances' has to be coming from the polluted spring water and he tried to find the source of the contamination.

The spring arises from a talus slope build up by Jurassic oolitic limestone rocks (Fig. 5.27). It was known that in connection with meadow irrigation or flooding in

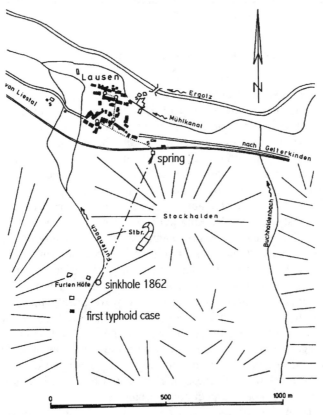

Figure 5.27. Topographic map of the village of Lausen near Basle, Switzerland, with the location of first typhoid case in 1872 and the spring involved.

Photo 5.24. One of the old village fountains supplied by the spring which was contaminated by typhoid bacteria in 1872 (photo: Käss 1996).

Photo 5.25. View from the Furlenbach valley down to the village of Lausen. On the right side is the old farmhouse, where the first typhoid sickness occurred in 1872. In the central part is the course of the Furlenbach, which sank at that time in a now closed sinkhole below the road (photo: Käss 1996).

the area of Furlenbach, a tributary creek of the main river Ergolz, as well as with the draining of that creek into a relatively new sinkhole south of the village, the discharge of the spring was increased. A detailed evaluation by Hägler showed that the first typhoid case had occurred in a remote farm house in this tributary valley (Photo 5.25) one month before the main epidemic event. The creek that flowed by the farm also carried away the sewage from the farm and percolated a short distance further downstream in the above mentioned sinkhole from the year 1862. The underground there is composed of a thick layer of oolitic limestone, which borders to the west on a fault against Liassic marls.

In order to prove this obvious source of the typhoid Hägler carried out successfully a salt tracer test in 1982, which was the first scientifically performed and semiquantitative recorded artificial tracer experiment (Käss 1992, Hötzl et al. 1991). For the tracer experiment 3000 l water with a salt content of about 900 kg from the nearby saltmine Schweizerhalle was injected in the sinkhole on September 7, 1872. After one day the water supply showed a strong reaction to chloride with silver nitrate.

5.7.2.3   Typhoid epidemic of Paderborn

The city of Paderborn was hit several times by typhoid and cholera epidemics in the last century. The city is located in the transition zone of the Egge mountain and the Münster basin in Westphalia, NW Germany. It dates back more than 1200 years, when Charles the Great founded one of his famous Pfalz castles just at the outlet of one of the karst springs forming the Pader River (Figs 5.28 and 5.29).

Over centuries people were taking out water just from the springs. In the 16th century the first pumping stations were installed to bring the water up to the distribution points in the town. In connection with a new pumping station most of the houses in the central part were connected to tapwater by 1888. After smaller typhoid and cholera epidemics in previous years more than 250 peoples fell ill with typhoid within two months in 1888. Detailed investigation first by the hygienist Gärtner (1915) proved that the reason for the epidemic was the polluted spring water. At that time it was still not fully accepted, that sickness can be caused from natural spring water. The hydrogeologic explanation was delivered by Stille (1903) in his famous contribution on the karst of Paderborn.

The Egge ridge is built of the strongly karstified Turonian Plänar limestone. The dipping of the bedding planes is orientated to the west towards the Münster basin (Fig. 5.29). On the border of the superposed Emscher marls (Upper Cretaceous) the Pader springs are located. Their catchment area on the so called Paderborner high plain shows different phenomena of karstification amongst them large sinkholes and numerous stream-sinks. The hydraulic connections of the swallow holes with the different spring groups in Paderborn had been shown already by Stille (1903) with tracing experiments. They were repeated later on from different places to define the relationship between the spring groups (Käss et al. 1996).

Two springs of the Börner Pader (Fig. 5.28) were used for the water supply at the end of the last century. They were selected due to their relatively constant temperature and discharge as well as their rare turbidity events. However the investigations of Gärtner (1915) revealed an increased number of germs even for those two springs after the typhoid epidemic. The conclusion that the typhoid bacteria were introduced

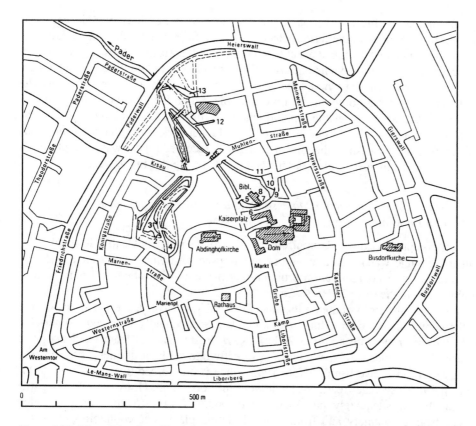

Figure 5.28. Map of the city of Paderborn, NW Germany (No. 1-13 are the main karst springs). The old central part is surrounding the main spring groups of the river Pader. The spring No. 4 is the Börnepader, where two outlets were used for the water supply of the city in the past (after Geyh & Michel 1979).

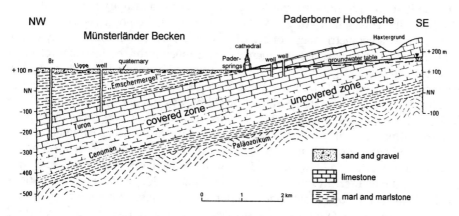

Figure 5.29. Schematic hydrogeologic section through the city of Paderborn and its surrounding (after Geyh & Michel 1979).

by one of the sinking rivers was suggested, because a previous typhoid case was known from a village in the farther catchment area. Stille (1903) did not exclude this assumption, but due to the constant behaviour of the springs showing less direct influence from stream-sinks he preferred the derivation of the pollution from the near surrounding of the spring where the centre of the city is only a few metres above the water level. The open joints of the limestone offered an ideal discharge opportunity for the sewage water. After the typhoid epidemic further use of the two springs was enabled by chlorinisation. After the second world war, when the general pollution of the springs was increasing, the city of Paderborn received a new water supply, with water coming from deep wells from the covered karst aquifer in the west.

5.7.3 *Aquifer contamination by domestic and industrial sewage waste, Central Kentucky* (Steven R.H. Worthington)

The Mississippian (Lower Carboniferous) limestone aquifer in Central Kentucky is a valuable ground water resource, and is widely used for domestic and industrial purposes. One major contamination problem that has now been alleviated is from contaminants from the town of Horse Cave.

Until 1912 the town obtained its water from Hidden River Cave, which is located in the centre of the town. Municipal water was subsequently obtained from wells, and more recently from a spring 26 km away. Waste disposal from residences and industry was by means of septic tanks or directly into wells or dolines. Hidden River Cave was a show cave from 1916 to 1943, when it was forced to close due to pollution. In 1964 a sewage treatment plant went into operation. However, the plant was not capable of adequately purifying the incoming waste stream, which included wastes from a creamery and from a metal-plating plant. The contamination of the show cave continued, and for many years the stench from the polluted water was noticeable through much of the centre of the town. In 1989 a new $13 million regional sewage treatment system went into operation, and this effectively ended the half century of gross contamination of the aquifer.

The unconfined Mississippian aquifer dips gently towards the north-west towards the Green River. Core samples indicate the limestone has a primary porosity of 3.3% and a hydraulic conductivity of $2 \cdot 10^{-11}$ m·s^{-1} (Brown & Lambert 1963). However, slug tests show a geometric mean of $5.7 \cdot 10^{-6}$ m·s^{-1}. Such enhancement of hydraulic conductivity as the scale of testing increases is a function of the inclusion of more permeable pathways at greater scales, and has been documented elsewhere in carbonate aquifers (Kiraly 1975, Sauter 1992, Rovey & Cherkauer 1995). The extent of groundwater contamination in the Horse Cave area was studied by Quinlan & Rowe (1977), who sampled 23 operating water wells between the sewage treatment plant and Green River, and also a number of springs close to the river. Quinlan & Ray (1981) carried out a comprehensive programme of tracer tests and water level measurements in domestic water wells to identify the water table, flow paths and flow velocities in the aquifer. Results are shown in Figure 5.30. Elevated concentrations of nickel, chromium, copper and zinc in a number of springs along the Green River were caused by contamination from the sewage effluent, and tracer testing confirmed the connections. The water table was found to be marked by a number of troughs,

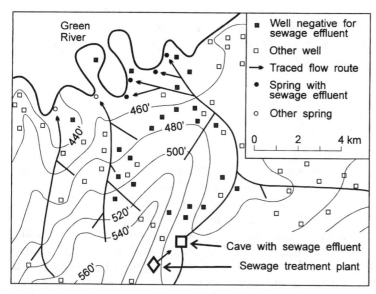

Figure 5.30. Contaminant pathways in the Mississippian limestone aquifer originating at the sewage treatment plant at Horse Cave, Kentucky (after Quinlan & Rowe 1977, Quinlan & Ray 1981). The 20 ft (6 m) spaced contours represent low flow water table elevations above sea level.

with each trough terminating in a downgradient direction at one or more springs along the Green River. Occasionally, open cave entrances allowed access to the major conduits which coincide with the water-table troughs, but most conduits appear to be below the water-table. The tracer testing demonstrated groundwater velocities of hundreds to thousands of metres per day.

The 23 sampled wells were all found to be uncontaminated, despite being located between the sewage treatment plant and the contaminated springs. The reason for this was convincingly demonstrated by the tracer testing. Flow in the aquifer is convergent to the major conduits, which offer a low-resistance pathway through the aquifer, and thus occupy troughs where hydraulic gradients are very low. The springs represent the output points for the conduits. The groundwater studies carried out in this area demonstrate a number of interesting points. Springs can be effective monitoring points in carbonate rocks, since they may discharge the integrated flow from considerable areas. Conversely, wells that are 'downgradient' from a facility are not reliable unless they can be shown by tracer tests to lie on the flow path from that facility (Quinlan 1990). Flow in most of the aquifer is convergent to major conduits, but distributaries are common close to the output point. In this case a total of 46 springs were found to be the distributary outputs from the single conduit draining the contaminated water.

The common practice in North America is to place three monitoring wells downgradient from a contaminant site as well as one upgradient well, and to sample these every three months. Such a monitoring programme is unlikely to provide adequate sampling of contaminant release where flow is convergent to conduits, as the downgradient conduit is most unlikely to be intercepted by a randomly-placed well. Furthermore, the rapid flow in conduits means that water quality sampling can be extremely variable in the short term, especially during major runoff events. The recommended protocol is to monitor springs or wells that have been shown by tracer testing to drain from the facility, and to sample intensively through major runoff events (Quinlan 1990).

5.7.4 *Pollution by surface water: The Reka River*
(Janja Kogovšek & Andrej Kranjc)

The Reka River is one of the biggest water flows of the Dinaric Karst and belongs to the Adriatic water basin. Its spring is in the Kvarner Bay recharge area (almost 20 km from the coast) below the carbonate massif of Sneznik Mt. (1796 m) at 784 m a.s.l. At the end of the river's 52 km surface course (Photos 5.26 and 5.27), it sinks at the contact with Kras (the 400 km^2 large region from where the international term 'Karst' derives) into the impressive Škocjanske Jame (underground canyon, 140 m deep, large underground chambers, one with more than 2 million m^3 volume) (Rojsek 1994, 1995). Inside the Kras the Reka waters are joined by other waters of the Kras aquifer that reappear at the surface, after 34.5 km at the Timavo Springs, close to the Adriatic coast at Devin (NW of Trieste, Italy) and are captured for town's water supply (Fig. 5.31) (Schmidl 1851, Boegan 1938, Habic et al. 1989, Civita et al. 1995).

According to a rough estimate the Reka water contributes some 10% at the most to the waters that reappear at Timava. Not only the rainfall, which immediately sinks into the karst interior but also the underground tributaries from the border contribute to the Kras aquifer; the Vipava river that flows on the northern side of the Kras and the Soca river flowing through the alluvial sediments of the Friuli Plain, are partly linked to the same aquifer.

The Reka has pluvio-nival fluvial regime, the primary maximum occurs in November and the primary minimum in August. The average discharge of the Reka is 8.35 m$^3 \cdot$s^{-1}, the minimum is 0.16 m$^3 \cdot$s^{-1} and maximum, according to gauging station some km upstream of Škocjanske Jame, 387 m$^3 \cdot$s^{-1}. On such occasions the Reka floods and the water level in Škocjanske Jame increases by about 100 m (Rojsek 1990, 1996).

The hydrological properties of the Reka are controlled by the geological structure and by climatic conditions. The Reka springs and its entire course up to 5 km before sinking into Škocjanske Jame are located on impermeable Eocene flysch (marl, mudstone, sandstone, calcarenite, breccia, conglomerate). From the right side come the karst tributaries (the most important is Bistrica at Ilirska Bistrica), from the left, from flysch, are flowing the superficial streams (the longest is the Pade stream). The climate in the Reka valley is of Mediterranean type, the rainfall maximum occurs in November. The last few kilometres of the lower Reka flow are on the Paleocene limestones, Škocjanske Jame and Kras are developed in Mesozoic (Cretaceous) limestones.

The Reka, as an important water artery on the border between two karst regions, Pivka and Istria, was always an important water source, used for power and water supply for people and animals. At the beginning of this century the Reka quality could be compared to Bistrica Spring; and the latter is still nowadays used for water

Photo 5.26. The big collapse structure at the entrance of the Škocjanske Jame (photo: Hötzl 1997).

Photo 5.27. The Reka River in the big collapse structure between Mahoricieva Jama and Škocjanske Jame. Due to the pollution thick foam mountains were formed below the waterfall (photo: Mihevc 1987).

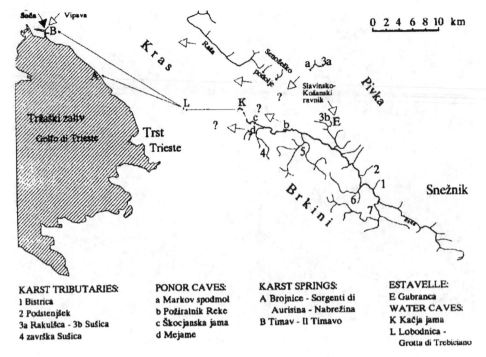

KARST TRIBUTARIES:
1 Bistrica
2 Podstenjšek
3a Rakulšca - 3b Sušica
4 završka Sušica

PONOR CAVES:
a Markov spodmol
b Požiralnik Reke
c Škocjanska jama
d Mejame

KARST SPRINGS:
A Brojnice - Sorgenti di
  Aurisina - Nabrežina
B Timav - Il Timavo

ESTAVELLE:
E Gabranca
WATER CAVES:
K Kačja jama
L Lobodnica -
  Grotta di Trebiciano

Figure 5.31. Map of the tributary net of Reka River, Slovenia, with underground connections to the springs of Timavo, Italy.

supply. Divaza was supplied by pumping the Reka water in its lower flow. There have been about 80 mills and saw-mills along the Reka. This also is the reason that industry settled in the Reka valley, most in the local centre Ilirska Bistrica (5000 inhabitants). Modern industry that started to develop after the Second World War included the wood industry (factories for veneer, wooden plates), the factory of organic acids, food-processing industry (agricultural and dairy products) and transportation (specialised organisation for transportation of fluids, oil and its derivatives). By the end of 1991 a strong motorised military unit was located at the outskirts of the town.

Due to progressive industrialisation and urbanisation the Reka had to receive more and more untreated waste waters (Gams & Habic 1987, Rojsek 1990, Kogovšek 1993, 1994). The decrease in its quality had been observed before 1960. The first serious warning occurred when in 1966 the Reka catchment for the Divaca water supply had to cease to operate. In 1972 the Reka was ranked in the third to fourth quality class (Peterlin 1972). In the years 1969-1979 systematic, continuos observations of Reka at Ilirska Bistrica, at Nova Sušica and at Matavun followed (Meja et al. 1983).

According to the measurements from 1969-1979 the ratio between chemical and biochemical oxygen demand (COD/BOD$_5$) at Ilirska Bistrica was 1.5; it means that the pollution consisted mainly of organic substances, very degradable and using all the oxygen transported by the Reka. At Matavun (just before the swallow-hole of the Reka) this ratio was 2.5 to 5; showing that easily degradable organic substances were, as far as Matavun, in general decomposed and the ratio of hardly degradable substances increased. At that time Reka was in the section between Ilirska Bistrica and Nova Sušica a virtual sewer where intensive anaerobic processes of decomposition took place. Such decay was accompanied by gas and easy volatizable products causing an odour in the valley of Reka and in Škocjanske Jame.

Particularly unfavourable effects on the Reka quality were caused by heavy short showers mobilising the seasonal increase of pollution and transporting the sedimentary particles from the bottom of the riverbed as well as the decay products resulting from the decomposition of the biomass. When water transport power decreased and conditions for progressive resedimentation of the suspended particles arose in the downstream direction (Škocjanske Jame and further) anaerobic processes started again. Mihevc (1984) reported on the methane gas in Kacna Jama, a re-appearance after 1.5 km subterranean flow of the Reka after Škocjanske Jame. Mihevc noticed the typical smell of methane gas in the lower level passages of the Kacna Jama and he linked it with piles of decomposed leaves and other organic material and brushwood accumulated on the bottom. This example shows that in the case of the excessive pollution received by the Reka, the solid impurities are to a great extent deposited on the bottom of the river bed. Under certain circumstances this pollution is carried by the stream, from the surface deep into the karst underground. It is not only the pollution of the underground with the impurities dissolved in the river, but also with solid organic substances which need for their decay much more oxygen and much more time. Comparable conclusions result from investigations of the polluted percolation water in Pivka Jama (Kogovšek 1987) and of the water of the Nanošica stream on the occasion of a spill of liquid manure (Kogovšek 1992).

In 1982 the daily quantity of pollution from the factory of the organic acids and wooden plates factory Lesonit from Ilirska Bistrica was reduced by one third due to various measures taken. Although the analyses of the Reka showed a decrease in all the measured pollutants by one third, the total pollution was far too high to be self-purified. Obviously the measures of pre-purification should be followed by the construction of common biological treatment plant.

The Karst Research Institute had analysed in the second half of 1981 and in 1982 the water of the Reka River five times (Kogovšek 1994, 1995). BOD_5 and dissolved oxygen at the swallow-hole to Škocjanske Jame were analysed. BOD_5 variations were from 2.9 to 8.5 $mg \cdot l^{-1}$ O_2, the level of dissolved oxygen was from 10 to 12.1 $mg \cdot l^{-1}$, respectively 78 to 103% of saturation, showing that the level of the dissolved oxygen was inversely proportional to the values of BOD_5. From July 1982 to June 1983 the Reka quality near Ribnica was measured using bimonthly sampling. The water temperature varied between 3 and 19°C, specific electric conductivity between 254 and 672 $\mu S \cdot cm^{-1}$, the chloride level between 3 to 30 $mg \cdot l^{-1}$, nitrate between 0.3 to 2.7 $mg \cdot l^{-1}$, phosphate below 0.35 $mg \cdot l^{-1}$, dissolved oxygen oscillated between 0 and 10.1 $mg \cdot l^{-1}$, BOD_5 was from 8.5 to 48 $mg \cdot l^{-1}$. Low values of dissolved oxygen and the highest BOD_5 occurred during low water levels. In October, November, December and March the oxygen level was higher, BOD_5 usually lower. The observed improvement of the Reka quality, though seasonal, was probably due to pre-purification in the factories Lesonit. But, from time to time the quality was extremely bad and it was suspected that the water was being retained and later released in bigger quantities from the industrial plants.

In autumn 1990 production in the organic acids factory (TOK) was stopped due to unfavourable economic conditions and very soon positive changes in the Reka River were seen. The abatement of pollution indicated what a burden were the waste waters from the factory of organic acids, in particular huge amount of non-degradable pollution. The Karst Research Institute analysed Reka water samples at the swallow-hole to Škocjanske Jame. On December 18, 1991 at low water level COD was 7.5 mg·l^{-1}, on June 6, 1992 6.1 with a BOD$_5$ of 2.3 mg·l^{-1} and 100% of O$_2$-saturation. On October 8, 1992, medium water COD was 6.9 mg·l^{-1}, and during high water level (October 23, 1992) only 3.3 mg·l^{-1}. Nitrate and chloride were at all measurements below 5 mg·l^{-1}, o-phosphate below 0.12 mg·l^{-1} (Table 5.3).

In the considerably less loaded river the self-purification processes started again. During extremely low water levels in July 1993 the water of the Reka at the swallow-hole to Škocjanske Jame was analysed again. COD was 12, BOD$_5$ 2 mg·l^{-1}. Chloride and nitrate levels were low, 5 mg·l^{-1}, comparable to 1992, phosphate was below 0.01 mg·l^{-1}. The ratio COD/ BOD$_5$ was 6, probably due to the lack of dilution. The decomposition of degradable substances was nevertheless successful but in respect of non-degradable components less obviously so. At high water levels in October 1993 the COD was 7, BOD$_5$ 1.3 mg·l^{-1}, the ratio COD/BOD$_5$ was 5.4. Rismal (et al. 1994) assessed that the Reka belonged in respect of most criteria to the second quality class in 1994. Seasonally COD, nitrate level and turbidity are surpassed, and more often there are mineral oils and bacteria of faecal origin present. Repeated analyses of the Reka in 1994 and 1995 indicate nitrate and chloride levels up to 6 mg·l^{-1}, phosphate just above 0.01 mg·l^{-1}, COD from 2 to 8.5 mg·l^{-1}. This, relatively high level of COD appeared in the conditions of low water and low temperatures at 100% saturation with oxygen indicating impeded degradation due to low temperatures.

The above mentioned results indicate a rather improved quality of the Reka River at the swallow-hole of Škocjanske Jame. Pollution flowing into the river is to some extent eliminated on its way to the swallow-hole. When the river flow increases after a long drought, the river bed material is disturbed and pollutants get transported underground. This pollution wave in turn is followed by major dilution which produces

Table 5.3. Reka River quality from 1974-1995.

			Il. Bistrica	Matavun
1974-1979	BOD$_5$	(mg O$_2$·l^{-1})	100-200 (400)	10-15
	COD	(mg O$_2$·l^{-1})	160-300 (700)	25-80
	Dissolved oxygen	(mg O$_2$·l^{-1})	0.1-0.5	5-8
	COD/BOD$_5$		1.5	2.5-5
1981-1982	BOD$_5$	(mg O$_2$·l^{-1})		2.9-8.5
	Dissolved oxygen	(mg O$_2$·l^{-1})		10.0-12.1
1991, 1992, 1993	COD	(mg O$_2$·l^{-1})	15	3.3-12
	BOD$_5$	(mg O$_2$·l^{-1})	2.8	1.3-2.3
	COD/BOD			2.7-6
1994-1995	COD	(mg O$_2$·l^{-1})		2.0-8.5
	BOD	(mg O$_2$·l^{-1})		2.0

a return to better water quality. However, the actual ratio COD/BOD$_5$ indicates the presence of non-degradable pollution. In any case the improvement is considerable but it is necessary to define the source of the non-degradable pollution.

5.7.5   *Remediation of a contamination by chlorinated hydrocarbons*
(Heinz Hötzl & Manfred Nahold)

5.7.5.1   *Introduction*
Among the dense nonaqueous phase liquids (DNAPL) volatile chlorinated hydrocarbons (CHC) are the most frequent group responsible for the pollution of the environment. As organic solvents they are widely and frequently used; for example in dry cleaning, to degrease installations, in metal processing and in slaughterhouses. The chlorinated hydrocarbons are now nearly everywhere in our environment. Due to their high density they contaminate mainly soil below the places where they have been used. Migrating down to the groundwater they can be dissolved and transported over long distances, because they are only slightly degradable. Because of their high vapour pressure and low adsorptivity within mineral soils they appear relatively easy to extract from unsaturated zone by soil air extraction. However, the process of cleaning the saturated zone especially in karstified rocks seems to be a very difficult task. A case study from South Germany, where finally remediation of a CHC contamination was successfully performed is given below (Böhler et al. 1990, Hötzl et al. 1990, Hötzl & Nahold 1992, Nahold & Hötzl 1993, Nahold 1996).

5.7.5.2   *Site description and contamination*
The source of the CHC contamination in the village of Sulzdorf half-way between Stuttgart and Nürnberg in South Germany was a slaughterhouse (Fig. 5.32) where animal food was produced and cadavers sorted out.

Mainly tetrachlorethylene (PCE) and in a minor amount trichlorethylene (TCE) was used in order to separate the fats. By way of leaking sewer systems as well as by penetrating through the concrete bottom of the building the solvents were migrating in phase into the underground over years. Figure 5.33 shows the distribution of the organic contaminants in the near-surface soil, water and air. The highest concentration with more than 10,000 mg·l^{-1} was found below the cellar and an adjacent seepage basin. In the lateral distribution a clear effect along the main collector sewer could be shown. In connection with the remediation work the whole main building was deconstructed and the contaminated soil was excavated down to a depth of 7 m below the surface.

The problem was that the contamination was not restricted to the uppermost soil layer, but had already migrated through the geologic sequence of the cover system down into the karstified limestone aquifer over the years (Fig. 5.34). As a part of a multistage exploration programme a total of 18 wells were drilled and selectively cased, allowing separate observation and treatment of the two main geologic sequences (Fig. 5.35). They revealed a deep contamination reaching down to about 50 m below the surface.

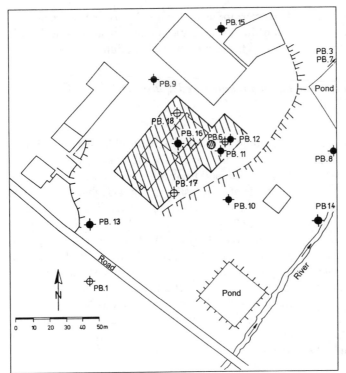

Figure 5.32. Map of the slaughterhouse in Sulzdorf, Baden-Württemberg, S Germany, source of a huge contamination of chlorinated hydrocarbons. The hatched part shows the area, which was excavated to a depth of 7 m; open circles are wells of the Keuper aquitard, closed circles wells of the Muschelkalk karst aquifer.

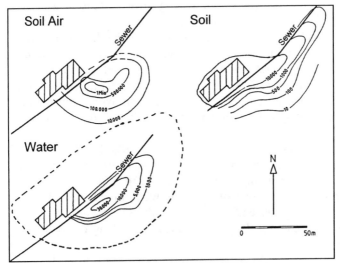

Figure 5.33. Results of the first monitoring of the chlorinated hydrocarbons in soil air, water (mg.m^{-3}) and soil (mg·kg^{-1}) close to the surface, slaughterhouse, Sulzdorf.

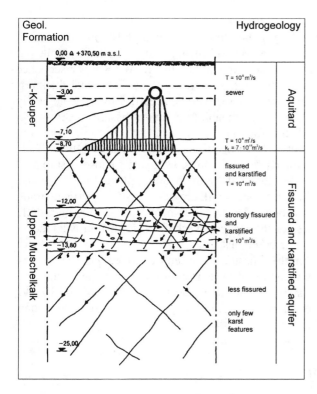

Figure 5.34. Schematic profile of the leaking sewer at Sulzdorf with spreading of the CHC contaminants down to the karstified and fissured aquifer.

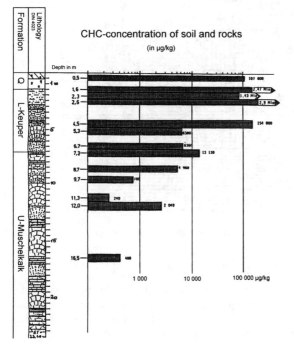

Figure 5.35. Distribution of the chlorinated hydrocarbons in soil and rock samples of well PB 16 from the centre of the contamination (CHC extraction by methylenglycol).

5.7.5.3   *Geologic and hydrogeologic setting*

The contamination site is situated on a flat slope dipping slightly towards a small natural drainage channel. The substrate consists of a 1-3 m thick loamy colluvial layer of Quaternary, below there are the remnants of the Lower Keuper sequence with clay, sandstone, marl and some dolomite layers (Fig. 5.36). The total thickness varies between 10 and 15 m. The Lower Keuper can be regarded hydrogeologically as a semi-permeable aquitard. The main permeability is due to fissures, as is shown by the remnants of the organic solvents found on fissure planes.

The lower stratigraphic member is formed by the Middle Triassic Muschelkalk formation. The limestones are regularly bedded and show a partly dense fissure network. Small solution cavities are developed along the bedding planes and the joints. The cavities are more frequent in some specific lithologies, causing a vertical stratification of the permeability, which shows also an enrichment of the contaminants (cf. Figs 5.34 and 5.35). The thickness of the Upper Muschelkalk formation is about 70 m, the base was reached in one of the boreholes.

The hydraulic conditions within the limestones are semi-confined. For the estimation of the hydraulic parameters with regard to remediation measures pumping tests where carried out. The transmissivity of the limestone lies in the range of 10^{-2} to 10^{-3} m$^2 \cdot$s^{-1}. Pumping test proved that with a yield of only 2-3 l$\cdot$s^{-1} the depletion cone more or less includes the main contaminated part.

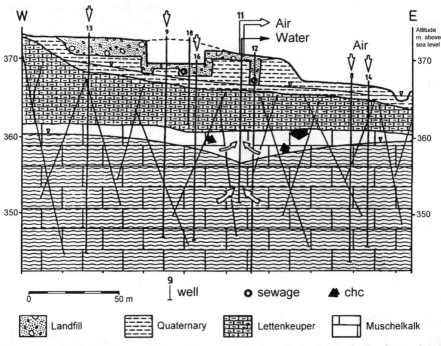

Figure 5.36. Geological cross section of the contamination site in Sulzdorf. The figure shows the effectiveness of the remediation procedure. By pumping from the karst aquifer an unsaturated zone is developed below the aquitard, causing the percolation of the CHC and its volatilisation, bubbling up into the unsaturated zone, from where it is extracted together with the soil air.

5.7.5.4 *Remediation technique*

As a first step the cleaning process started with groundwater pumping in order to prevent further subsoil spreading of the CHC. The CHC compounds of the discharged water were recovered by on site flushing. In the second stage of cleaning, the highly CHC penetrated central area of contamination of about 600 m^3 within the Quaternary sediments and the strongly weathered Lower Keuper, was excavated and treated off site. The original plan to clean up the area of the cohesive sediments using vacuum pumps and hot air injections failed due to the low permeability of these sediments.

In the third stage cleaning was continued by using an improved technique which was efficient for the carbonate aquifer. For this purpose, a modified soil air extraction was examined in two pilot tests (Hötzl et al. 1990). In this modified method, the groundwater level in the deeper fissured and karstified aquifer was lowered and the air was extracted from the then newly unsaturated zone by means of a lateral channel compressor. Figure 5.36 shows the geological and hydrogeological situation, the arrangement of certain measuring points and the schematic function of the cleaning equipment. The artificially aerated zone lies underneath the saturated Keuper aquitard. Air recharge arrived from the surface via individual drillholes. The selective air flow through contaminated areas could be controlled by bars at the top of the drillholes. Some of the results of different remediation procedures are shown in Figure 5.37.

This method of air flushing caused the highly contaminated contact area at the base of the semipermeable aquitard to be cleaned. In addition, due to the pressure applied, an increased hydraulic gradient was induced in the upper highly contaminated aquitard. As shown in Figure 5.36 the highly volatile compounds in the water then recharged into the karstified zone vaporised as soon as they reached the unsaturated zone and were then flushed out immediately with the air stream.

In the fourth and last stage of the cleaning programme and groundwater treatment was intensified in the fissured and karstified aquifer after the pollutant load in the covering system had subsided. As a proven technique the air sparging method together with pump and treat was used in a similar way as is shown in Figure 5.36.

5.7.5.5 *Conclusions*

Contaminated subsoil and karst groundwater can be more or less successfully treated with different combinations of in situ and off site techniques even under complex hydrogeologic conditions. In case of the CHC contamination of a karst aquifer with a semipermeable cover system a combination of hydraulic and pneumatic in situ techniques were applied after excavation of the uppermost highly contaminated soil. The combination of pumping and lowering of the water table together with air sparging into the saturated zone and air extraction from the artificially unsaturated part has proved to be the most effective technique. Their application, however, requires a detailed exploration and an adaptation of the cleaning process to each individual site.

In general it has proved that low pressure and low airing rates rather than powerful ventilation permit water flow through large rock masses, whereas higher pressure and airing rates do not enlarge the radius of action, but cause breakthroughs close to the injection wells. The upward movement of the air bubbles and the differences in density cause complex flow conditions in the fissured and karstified aquifer. This can

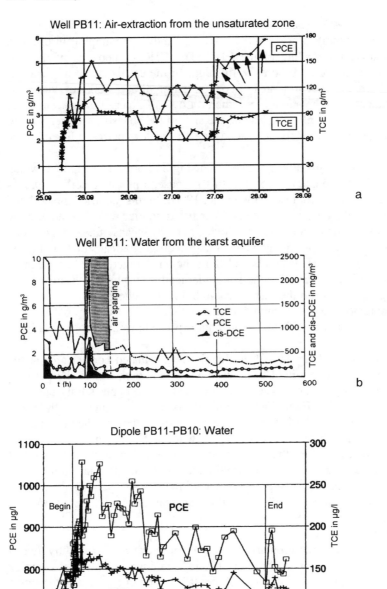

Figure 5.37. Results of the remediation procedure at Sulzdorf: a) Extraction of CHC by soil air ventilation in one of the suck wells, the second rise (black arrows) after the beginning of the tailing is due to the change of pressure conditions, b) CHC concentration in the karst water from beginning of pumping, including air sparging phase (10-150 hours), and c) Extraction of CHC from the karst water with a closed symmetric dipole configuration between two wells.

positively affect the cleaning exceeding the directly aired zone, but may also cause a shifting of pollutants into uncontaminated areas of the aquifer. Therefore detailed studies with hydraulic, pneumatic and tracer tests are recommended before the final design of the remediation work in a karst aquifer.

5.7.6   Contamination of a karst aquifer by a sanitary landfill, SW Germany
(Matthias Eiswirth, Heinz Hötzl, Georg Jentsch & Bernd Krauthausen)

5.7.6.1   Development of the sanitary landfill
This case study deals with the sanitary landfill of Grötzingen near Karlsruhe, Southwest Germany (Fig. 5.38). It was laid out in an old limestone quarry at the top of the escarpment delineating the Upper Rhine Graben east of Karlsruhe. The quarry dates back more than 50 years. It was operated to gain limestones for construction and road material, for which the limestone rocks at the top of the escarpment offered a good quality. Looking for a proper repository for garbage and industrial waste in the early seventies the closed quarry seemed to be a reasonable place restoring the landscape at the same time. The plan finally even got official geological consent from the state authorities, underestimating the flow and transport condition in a karst aquifer. Filling up of waste in the quarry started in 1973 without any special liner system and was operated until 1997 (Photo 5.28). In total 700,000 m³ of garbage and waste was deposited.

After deposition of more than 400,000 m³ of waste a more detailed study of the possible impact was done in 1985. The collector system for the leachate extracted only 450 m³ which was less than 5% of the estimated total amount. In order to avoid a further leaking of the leachates a intermediate mineral sealing system was built covering the already deposited waste and forming the base liner for the second stage of filling up. Figure 5.39 shows the increasing discharge of the effluents from the new collector system above the intercalated liner in connection with the extension of the waste site from 1990 to 1996. The calculation of the water balance with regard to precipitation and evaporation showed a good agreement with the volume of extracted leachate. The control of the leachate from the lower part of the landfill shows a reduction of the accessible leachate from 430 to 25 m³·a⁻¹ (Fig. 5.39). Though this doesn't mean that there is no additional loss into the underground, one can assume that the seeping part of the leachate is reduced in the same ratio. It can be expected that the lower waste body is slowly running dry; that however may take additional 5 to 10 years.

5.7.6.2   Geological and hydrogeological situation
The Grötzingen sanitary landfill is located at a distance of only 500 m from the eastern Rhine Graben master fault. The vertical displacement on this Tertiary fault system amounts more than 2000 m. The uplifted escarpment block is built up by Lower and Middle Triassic formations of the Bunter and Muschelkalk (Fig. 5.40).

The terrestrial sandstone sequence of the Bunter is topped by a 6 m thick layer of clay- and siltstones, which acts as a main aquiclude. The hanging wall is then formed by the marine carbonate sequence. The lower part consists of 60 m of thin-bedded marls, limestone and dolomites. There permeability is rather low and connected to the intensive jointing. The Middle Muschelkalk consists now mainly of dolomites

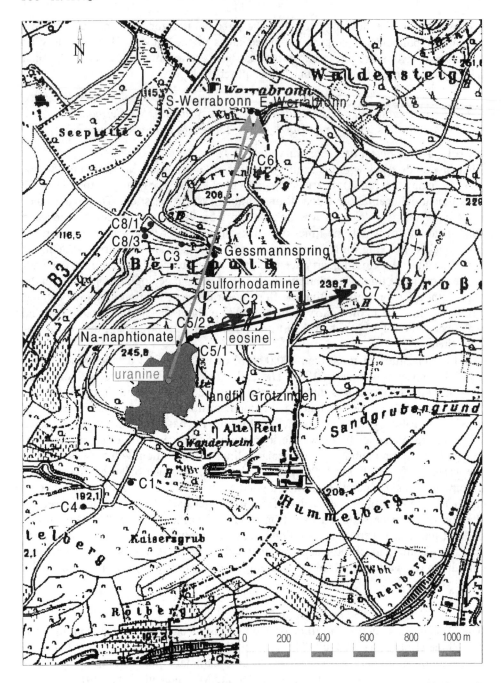

Figure 5.38. Map of the sanitary landfill Grötzingen near Karlsruhe, SW Germany, with the position of observation wells (C1 to C8) and the results of tracing experiments.

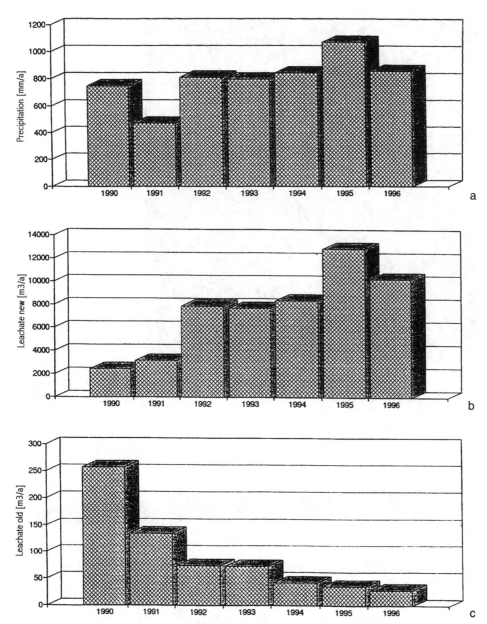

Figure 5.39. Development of leachate yield in the years 1990 to 1996: a) Precipitation, b) Leachate from top of the intercalated mineral liner (new leachate), and c) Leachate from the part below the liner (old leachate).

Photo 5.28. Aerial view of the landfill Grötzingen (viewed from North to South). The southern part shows the intermediate liner and the drainage layer prepared for further waste deposition (photo: Eiswirth 1993).

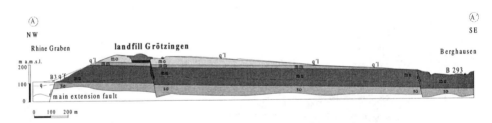

Figure 5.40. Geological cross section (q, q´f, q ´l = Quaternary sediments, mo = Upper Muschelkalk, mm = Middle Muschelkalk, mu = Lower Muschelkalk, so = Upper Bunter).

with intercalations of some limestone beds and some residual clay layers, which are the result of the solution processes of the originally intercalated thick salt and anhydrite formation. The Upper Muschelkalk limestones quarried at Grötzingen consists of micritic or fossiliferous limestones. The up to 60 m thick sequence is strongly fractured with pronounced karstification along the joints.

The bedding conditions are marked by the general flat dipping of the sequence to the east (Fig. 5.40), though some of the blocks like that of the sanitary landfill show certain deviations dipping toward the N or NE. A main characteristic of this area is block faulting and strong fracturing due to the graben tectonics. Several faults exist in the near surrounding of the landfill partly running parallel to the Rhine Graben system (NNE), partly crossing it (NW). One of the main faults even runs through the landfill area.

The transmissivities determined in the wells by hydraulic tests are in the range between $6 \cdot 10^{-4}$ and $3 \cdot 10^{-7}$ $m^2 \cdot s^{-1}$. The water table of the upper aquifer lies about 30 m below the bottom of the sanitary landfill. The water table fluctuation shows significant seasonal variations with high levels in February and March. The most important spring is the Werrabronn, 1.5 km N of the landfill. The discharge varies between 10 and 15 $l \cdot s^{-1}$ and shows for a karst spring surprisingly small variations. The Gessmann Spring half on the way to the landfill has a discharge of less than 1 $l \cdot s^{-1}$ and is mainly of local origin. The next spring to the east is located 1.7 km distant from the landfill and has an average discharge of 7 $l \cdot s^{-1}$.

5.7.6.3 *Impact of the sanitary landfill on the karst aquifer*
Though pollution of the karst aquifer has taken place since the seventies, no clear observation on negative effects were recorded in the first years. The main reason was that in the near surrounding no monitoring wells existed. The only captured spring in the vicinity, the Werrabronn was abandoned after pollution first occurred. After realising from the small amount of pumped leachate that strong seeping into the underground must have taken place a detailed investigation program was started including hydrogeologic and geophysical surveying, drilling works, hydraulic tests, hydrochemical studies and tracer tests. The aim was to prove the seeping of the leachate to determine the discharge behaviour in the underground and to quantify the effects on the karst water. Subsequently three aspects of the reconnaissance program will be discussed.

Soil gas screening for the detection of preferential flow paths
The aerobic biodegradation of leachate compounds in the unsaturated zone decreases O_2 and increases CO_2 in the soil air and in the groundwater (Hendry et al. 1992, Deyo et al. 1993). The degradation of leachate compounds on small preferential flow paths, like fault zones, leads to relevant changes. Soil gas screenings have been carried out. Beside CO_2 and O_2, CH_4 and H_2S as well as radon were surveyed along profiles crossing the assumed fault zones. To get additional information geophysical measurements was carried out along the same sections. The work was focused on electromagnetic sounding, with horizontal and vertical coplanar coil configurations with spacings of 10, 20 and 40 m leading to a various depth penetrations along the cross sections.

The main results (Eiswirth 1995, Eiswirth & Hötzl 1997) are summarised by reference to one of the sections (Fig. 5.41). During the soil gas investigation on this profile soil gas CO_2-maxima were detected at location 73 m and 85 m, corresponding clearly with detected minima in soil gas O_2. Both anomalies refer to contaminant transport on the extensional fault zone. The elevated CO_2 and depressed O_2 concentrations are caused by biodegradation of leachate in a discrete flow paths within the karstified limestones. The results of the ^{222}Rn investigations indicate also a main anomaly at location 73 m and three minor anomalies. The first refers to the main fault zone distinguished by high radon concentrations. The results of the electromagnetic soundings showed a broad zone below 35 m and two areas with low ap-

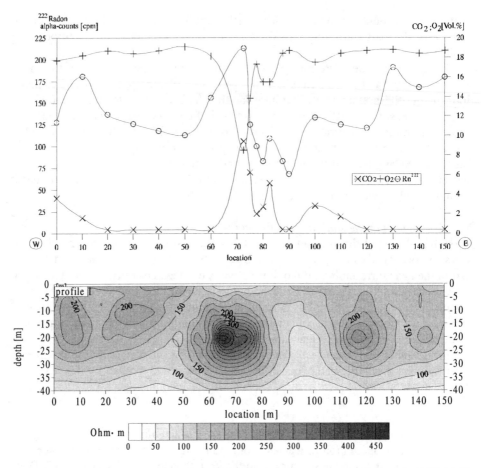

Figure 5.41. Results of the soil gas screenings and electromagnetic soundings on a E-W profile north of the leaky landfill (for explanation compare the text).

parent electrical resistivities. The zone below 35 m depth refers to the groundwater of the karst aquifer. These very low resistivities (<50 Ω·m) possibly indicate saturated conditions (water, leachate) within the limestones. Similar results were received from the parallel sections so that one of the main water path for the contaminant flow could be determined and followed to the North. The results were used also for the location of the deep boreholes.

Hydrochemical studies
The leachate from the lower part of the sanitary landfill has shown relatively constant concentrations for several years. It is characterised by a high general mineralisation, expressed by the electric conductivity there are more than 10,000 μS·cm^{-1}. From the high content of organic substances result values of DOC >300 mg·l^{-1}, AOX >1500 μg·l^{-1} and ammonia concentrations up to 900 mg·l^{-1}. Significant

parameters for the leachate are chloride (500 mg·l^{-1}), boron (8 mg·l^{-1}), the pesticide Mecaprop (10 µg·l^{-1}), tritium (400 T.U.) and some heavy metals (Al, Cr, Ni, As).

Direct evidence of the leachate in the different sampling points was only given for the groundwater monitoring well C2, which was drilled in 1989. Because the installation of the well coincided with the completion of the intermediate sealing system in the landfill, the curve of the concentration between 1989 and 1997 (Fig. 5.42) reflects the partial restoration of the old water quality, only ammonia showed a delayed breakthrough of the maximum in 1992, from then the curve has dropped away.

For the Werrabronn spring an influence of the landfill was assumed, since the spring had to be cut off from the water supply due to the increasing pollution. The dis-

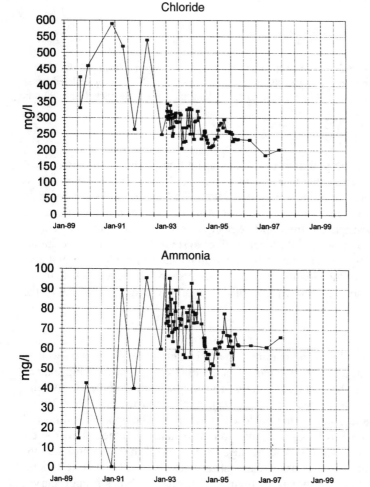

Figure 5.42. Chloride and Ammonia concentration in the observation well C2 from 1989-1997. The increasing concentrations are due to the leachate from the sanitary landfill, the reduction are caused by the construction of a intercalated cover system.

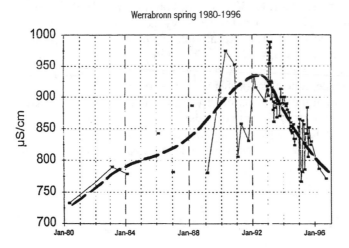

Figure 5.43. Graph of the development of the mineralisation (shown in the values of the electric conductivity) of the Werrabronn spring in the period from 1980 to 1996. The intercalated cover system, reducing drastically the leachate in the lower leaking part, was constructed in 1990 and 1991.

cussion on the origin of the pollution could be clarified only after detailed chemical investigations. They revealed that two different sources of contamination are responsible for the total amount of pollutants in the water. The long-term observation data of this spring (Fig. 5.43) demonstrate impressively the increase of pollution since the start of the waste deposition and the improvement after the reduction of the leachate losses by the construction of the intermediate liner system. Beside these two monitoring stations, weak signs of an impact by the sanitary landfill could be found also in C 5/1 and C 5/2 and at the Gessmann Spring. Because of the location close to the landfill these can be interpreted as a result of the inhomogenious distribution of the karstified water paths.

Tracer tests
In order to prove the results of the chemical analysis and data collection as well as to get some more quantitative parameters of contaminant transport, tracing experiments were carried out. The selected tracers, uranine, sulforhodamine B, sodium-naphtionate and eosine were checked before their application for possible interference with the leachate or some of the enclosed substances. The applied volumes of the tracers were between 5 to 15 kg depending on the distance and the assumed preferential water paths. Up to 45 observation sites were included in the control stations either by direct sampling of water or by charcoal sampling bags. The main results regarding the flow direction are shown on Figure 5.38, selected breakthrough curves are given in Figure 5.44.

The flow distribution confirms the essential influence of the structural setting on the underground discharge. The main direction (in Fig. 5.38 characterised by the spreading of uranin) follows the fault zone in the Rhenish direction to NNW. For the

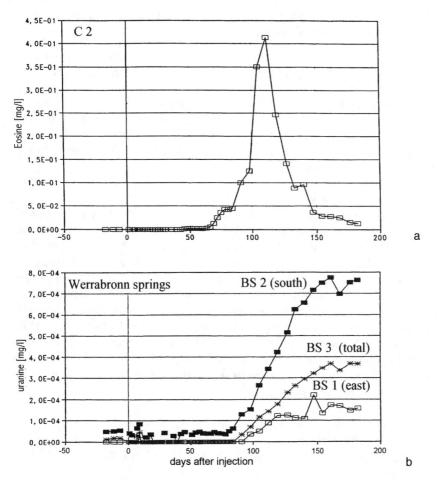

Figure 5.44. Results of the tracing experiments in the sanitary landfill of Grötzingen. The figure show the breakthrough curve of: a) Eosine in the well C2, and b) Uranine in the Werrabronn Springs (east: BS1, south: BS2, total: BS3).

second main direction (discharge of eosine) the dipping of the bedding planes is re-sponsible. Though in karst aquifers the dipping of the beds does not have so strong an influence, this can be explained for the area around Grötzingen by the fact that the karst water table below the escarpment crest comes is in the Middle Muschelkalk where some residual clay layers are nearly impervious so that the percolating water in the unsaturated zone follows the dipping of these beds. The breakthrough times of the different tracers offer interesting information on the complex underground flow dynamics (Fig. 5.44). From the first appearance of the tracers average flow velocities of 10 m·day^{-1} were calculated. These are rather slow velocities compared with other karst areas. It shows that karstification in the deeper part of the Muschelkalk se-quence is not very well developed. Both the velocities as well as the long tailings of the breakthrough curves are rather typical for fissured aquifers.

5.7.7   *Impact of major tunnels: Excavation of the Mont d'Or railway tunnel*
(Jacques Mudry & Pierre Chauve)

At the beginning of the twentieth century, the opening of the Simplon Tunnel between Switzerland and Italy required the improvement of passage through the Jura Mountains to shorten the distance to Paris. The new railway project between Paris and Milan had to pass through the Mont d'Or with a single slope 6099 m long tunnel. During 1912, digging was troubled by water intrusion into the tunnel; at the same time several springs which are situated up to 5 km distant as the crow flies, completely or partly dried up. The sealing of the karstic conduits with concrete restored part of previous discharge of the spring.

5.7.7.1   *Topographical and geological context*
The Mont d'Or is a 1461 m high range, which is situated between Pontarlier (Doubs, France) and Vallorbe (Vaud, Switzerland). It looks down upon the valleys of Lac de Joux (elevation 800 m) and Metabief (950 m). Geologically, as shown on Figure 5.45 (Fournier 1919), Mont d'Or is a Jurassic asymmetric anticline (Risoux) which is bordered by two Cretaceous synclines (Vallorbe and Metabief, Fig. 5.46). The core of the anticline is composed of Dogger clayey limestones (δ_{1-11}), covered by thick Oxfordian marls (j_{1-2}). The top of the structure belongs to the Malm limestone series (j_{3-6}). Both synclines include limestone and marl of Lower Cretaceous age. The main feature of the Upper Jurassic structure (Fig. 5.48) is the Pontarlier fault. This 50 km long basement fracture cuts the folded area into 2 independent divisions. Although it looks a sinistral strike slip fault the number and geometry of folds are different on both compartments of this fracture.

5.7.7.2   *Hydrological disturbances induced by digging the tunnel*
As the tunnel has only a single slope tilted towards Switzerland, digging on the French side had to be stopped at 1006 m, when a 30 l·s^{-1} spring was added to a

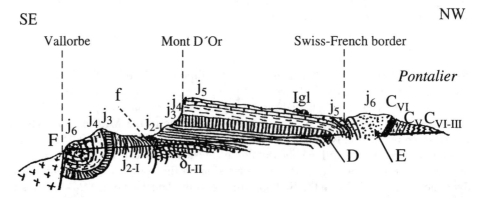

Figure 5.45. Geological cross-section through the Mont d'Or along the railway tunnel. Legend: δ_{I-II} = Dogger, j_{1-6} = Malm, C_{I-VI} = Cretaceous, F = fault, f = Thrust fault, D and E = main entrance of water (after Fournier 1919).

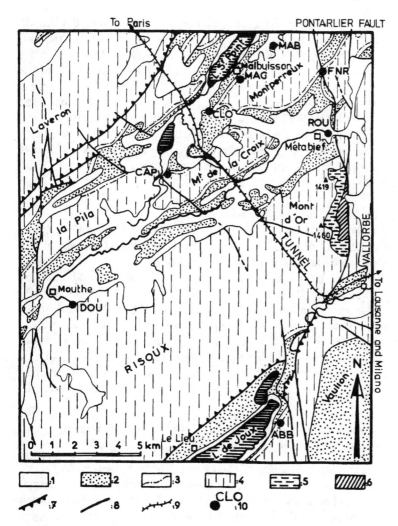

Figure 5.46. Geological map of the Mont d'Or area, south of Pontarlier, Jura Mountains. (Chauve & Mudry 1980). Legend: 1 = Quaternary, 2 = Cretaceous and Tertiary, 3 = Limit Cretaceous – 'Purbeckian', 4 = Malm limestones, 5 = Argovian marls, 6 = Dogger, 7 = Overthrusting, 8 = strike-slip fault, 9 = railway Paris – Lausanne, 10 = Outlet. Springs: ABB = Abbaye, CAP = Capucins, CLO = Clouterie, DOU = Doubs, FNR = Fontaine Ronde (periodic), MAB = Source Bleue, MAG = Grande Source, ROU = Bief Rouge.

33 l·s^{-1} discharge coming from the Lower Cretaceous. Digging continued on the Swiss side, passing through the Dogger limestones (1450 to 2175 m), and the less in-clined Oxfordian marls (2175 to 4200 m) and again Malm.

On the Swiss side, the first 650 m were dry. On the eastern crossing of the Oxfordian only 12 to 15 l·s^{-1} were found, rich in sulphate. When they reached 4723 m, a frac-ture unclogged, and produced 1800 l·s^{-1} in the tunnel. With respect to rainfall the discharge fluctuated between 1000 and 5000 l·s^{-1} and flooded the tunnel to 50 cm depth with a cascade at the entrance. As a consequence, The Bief Rouge, the Grande Source at Malbuisson and the Fontaine Ronde periodic spring completely dried up, and the discharge of the other springs, the Source Bleue and Sources Martin, also decreased. Later the fracture was concreted and equipped with a manometer and a water gate. 45 h after it was closed, the pressure reached 80 m, after 48 to 52 h the 3 springs of Bief Rouge started to overflow once again. One day later, the discharge was 300 l·s^{-1} and three days later 477 l·s^{-1}. A new experiment was performed on February 21st, 1913 and the water gate was reopened. Between 3 and 16 h later, the 3 gryphons of Bief Rouge dried up, after 36 h the Grande Source dried up, and 2 days later, Fontaine Ronde periodic spring too, at the same time the discharge of the sources Martin was reduced by the factor 2.5.

5.7.7.3   *Hydrogeological results of this experiment*
The existence of a siphon could be proved between the tunnel and the Bief Rouge springs, which are situated 84 m higher than the fractured aquifers. Evidence of this siphon is shown by pressure variations on the manometers.

A part of the Pontarlier fault is in connection with Bief Rouge and Fontaine Ronde springs of Mont d'Or. Evidence of a direct connection between Grande Source and Mont d'Or was found (Jeanblanc & Schneider 1981). The possibilities for sealing the intrusion and restoring the original karst water conditions were examined. The practical results of that water intrusion were the erosion of about 17,000 m^3 rock and soil material blocking up the Swiss exit and the destruction of two roads. It caused a 14 months of delay in putting the track into service. The overall cost was almost 100 million francs instead of 17 million which had been forecast.

5.7.8   *Impact by man-induced infiltration of surface water on karst processes, examples from Russia* (Vladimir S. Kovalevsky)

5.7.8.1   *Reactivation and intensification of karst process by water reservoirs*
Making reservoirs in regions subjected to karstification may bring about the formation of new karst forms and processes. A rise in the level of surface water and related to it, of groundwater levels, provides conditions for forming karst both in the reservoir bank zone and due to leakage from the reservoir to the adjacent valleys downstream around the dams. Karst is most widely developed in zones of seasonal fluctuations of reservoir levels, in its marginal part and in the zone of groundwater fluctuations. Thus for example, formation of new karst sinkholes and voids is observed both in carbonate and sulphate rocks in the Vierkhne-Kamsk and Bratsk reservoirs. A zone of previously unknown sinkholes and new collapses was apparent up to 3-5 km from the reservoir banks after the reservoir was filled. The intensity of karstification and the depth of the affected zone depends on the amplitude of seasonal fluctuations in the reservoir levels. Thus 130-140 sinkholes per km^2 at a distance from a reservoir shore toward the watershed are observed when the amplitude of the reservoir level fluctuation reaches 10-80 m. The intensity of karstification increases sharply compared with natural conditions, contrary to the earlier idea that the process develops slowly. Thus during the summer of 1966 more than 100 new sinkholes with diameters of 2 to 10 m were observed in a distance of 10 km along the Kama Valley.

When karstic deposits are covered with loose alluvial materials the karst suffusion process is accompanied by surface subsidence at a rate of 13-20 cm·a^{-1}. Karstification is particularly intensive in sulphate and halide deposits. The rate of dissolution of the former can be 10 times, and the latter 1000 times higher, than that of carbonate deposits.

5.7.8.2   *Karst development due to leakage from water pipes*

Groundwater level rises and the waterlogging of urban areas are characteristic after-effects of urbanisation. There are some hundred towns in Russia where waterlogging processes have already manifested themselves. Leakage from water conduits and canalisation as well as street cleaning and moisture condensation under buildings etc. are the main reasons for waterlogging. Water losses from conduits often amounts to 7-10% of the total water consumption in cities. It is about 5 m^3·sec^{-1} in Moscow. Groundwater level rises in some places reach 10-30 cm·a^{-1} and groundwater infiltration recharge increases in this case 5-10 times, equivalent to 500 mm and more per year with a mean regional norm of about 70 mm·a^{-1}. Surface water from reservoirs with free carbonic acid concentrations of 20-25 mg·l^{-1} are used in Moscow for water supply. Infiltration of this aggressive water into areas liable to karstified aquifers furthers karst process activation.

Mapping such zones of intensive leakage and related karst activation is a complicated problem in urbanised area where intensive ground water exploitation is also taking place. Indirectly these zones can be revealed through helium anomalies, its low concentrations can only result from surface water infiltration. The map of helium concentration in the Middle Carboniferous aquifer beneath the city of Moscow is given in Figure 5.47. There is a zone on the map of helium low concentrations

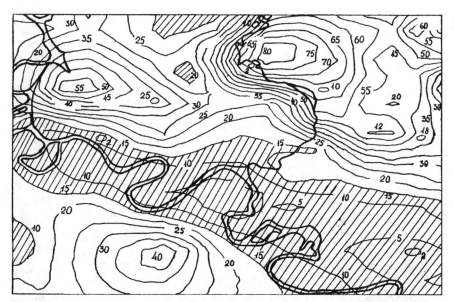

Figure 5.47. Distribution of helium concentration in the shallow aquifer beneath Moscow (not to scale). The concentrations are given in 10^{-5} mg·l^{-1}. The hatched parts represent low concentration areas.

($<15 \cdot 10^{-5}$ mg·l^{-1}) along the river valley due to the induced river water into the aquifer under the impact of groundwater pumping. Separate irregular localised helium anomalies with concentrations of $2\text{-}20 \cdot 10^{-5}$ mg·l^{-1} are associated with localities of downward infiltration of conduit and unconfined groundwater. These helium anomalies comparable with pH, CO_2 and O_2 anomalies in the confined groundwater of an aquifer subject to karstification. Karst sinkholes on the land surface, some of them causing failure of buildings, are related to the most intense helium anomalies ($2 \cdot 10^{-5}$ mg·l^{-1}).

5.7.9 *Pamukkale travertine area* (Sakir Simsek)

The travertine area of Pamukkale in western Turkey is one of the world's most important natural heritage sites (Photo 5.29). It has partly overwhelmed the Roman City and the Necropolis of Hierapolis. The main aquifer supplying hot water to the Pamukkale thermal springs is the karstic marble of Paleozoic age (Simsek 1990, Yesertener & Elhatip 1997, Günay et al. 1997). Recharge of the aquifer is provided by infiltration from precipitation through the outcrops of marble and limestones which have a high secondary porosity (Fig. 5.48). Four major springs exist at the edge of the travertine area. All these outlets are located along a major fault line. The total discharge of the springs is about 385 l·s^{-1} on average. This figure does not change significantly throughout the year. Recession curve analysis has revealed that the annual active reservoir capacity of the aquifer is about 16 Mm^3. A number of tracing experiments and pumping tests were carried out to find out hydraulic connections, the flow direction and interactions among the major outlets of the hot water. Chemi-

Photo 5.29. Sinter terraces from the travertine area of Pamukkale, Turkey (photo: Hötzl 1992).

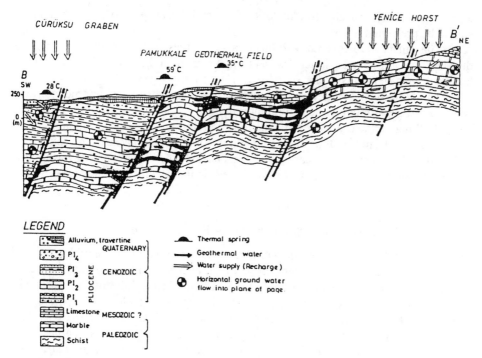

Figure 5.48. Geological cross section of Pamukkale geothermal field, Turkey (Günay et al. 1997).

cal and environmental isotope investigations revealed that hot waters are mainly of meteoric origin with a turnover time of some decades. The chemical composition of the waters is almost constant.

The precipitation of calcium carbonate is found to be enhanced by water turbulence. Travertine deposition is maximum at the rim of the terraces where outgassing of CO_2 is enhanced by a turbulent flow. However, deposition is also significantly high when the water flows in high velocity and the depth of flow is low. These hydraulic controls were investigated using the relation between pH-value and saturation index and the results were used in designing the canals conveying hot water on to the travertine area.

The white travertines are becoming darker, yellowish and brownish after the establishment of touristic sites and hotels in the area. The hotels take the hot water directly from the outlets to the swimming pools and afterwards they release it on to the travertine. Organic materials from the swimming pools cause a rapid growth of algae which change the colour of the travertine. The lack of a sewage system and leaky septic tanks are other sources of pollution on the travertine. Finally walking on the travertines and the direct road to the site are additional points for negative influences on the travertine cascades.

Based on the results obtained from the above mentioned interdisciplinary investigations a conservation strategy was established (Günay et al. 1997):
– New concrete canals were constructed to protect the travertine depositions

against pollution, and any further loss of water by leakage along the water path should be prevented.

— The canals were covered with lids to avoid outgassing before the water reaches the travertine area. Intake structures were implemented at the outlets for an advanced hot water management system.

— The old asphalt road crossing the travertines was abandoned and new terraces were constructed imitating the natural morphology.

— Walking on the travertines and the traffic in the site is strictly prohibited.

— Tourist activities will be decreased and in some fragile places even prohibited.

— Special regulations by law should be enacted to conserve the travertine area and therewith the outflowing karst water based on the scientific findings and evaluations.

5.7.10 *The Plitvice Lakes* (Srecko Bozicevic & Bozidar Biondic)

The Plitvice Lakes are situated in central Croatia, in Likia, in the central part of the Dinaric Karst (Herak 1965). When full of water, as it happens every year during the wet season, these lakes display a rarely seen natural beauty. The Plitvice Lakes with their 16 lakes separated by cascades, and with specific geological, hydrological, vegetal and faunal contents are of global significance (Pevalek 1938, Roglic 1951, Fritz et al. 1984). This was the reason why the Croatian Parliament proclaimed them as a National Park. The natural scenery of the lakes led to very great tourist interest resulting in the visits of about one million tourists each year during the 1980's. A number of small wooden bridges and narrow paths enable the visitors to cross directly under the waterfalls. It is not surprising that United Nations, in accordance with their convention for the protection of world cultural and natural inheritance, included the Plitvice Lakes into their list of World Natural Heritage Sites (Photos 5.30 and 5.31). Unfortunately, after the collapse of the former Yugoslavia, the lakes were occupied for four years till to early August of 1995, and during that long time heavily neglected and devastated. Fortunately, this only happened to the paths and some constructions, while the basic natural phenomena have remained almost intact. Therefore it is possible to arrange nowadays properly the whole area and to emphasise more the protection of the natural phenomena and to control efficiently the touristic impact to the environment.

The Plitvice Lakes occur within a discharge zone at the northern side of the carbonate massif of Lika, where the Mala Kapela mountain dominates (Herak 1972, Roglic 1951, 1981). The cascade of the lakes is 150 m high and the water forms the beginning of the Korana River. The springs, which feed the lakes, emerge at the boundary of a very permeable mostly calcareous massif and low permeable Triassic dolomites (Figs 5.49 and 5.50). Large masses of dolomites in the Dinarides usually function as barriers to groundwater flow. The upper lakes have low-permeable dolomites at their base, and they are larger and deeper than the lower ones. The latter are located in a canyon which cuts the thick and very permeable Upper Cretaceous limestones. A part of the lake water and of the upper Korana River, which originates from the last lake, is seeping down to the limestones. Especially during dry summer periods river water disappears completely along several kilometres downstream the lakes. Tourism at the lakes may have a possible impact on the karst water down-

Photo 5.30. The lake of Gradina with the Burgeta travertine barriers and the Kozjak lake, Plitvice, Croatia (photo: Bozicevic).

Photo 5.31. High waterfall of the Plitvice creek, the Novakovica Brod lake and the Sastavci, under which is the river Korana spring, Plitvice, Croatia (photo: Bozicevic).

stream. One part of the water flows through the karst underground to the catchment area of another river, the river Una, which belongs also to the Black Sea drainage basin, the other part reappears downstream in the river Korana and at the source of river Slunjcica.

The upper lakes are recharged by the water flowing from karst springs of the small rivers Crna Rijeka and Bijela Rijeka and from some creeks draining the neighbouring karst massif. The Plitvice creek enters the lower lakes through a high waterfall (Pevalek 1938, Roglic 1951, 1981). Maximum inflows are during spring and autumn rain periods and particularly during late spring when the snow melts. During winter the waterfalls are partly frozen offering outstanding visual effects. The inflow is substantially smaller in the dry summer period, but some 400,000 m^3 of water are still retained in storage. The lakes are separated by travertine barriers, and they are in a constant biogenetic process of growth. The creation and growth of travertine is controlled by these processes, together with the physico-chemical composition of the water, the climatic conditions and the development of organisms (the moss of Bryum and Cratoneuron genera). Most of the lakes are shallow except for some of the upper ones which exceed about 20 m in depth. The travertine barriers are of various heights, creating a series of cascades (Figs 5.49 and 5.50).

The protection of the National Park of Plitvice Lakes is rather a complex task. The tourism includes several-day stays of visitors including hotel accommodation, which generate a relatively high amount of waste water and crude waste as well as intensive transport through the centre of the National Park. Due to such anthropogenic impacts during the last decade a considerable increase in eutrophic processes has been recorded in the lakes. This should be stopped or at least decreased in the near future by proper technical measures. Such changes of water quality in the Plivice lakes spoil the karst groundwater further downstream, where the river Korana infiltrates partly into the karst system (Fig. 5.50). The present knowledge of the hydro-

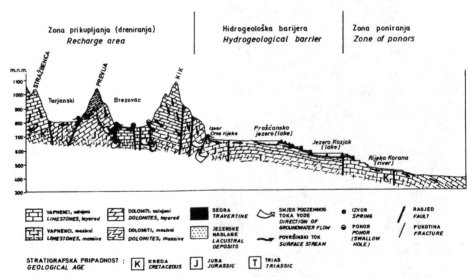

Figure 5.49. Hydrogeological cross section (not to scale) through the Plitvice lake district, Croatia.

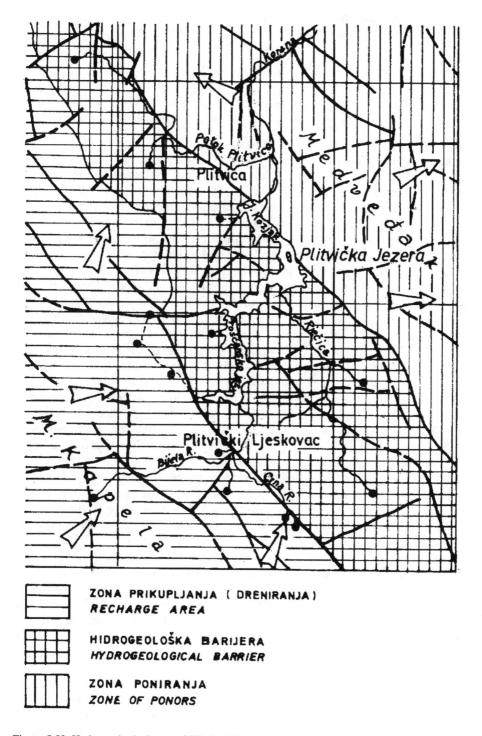

Figure 5.50. Hydrogeological map of Plitvice lake district, Croatia.

geological conditions indicates that the most environmentally hazardous matter (mainly solid and liquid wastes) should be removed from the area.

5.7.11 *Human influence on the karst water of the Nerja cave, Malaga, Southern Spain*
(Francisco Carrasco-Cantos, Bartolomé Andreo-Navarro, Juan José Durán-Valsero, Iñaki Vadillo-Pérez & Cristina Liñán-Baena)

5.7.11.1 *Site description and hydrogeologic settings*
The karstic underworld is a domain which has been traditionally studied from a speleological or scientific approach (Trombe 1952) but in the last few years the interest is also economic (Mangin & d´Hulst 1995, Huppert et al. 1993). Everyday, many cavities are visited by the public and constitute important economic activities in the region where they are situated. The human visits produce impacts in the underground environment: variations of the climatic parameters (temperature, humidity, CO_2), contamination and changes of the physical parameters of the groundwater and rock alteration (Cigna 1993, Pulido-Bosch et al. 1997). This contribution deals with the human impact on the environmental parameters and on the drip water in the Nerja cave, a touristic cave situated in the south of Spain (Andalusia), 50 km to the east of Malaga city and less than 1 km from the Mediterranean Sea (Fig. 5.51). The Nerja cave extends through a series of chambers and galleries to a total of almost 5 km, with a difference of height of 70 m and occupying a volume of over 300,000 m^3. Its shape is elongated in a more or less N-S direction and practically horizontal. The cave entrance is situated at 158 m altitude. From a geological standpoint, the cave is situated in the Almijara unit belonging to the Alpujarride complex of the Internal Zone of the Betic Cordillera (Avidad & García-Dueñas 1980, Sanz de Galdeano 1986). This complex has two lithological formations, represented in the geological sketch of Figure 5.51. The lower one is made up of metapelites (schists and quarzites), attributed to Paleozoic age. The upper formation is carbonated: white dolomitic marbles towards the base and blue calcareous marbles towards the top, with discontinuous metapelitic intercalation, dating from the middle to upper Triassic. The cave is developed over the dolomitic marbles which have a dense fissuration. Outside the cave, detrital Neogene deposits outcrop discordantly over the Alpujarride materials (Fig. 5.51). Karstic forms (karren, dolines, sinkholes) hardly exist in these carbonate materials. The Nerja cave is a major exception. The karstification process which gave rise to the cave occurred throughout the Pliocene and the Pleistocene. During the temperate and hot periods of the Quaternary age enormous quantities of speleothems were generated (Durán et al. 1993). The cave is actually in the unsaturated zone of the aquifer, several metres above the water table(Andreo & Carrasco 1993a), because of the neotectonic activity and uplifting along of the faults which limit the aquifer at the south.

5.7.11.2 *Tourism*
Since its tourist habilitation in 1960 the cave is one of the most visited natural sites in Spain, with 500,000 visitors per year throughout the period between 1988-1996. The monthly distribution is very similar for the different years (Fig. 5.52). The numbers of visitors vary between 200 and 3500 daily; some days increasing to 5000. The

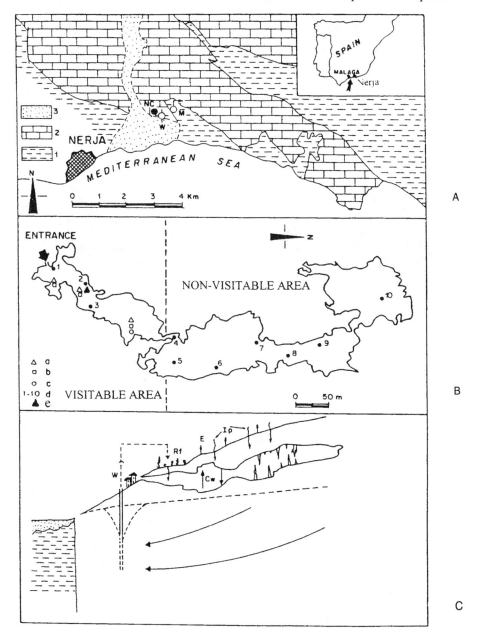

Figure 5.51. Situation and geological sketch of the Nerja cave (modified from Carrasco et al. 1995). A) 1 = metapelites, 2 = carbonates, 3 = Pliocene and Quaternary deposits, NC = cave entrance, M = maro spring, W = cave well. B) Location of the monitoring points in the A, B and C halls: a = air temperature, b = humidity, c = air CO_2 content, d = drip water sampling point, e = rock temperature sensor. C) Cross-sectional sketch (not to scale) showing the main processes influencing the water chemistry inside the cave: E = evapotranspiration, Ip = infiltration of the rain, Cw = water condensation, W = groundwater from the well located near the cave, Rf = return of irrigation on the garden.

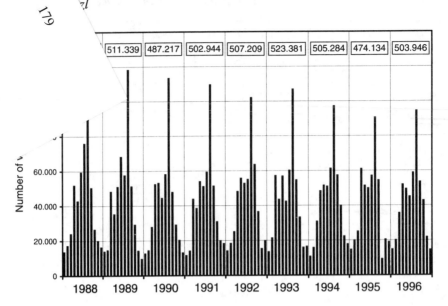

511.339 487.217 502.944 507.209 523.381 505.284 474.134 503.946

Figure 5.52. Data about the visitors to Nerja cave (1988-1996 period).

stay of the visitors in the cave changes its environmental conditions. During a visit of one hour a person contributes 60 calories and 40 g of water vapour, and exhales around 20 l CO_2, thus increasing temperature, humidity and CO_2.

The changes produced are a function of the number of visitors and the mean stay time. Others factors, inherent to the cave, are also important, i.e. cave volume and natural capacity of ventilation. The main changes in environmental parameters are the following: cave air temperature rises 0.3°C by the lighting and 0.1°C by 1000 visitors/day, temperature inside the rock rises between 0.05°C and 0.15°C in a day; relative air humidity increases to 2-3% in a day, nearly reaching saturation on some summer days. The carbon dioxide concentration can reach values 8 times greater than the natural mean value. In a period of few visitors, the minimum values are near 300 ppm, according to the natural value of the atmosphere in the exterior of the cave. In a day, the increase of concentration can be 500-700 ppm, but during the night it decreases to minimum values. In the periods with major visitors flux concentration reaches 2300 ppm.

5.7.11.3 Anthropogenic impact on the water

Hydrochemical monitoring of two types of groundwater has been carried out since October 1991:

    1. The drip water inside the cave,
    2. The water of the saturated zone near the cave.

The first water type was collected at 10 points (Fig. 5.51 and Table 5.5). The second water type was sampled in the near Maro spring (M in Fig. 5.51) and from a well (W) that forms part of the installations of the cave, used for watering the gardens and cleaning. The sampling periodicity can be considered monthly, in general.

Table 5.5. Statistical summary of the physical-chemical data registered in the monitoring network.

		pH	t	EC	Ca^{2+}	Mg^{2+}	Na^+	K^+	Cl^-	SO_4^{2-}	ALK	NO_3^-	SiO_2	pCO_2	SIc	SId
P	n	99	3	114	74	72	125	125	111	47	80	103	78			
	m	6.92	16.3	87	3.5	2.5	4.6	2.8	10.7	11.8	18.9	1.8	3.6			
	V(%)	5.91	18.9	91	145.9	89.4	127.6	242.7	78.3	96.2	97.7	276.2	387.2			
1	n	66	50	66	67	67	67	67	67	66	67	67	65	66	66	66
	m	7.88	18.9	1148	135.9	65.3	32.6	2.3	83.8	296.5	371.8	18.2	14.8	0.43	0.86	1.64
	V(%)	2.16	4.3	9	14.7	17.5	31.5	147.7	19.1	14.5	9.3	73.5	13.5	43.17	19.31	20.72
2	n	13	13	13	14	14	14	14	14	14	14	14	14	13	13	13
	m	8.35	18.7	531	41.4	41.6	7.7	2.2	25.1	14.7	300.0	2.3	7.2	0.13	0.85	1.92
	V(%)	2.65	6.6	11	28.9	15.3	18.8	121.1	19.0	24.0	13.7	39.4	16.0	71.24	30.35	24.63
3	n	136	121	131	131	130	138	138	133	118	135	136	123	132	126	125
	m	8.32	19.1	484	31.5	44.4	10.9	3.6	29.1	21.1	283.4	3.2	9.2	0.14	0.65	1.70
	V(%)	2.71	3.8	11	30.0	11.0	30.7	289.6	51.0	42.4	10.5	106.1	50.2	94.10	32.89	24.38
4	n	17	17	17	17	17	17	17	17	17	17	17	17	17	17	17
	m	8.31	20.4	529	44.9	42.3	8.5	2.8	29.7	17.4	304.8	6.1	10.8	0.18	0.85	1.93
	V(%)	3.27	0.9	5	16.4	11.1	9.0	168.2	14.5	14.9	6.2	26.9	11.7	133.88	30.76	26.49
5	n	8	8	8	8	8	8	8	8	8	8	8	8	8	8	8
	m	8.49	19.1	522	40.1	42.9	9.4	1.1	32.8	21.0	298.0	3.7	8.6	0.08	0.95	2.17
	V(%)	1.45	4.1	4	16.7	7.3	9.5	67.9	1.9	12.9	2.4	55.9	22.2	24.57	7.21	8.66
6	n	15	15	15	14	14	14	14	14	14	14	14	14	14	14	14
	m	8.37	19.1	504	48.3	40.3	5.8	1.7	21.2	11.0	317.8	8.9	8.6	0.13	0.93	2.03
	V(%)	1.77	2.1	9	23.9	13.6	8.5	84.8	13.0	27.0	7.5	37.3	25.1	37.86	16.10	12.52
7	n	60	45	60	60	60	60	60	60	59	60	60	60	59	59	59
	m	8.29	19.1	472	41.9	36.7	6.3	3.2	27.3	19.4	277.2	5.5	7.5	0.14	0.76	1.69
	V(%)	2.57	4.1	18	16.3	19.4	22.6	413.5	82.4	36.8	5.7	83.3	17.4	58.92	25.74	25.37
8	n	14	14	14	14	14	14	14	14	14	14	14	14	14	14	14
	m	8.66	19.3	392	19.0	42.5	8.4	2.3	26.6	16.2	233.2	3.1	9.5	0.05	0.70	2.00
	V(%)	1.48	1.1	11	18.1	5.3	7.7	147.5	12.5	16.3	2.9	20.6	10.3	32.26	13.17	9.20
9	n	10	10	10	10	10	10	10	10	10	10	10	10	10	10	10
	m	8.55	19.5	418	22.3	42.8	6.9	3.1	26.9	13.5	251.2	2.4	7.2	0.06	0.71	1.95
	V(%)	0.93	1.2	6	13.0	3.2	13.3	140.7	13.5	26.8	3.9	25.0	15.8	18.57	11.31	8.51
10	n	11	11	11	10	10	10	10	10	10	10	10	10	10	10	10
	m	8.39	19.6	483	44.0	38.5	5.5	1.3	20.6	13.7	304.7	2.2	7.1	0.10	0.95	2.09
	V(%)	1.46	3.0	7	13.0	6.8	18.2	84.6	4.5	27.8	7.3	40.1	11.5	26.60	4.43	4.67
M	n	87	92	97	54	54	53	53	53	54	53	53	53	51	51	51
	m	7.61	18.9	624	100.9	23.9	9.1	2.1	65.8	197.1	169.9	0.6	15.1	0.29	0.12	-0.19
	V(%)	2.49	3.1	20	25.6	44.0	40.5	32.0	113.3	38.8	50.5	95.7	64.5	67.72	507.65	-679.1
W	n			50	29	29	29	29	29	28	29	29				
	m			651	77.9	35.7	13.8	2.8	33.3	89.2	283.0	10.9				
	V(%)			8	10.4	13.7	30.3	38.9	16.5	15.4	9.6	73.1				

Legend: P = rainwater data, n = number of measurements taken, m = arithmetic mean, V = coefficient of variation. Chemical contents in $(mg \cdot l^{-1})$, pCO_2 $(atm \cdot 10^2)$, temperature ($^{\circ}$C) and EC $(\mu S \cdot cm^{-1})$.

The majority of the drip water have low conductivity values (392 to 531 $\mu S \cdot cm^{-1}$), with low contents in the major components (Mg^{2+} around 40 $mg \cdot l^{-1}$, Ca^{2+} 30-40 $mg \cdot l^{-1}$, alkalinity around 300 $mg \cdot l^{-1}$ of HCO_3^- and SO_4^{2-} 15-20 $mg \cdot l^{-1}$). This water is HCO_3 Mg-Ca type (Andreo & Carrasco 1993b, Carrasco et al. 1995, 1996). In a previous work Romero et al. (1991) indicate that this water shows microbiological contamination, which rises in summer when the number of visitors increase.

The samples taken from the saturated zone of the aquifer are of a HCO_3-SO_4-Ca type (M) and HCO_3 Ca-Mg type (W). The electrical conductivity is slightly higher than the above mentioned one (Table 5.5): 624 and 651 $\mu S \cdot cm^{-1}$, respectively. The well water is used for supplying the installations of the cave. Recently, point microbiological contamination has been detected, owing to an waste water leak in the services of the installations, which has been corrected.

Nevertheless, the most mineralised water is a drip water collected inside the cave near the entrance (point 1, Fig. 5.51), which is HCO_3-SO_4-Ca-Mg type. Its electrical conductivity ranges between 895 and 1453 $\mu S \cdot cm^{-1}$, because of the higher content in all the components analysed. This water corresponds to the water of the well (W) which is used for watering the gardens, passes through the anthropogenic soil prepared to cultivate the gardens, and then drips into the cave. The concentration factor of the well water by infiltration and percolation is calculated at approximately 1.5 for the electrical conductivity and carbonate parameters (Alkalinity, Ca^{2+} and Mg^{2+}), whereas this factor is superior to 2.0 for Cl^- and Na^+ and reaches 3.2 in the SO_4^{2-} content. The flow towards the interior of the cave produces an increase in all the major components, particularly in SO_4^{2-} (297 $mg \cdot l^{-1}$ as mean value), Cl^- (84 $mg \cdot l^{-1}$) and Na^+ (33 $mg \cdot l^{-1}$), which come from the anthropogenic soil. In the external part of the cave there are several points, not considered here, which present the same hydrochemical characteristics.

The partial equilibrium pressure of CO_2 (pCO_2) and the saturation indexes for calcite (SIc) and dolomite (SId) were calculated with the available data, using the program SOLUTEQ (Bakalowicz 1984). Normally the sampled drip water is oversaturated in calcite (SIc = 0.65-0.95) and dolomite (SId = 1.64-2.17) and presents pCO_2 values varying 0.05-0.18 $\cdot$ 10^{-2} atm. The water from point 1 presents higher values in these parameters SIc = 0.86, SId = 1.64, pCO_2= 0.43 $\cdot$ 10^{-2}, because it comes from the well and passes through the garden soil where relatively important biological activity takes place, producing more CO_2 than in the soil which exists above the rest of the cavity.

There are differences in the calco-carbonic parameters of the drip water corresponding to certain points, due to the situation inside the cave. Thus, generally, the drip water sampled in the area open to the public or near this (points 2-4), presents higher values of pCO_2 average and greater coefficient of variation than the rest of the cavity (Table 5.5). These differences may be related with the CO_2 coming from the human visits. The seasonal evolution of pCO_2 in the drip water (Fig. 5.53) shows, in general, the natural fluctuations of the soil CO_2 content in temperate climates (Atkinson 1977a, Troester & White 1984), with low values in winter and high values in summer. Nevertheless, the temporal evolution of pCO_2 also shows that in the summer this parameter is frequently higher in the area open to the public than in the non visited area (Fig. 5.53).

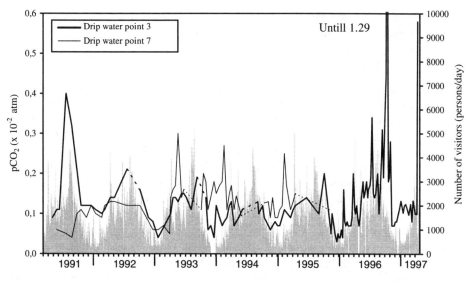

Figure 5.53. pCO_2 variation of drip water in the visitable zone (drip water point number 3) and the non visitable zone (point 7).

5.7.11.4 *Conclusions*

In caves open to the public, such as the Nerja cave, the considerable number of visitors that go in, produce an impact on the environmental parameters inside the cave and on several hydrochemical characteristics of the drip water.

The majority of drip water has a low mineralisation, it is HCO_3 Mg-Ca type and presents microbiological contamination, specially in summer, when the cave is most visited. In addition, in the tourist area of the cave, the sampled drip water has pCO_2 higher than the monitored water in the non-visitable area, also in summer, due to CO_2 from by the visitors. This CO_2 lessens the degree of saturation to calcite, which may originate alteration processes in the speleothems. Condensation processes may occur, because during some summer days the humidity in the cave atmosphere reaches 95%, which could generate unsaturated water also capable of deteriorating the speleothems.

In the external part of the visited zone, the drip water sampled is very mineralised and HCO_3-SO_4-Ca-Mg type, because it is water pumped out of a well, used for watering the garden and then infiltrated through the fissures of marbles above the cave. So this drip water has high concentration in SO_4^{2-}, Cl^- and Na^+, originated by dissolution of these components in the anthropogenic soil prepared for the garden cultivation. In addition, point contamination process has been detected in the well groundwater because of an urban waste water leak in a pipe line.

Acknowledgements. This article has been financed by the Nerja Cave Management and the Research Group 4021 of the Junta de Andalucía and it is a contribution to the UNESCO project IGCP 379 'Karst processes and the Carbon cycle'.

5.7.12   *Possible impacts of climate change on the karst of the Russian platform*
(Vladimir S. Kovalevsky)

Man made climate changes as predicted by climatologists can affect karst process development but also karst water resources and that is of practical importance. The supposed doubling of carbon dioxide content in the atmosphere in the first decades of the 21st century will cause a global warming and an increase in precipitation and evaporation. These changes will not be very significant in the equatorial zone, but in high latitudes they will be very noticeable. Thus, according to Budyko (1988) it is possible that summer temperatures will increase by 2-5°C and in winter by 10-15°C in the Russian Platform by the year 2020. The amount of precipitation in this instance is projected to increase by 200-400 mm·a^{-1} in the north and by 50-100 mm·a^{-1} in the south of the territory. Potential evaporation will also increase by 200-300 mm·a^{-1} in the south and the rise in actual evapotranspiration is expected to be 10-200 mm·a^{-1}. The amount of effective precipitation in the northern part of the Russian Platform will increase to 300 mm·a^{-1}. Taking into account the fact that mean annual precipitation for this area amounts to 600-800 mm·a^{-1} and that karst water recharge is about 30-50% of that, it can be concluded that karst water recharge will greatly increase in the north and may decrease in the south. Similar tendencies in karst water resources changes may occur in other regions of the world.

Similar trends in groundwater levels and discharges in karst areas of the Russian Platform are already being manifested independent of whether the reasons are natural or artificial. Increases in total precipitation, annual runoff and groundwater level have been observed in most parts of the Russian Platform for the last 2-3 decades. Whilst in the southern, dry part of the platform things are reversed.

The observed climatic changes are being used to predict possible changes in karst water resources in future. Predictions can be made using different methods depending on the initial data availability and the required detail of assessment. There is an extensive literature of all these methods available (Bredehoeft 1982, Kovalevsky 1976, Anderson & Burt 1988). Due to calculations of mean annual and winter baseflow an increase of karst water resources of up to 5-30% is expected in the Russian Platform compared with the present-day mean annual norms. Comparison of probable changes in the mean annual and low water dry season discharge of karst water indicates greater considerable changes in the latter. Thus karst low water discharge may double on the Izhorsk plateau, whilst in the southern part the increase will only be 10-35%. The predicted changes in karst water discharges will cause not only changes in mean and low water discharge but also changes in the whole karst water regime.

Due to the fact that the intensity of solution is directly proportional to the discharge of the groundwater filtering through it, a change of karst water recharge will bring about a change in the karst process development. In this case the most intensive accentuation of karst process may be expected in the northern areas of the Russian Platform. Karst process intensity will be amplified by the increasing corrosiveness of precipitation due to the predicted doubling of the carbonic acid content in the atmosphere. The greatest increases in karst process intensity should be expected in the Baltic area where, apart from the predicted increase in karst water recharge, processes of karstification will be amplified due to increasing karst water exploitation and mine water pumpage. The intensification of karst processes will have negative effects on the environment (new sinkholes, loss of surface runoff etc.). In certain parts it will promote the recharge of karst water and improve the conditions for water supply in the areas. However, the increase in temperatures and potential evaporation will lead in other parts (in conjunction with the increased discharge into the karstic underground and the storage capacity of the karst system) to drying up of soils and an increased frequency of droughts. Taking into account the very limited karst water resources, it is reasonable to expect a deterioration the of water supply to these areas.

The predicted values show that the possible effects of climate changes caused by human activities on the karst water regime and karst water resources, as well as on the intensity of karst processes can be significant and may vary between regions.

Acknowledgements. The research described in this contribution was made possible in part by Grant N J 93100 from the International Science Foundation and Russian Government.

CHAPTER 6

Extractive industries impact

JOHN W. HESS (co-ordinator)
Water Resources Center, Desert Research Institute, Las Vegas, USA

LINDA D. SLATTERY (co-ordinator)
Ohio Environmental Protection Agency, Division of Emergency and Remedial Response, Columbus, Ohio, USA

6.1 OVERVIEW OF EXTRACTIVE INDUSTRIES IMPACT
(John W. Hess & Linda D. Slattery)

6.1.1 *Introduction*

The minerals associated with karst terrains have been exploited for many years as a valuable natural resource. This exploitation has severely impacted these areas both physically and chemically, altering landforms and hydrological and geomorphological processes. The types of extractive industries that adversely impact karst sites are: 1) subsurface mining, 2) surface mining (quarrying), and 3) oil and natural gas exploration. Impacts from these extractive industries may not be limited to the local area, and can impact water quality and quantity at significant distances from the extraction site.

Carbonate rocks host many ore deposits such as lead, zinc coal, iron, and gold, with about 10% (Ford & Williams 1989) of bauxite mined from karst sites (primarily in Jamaica and Hungary, Photo 6.1). Exotic minerals such as antimony, copper, uranium and vanadium are formed as thermal precipitates in caves and collapse breccias. These are also extracted from karst areas but not mined extensively.

Limestone and dolomite have more industrial uses than any other rock. Limestone is commonly used as a building material and an aggregate for road beds. Lime (a product of limestone) and magnesium (obtained from dolomite) is used for agricultural applications (fertilizers). Gypsum is extracted from karst areas for use in plaster of Paris and drywall production.

Approximately 50% of the current production and known oil and gas reservoirs are found in karst areas (Ford & Williams 1989). The world's most productive basin is the Persian Gulf, which is primarily limestone. In contrast, most carbonate oil and gas fields in North America are found in dolomites (Craig 1987).

6.1.2 *Physical processes and impacts*

Exploitation of natural resources impact karst areas both physically and chemically.

Physical impacts include aquifer dewatering, lowering of local and regional water tables resulting in land subsidence and sinkhole development, and irreparable land destruction from quarrying.

Aquifer dewatering

Substantial literature exists on the destructive impacts to karst terranes from subsurface mining. Two major themes emerge from this literature: the impacts of inrushes of groundwater to excavated galleries, and the impacts of dewatering operations designed to reduce these inrushes. During mining operations, the water table must be lowered to create dry working conditions and to avoid catastrophic mine floods which could damage equipment and claim lives (Bosak & Koroe 1991). In mature karst aquifers, the secondary permeability can be significant, resulting in high yields and appreciable aquifer dewatering. For example, Yuan (1992) records over 1000 instances of sudden flooding in karst-related coal mines in China, and Adamczyk et al. (1988) record similar groundwater inrushes in lead-zinc mines associated with a karstic aquifer in Poland. In the Transdanubian Mountains of Hungary, pumping operations to dewater large areas of limestone for coal and bauxite mining have occurred for over 80 years (Böcker & Hegyi-Hovanyi 1983). This has lowered the water table by 15 to 150 m, producing a widespread cone of depression. As a consequence of this, springs have ceased to flow and there is concern that the famous thermal springs of Budapest may be at risk (Alföldi 1984). In South Africa, there is a long history of problems associated with karst groundwater inflows to the gold-mines of the Far West Rand as a result of dewatering operations (Kleywegt & Pike 1982, Vegter & Foster 1992). The dewatering is also responsible for widespread ground subsidence and sinkhole formation. These problems have a particular relevance to sub-water table limestone quarrying. Grimmelmann (Section 6.4) examines the influence of underground mining on the karst in Mansfeld, Germany including accelerated dissolution rates and increased land subsidence.

Land subsidence and sinkhole development

In carbonate rocks, the water table is controlled by local factors such as its secondary permeability, topography and precipitation. Dewatering of a karst aquifer over time may result in the lowering of the local water table. This has been reported to have adverse effects on local springs, streams and lakes. As water is pumped to dewater a mine, a large cone of depression is formed, greatly expanding the affected area. Alföldi (1984) described a karst dolomite region in Hungary where the water table was lowered 15 to 20 m from mining activity and caused most of the lukewarm streams to dry up. Böcker (1984) identified a spring-fed lake in Hungary where discharge from the Heziv Spring has been significantly reduced from lowering the water table by approximately 80 m. The Oberholzer Spring, near Carletonville, South Africa, dried up after the water level in the immediate vicinity of the gold mining district declined by more than 50 m (Foose 1967). Drastic lowering of the water table by excessive aquifer dewatering can, in some instances, result in land subsidence and sinkhole and doline development. In most cases land subsidence will only occur if support to the overlying unconsolidated materials is removed (Foose 1967). When the water table is lowered, a new level of subsurface erosion is initiated. By remov-

Photo 6.1. Exposure of palaeokarst features in a bauxite mine, Western Hungary (photo: Hötzl 1978).

Photo 6.2. Interception of active karst conduit features by quarry activity (photo:Hötzl 1992).

ing the support and allowing the materials to desiccate, the sediments can move downward into the cavities resulting in subsidence or sinkhole development.

Some catastrophic sinkhole formation has been reported in Far West Rand, Transvaal Province, South Africa. The dolomite in this gold mining district has been dewatered by more than 300 m since 1960, resulting in sinkholes a maximum of a few hundred metres wide and 50 m deep (LaMoreaux & Warren 1973). Brink (1979) described a three-story building that was swallowed into a sinkhole near the West Driefontein Mine in 1962, permanently burying the building and all 29 occupants. Quinlan (1974) reported that sudden collapses in this area have resulted in 34 deaths and 35 million dollars in expenditures on rebuilding, safety measures, research, and compensation for damages including loss of water supplies. Bredenkamp (Section 6.5) discusses these issues associated with the gold mining in South Africa.

Quarrying
Quarrying is one of the oldest extraction methods for limestone. Removal of the carbonate rock cover of an aquifer may lead to contamination of the aquifer (Ekmekci 1993). The quarry acts as a sinkhole which will rapidly carry surface water to the groundwater system. This is particularly a problem if the quarry is downstream of a village or town. Blasting in quarries may disrupt or change groundwater flow paths and may alter the quantity of water flowing through the karst system. This can drastically affect water supply systems downstream. Ekmekci (1993) described a site at Beytepe, Ankara, Turkey where springs downstream from a quarry were found to be microbiologically polluted. Studies showed that this might be from rapid infiltration of surface runoff containing animal waste through the quarry and consequently affecting the groundwater.

The destruction of both relict and active caves by quarrying has been recorded in many countries (e.g. Gillieson 1989, Stanton 1990), and the wider geomorphological impacts of limestone quarrying and possible techniques for their amelioration have been described by Gagen & Gunn (1987) and Gunn & Bailey (1993). Although the focus is primarily geomorphological, there are hydrogeological implications, particularly as both Gunn & Gagen (1987) and Kiernan (1989) note accelerated growth of dolines as a result of quarrying activities. Other authors who have considered the hydrogeological aspects of limestone quarrying include Michel (1988) who examined the conflict between groundwater exploitation and limestone extraction in Germany, and Ekmekci (1993) who briefly outlined some impacts of extraction on karst groundwater systems, with particular reference to quarries near Ankara, Turkey. In Britain, the main focus of research into the hydrogeological impacts of quarrying has been the Mendip Hills (Atkinson et al. 1973, Stanton 1977, Harrison et al. 1992). Here, the local planning authority adopted an approach whereby preference was given to working downwards to create 'invisible' holes, rather than lateral extraction scarring the landscape. In addition, a 'sacrifice' area was proposed in order to save more aesthetically pleasing and geomorphologically distinct areas. At this time, it was suggested by Stanton (1966) that stone reserves be maximised by deep sub-water table working. More recently however, Stanton (1989, 1990) has argued that limestone is more valuable in situ as a water resource and for its amenity value. Much of the research on sub-water table quarrying in limestone is being undertaken

Photo 6.3. Quarrying of Upper Jurassic limestone above the karst water table, Solnhofen, Bavaria, Germany. The removal of the natural cover system and the opening of karst shafts as well as the extension of existing joints favours the unfiltered seepage of surface water into the underground (photo: Hötzl 1982).

Photo 6.4. Quarrying of Devonian limestone below the karst water table, Warstein Sauerland, West Germany. The exposure of the karst water presents a direct pollution risk. For the quarrying to continue the water level has to be lowered thus significantly reducing the storage (photo: Hötzl 1986).

by environmental consultants working both for quarrying companies and regulatory authorities. Gunn and Hobbs (Section 6.2) discuss the impacts of quarrying and Gillieson & Houshold (Section 6.3) review quarry rehabilitation on carbonate terrains.

6.1.3 *Chemical processes and impacts*

Water, land and air pollution, and increased karst dissolution rates from acid mine drainage are the primary chemical impacts to karst terrains. Acid mine drainage into carbonate aquifers can severely affect water quality by altering the chemistry and pH. Karst waters affected by acid drainage display a low pH, high sulphate, and negligible alkalinity (Sasowsky & White 1993). Low water pH may result in high metals concentrations, such as aluminum and iron, which is a health concern. The high sulphate and low alkalinity concentrations create a detrimental environment for carbonate rocks, significantly increasing dissolution rates (Wicks & Groves 1993). This may alter or create flow paths which can reduce the output of local springs. During low flows, the acid water may significantly downcut the conduit flow path, resulting in a lowering of the local water table. In addition to the chemical problems introduced (increased concentrations of metals), increased carbonate dissolution rates (resulting in altering, creating, or downcutting flow paths) can create problems for local residents who depend on a water supply for drinking or irrigation.

6.1.4 *Summary*

Environmental impacts on karst terrains from extractive industries have been severe and in some cases widespread on land and air quality, and water quality and quantity. Impacts include changes to landforms and disruption of hydrological and geomorphological processes. New ideas in mining techniques, aquifer protection and rehabilitation are concentrating efforts toward effective environmental management. The difficulty and expense of rehabilitating complex carbonate aquifers indicate that changes in mining techniques can often be more effective than post mining efforts.

The following six contributions highlight the environmental impacts of extractive industries. Gunn and Hobbs discuss the hydrogeological impacts of limestone quarrying (Section 6.2). Grimmelmann reviews subsurface mining in Germany (Section 6.4), while Gillieson and Houshold examine the rehabilitation of a quarry in Australia (Section 6.3). The influence of gold-mining on water quality in South Africa is discussed by Bredenkamp (Section 6.5) and that of lead-zinc mining on water and land subsidence in Silesia by Tyc (Section 6.6). The final contribution by Johnson examines the special conditions that pertain in highly soluble salt (halite) karst, with examples from the USA.

6.2   LIMESTONE QUARRYING: HYDROGEOLOGICAL IMPACTS, CONSEQUENCES, IMPLICATIONS (John Gunn & Steve Hobbs)

6.2.1 *Hydrogeological aspects of stone extraction*

In many industrial countries, the quarrying of limestone from open-pits represents the most visually obvious and the most dramatic anthropogenic impact on karst terrains affecting both landforms and geomorphological and hydrogeological processes. The impacts on karst landforms, many of which are located in areas of high scenic value and are of considerable scientific interest, are immediate and obvious, but changes in groundwater quality and quantity may be manifest some distance from the actual extraction site. Surprisingly little attention has previously been given to these impacts.

The extraction of limestone has a long history going back quite literally to the Stone Age, but the pace of the activity has accelerated markedly in recent years as a result of a worldwide increase both in the demand for limestone and in the technical ability to excavate, transport and process large tonnages of rock. In Britain, Stanton (1977) suggests that annual removal of limestone from the Mendip Hills by quarrying first exceeded that removed naturally in solution at some stage in the 1700s or early 1800s. Similarly, Gunn and Gagen (1989) estimated that quarrying will have removed more limestone from the Peak District by the end of this century than natural processes over the whole of the Holocene. Overall, UK limestone extraction increased from 14 million tonnes (Mt) in 1895 to 28 Mt by 1935, and to 125 Mt in 1991. Similar production trends are apparent in other countries, and have been paralleled by a decrease in the number of quarries but a substantial increase in their size, resulting in a greater impact. The increased output of limestone has only been made possible by technological advances in extraction techniques, particularly the use of explosives.

The earliest and simplest methods of extraction involved the use of human muscle to remove limestone from free faces. This form of quarrying has minimal hydrological impacts although it does result in significant landform modification. Beginning in the 19th Century manual extraction was largely superseded by the use of explosives in most developed countries. Mechanical excavators are used to remove some weaker limestones such as the English Chalk and mechanical saws are used to remove harder, decorative limestones such as the Globerigina limestone on Malta, and fine Italian marbles. In contrast to manual techniques, these methods of quarrying require relatively low faces, and the quarry is more likely to be a hole in the ground rather than cutting back into the side of a valley or hill. As such there is greater potential for hydrological impacts, particularly if the quarry is excavated to, or beneath the water table.

In order to excavate large tonnages of limestone efficiently, blasting is necessary to fragment the rock for further handling and processing. In most countries, low explosives (black powder) were used from the 1800s until the middle of the present century. Subsequently, they have been progressively replaced by high explosives including TNT, blasting gelignite, and slurry and emulsion explosives which may be used either alone or in combination with ANFO (a mixture of Ammonium Nitrate and Fuel Oil). Differences in explosive properties determine the amount of rock liberated and the shape and size of material in the resulting blast pile.

Hydrogeologically, two aspects are important. Firstly, the introduction of explosives allowed quarries to expand in both area and depth. In general, use of black powder for blasting in quarries limited their location to the sides of valleys and hills and they rarely exceeded 50 m in depth. However, the greater power of high explosives, together with advances in drilling and blasting technology, means that modern quarries can be excavated in almost any topographical situation and to depths in excess of 100 m. Secondly, both the type of explosive used on the final blast on a particular quarry face and the blast design will influence the amount of blast-induced fracturing of the remaining unexcavated rock, and hence its permeability, porosity and hydraulic conductivity. For example, Smart et al. (1991) suggest that in sub-water table quarries the blast zone beneath the quarry floor may be considered as a separate aquifer characterised by high fracture density, low primary porosity, and negligible conduit development. Ekmekci (1993) further suggests that this may lead to changes in the direction of underground drainage such that more water may move in one direction and less in another. Explosive type and blast design also influence the form, and future development, of the blasted rock face (Gagen & Gunn 1987, Gunn & Bailey 1993). Increased permeability as a result of blast-induced fracturing may accelerate drainage towards the quarry face, and enhance the development of both subsidence and solutional dolines (Gunn & Gagen 1987).

6.2.2 *Potential impacts upon the unsaturated zone*

It has generally been considered that the unsaturated zone only contains a small percentage of the total volume of storage in a karst aquifer. For example, Atkinson (1977b) calculated that only 12% of water in the Cheddar Spring Catchment, England is stored within the unsaturated zone. However, Smart & Friederich (1986), working in the same area, suggested that storage in the unsaturated zone is comparable to that in the saturated zone. In the Causse Comtal, France, over 60% of storage is in the unsaturated zone (Dodge 1984), and the unsaturated zone accounts for almost all storage in two catchments near Waitomo, New Zealand (Gunn 1986). The differences may be accounted for by lithology, hydrogeology and drainage basin size. Where the unsaturated zone is thin or has a poorly-developed subcutaneous zone and/or fracture/fissure network, storage will usually be low and the impact of quarrying on groundwater quantity will be minimal. However, where a thick, well developed unsaturated zone is present then storage, and impacts may be much greater. In both cases quarrying will substantially modify the routing of recharge through the unsaturated zone, and deterioration of groundwater quality may occur.

Although some limestone areas have little or no soil cover, in most there is sufficient to support grass/shrub vegetation as a minimum. In these areas, the first impact of quarrying is the removal of soil and vegetation, thereby reducing evapotranspiration losses and increasing effective rainfall. In England, a change from grass to bare ground is likely to increase annual effective rainfall by about 8% in the north and by up to 40% in the Mendip Hills (Harrison et al. 1992); in warmer countries the increase will be higher. This additional water generally forms 'run-in' to the quarry, mostly as surface flow since the quarry floor becomes partly sealed due to packing of fractures, fissures and joints with dust and rock chippings compacted by vehicle movements. Runoff is often turbid and is usually channelled to some form of pond (sump). From here some water is lost by evaporation, partly negating the results of vegetation removal, some may infiltrate into the quarry floor, and the remainder may either be discharged into a convenient sink point or pumped out of the quarry. Unless great care is taken to settle out the majority of the fine material, deterioration of water quality is likely. For example, in the Peak District, water is discharged from

Eldon Hill Quarry into a closed depression. This in turn feeds a cave where there has been a significant accumulation of fines.

In addition to point-source pollution, the removal of soil increases the potential for diffuse contaminant inputs because the soil is normally an important zone of filtration and water purification. In the longer term, there will be changes in the amount and locus of solutional erosion, as in most karsts the soil is the major source of CO_2, and the majority of solution takes place in the subcutaneous zone representing not only a major store in the unsaturated zone, but also controlling water movement (Williams 1983). During periods of heavy rainfall the subcutaneous zone becomes saturated and water flows laterally towards zones with enhanced vertical permeability(subcutaneous shafts, transmitting water rapidly to the saturated zone (Gunn 1981, Friederich & Smart 1982). Thus, if the subcutaneous zone is removed during quarrying the hydrologic function of the unsaturated zone is partly destroyed.

The overall result of these near surface changes is likely to be an increase in annual and peak discharges but a decrease in base flows. For example, Stanton (1990) has recorded increased winter and decreased summer discharges from springs draining quarried areas in the Mendip Hills. However, quarrying through the unsaturated zone may also intersect vadose cave streams, resulting in increased surface runoff in the quarry and loss of water from the natural groundwater regime with the potential to reduce spring discharges.

6.2.3 *Potential impacts upon the saturated zone*

Once a quarry has reached the lateral limits of its planning boundary and extracted all the stone possible from the unsaturated zone it has to work beneath the water table to win more reserves. Where the aquifer in question has a large seasonal fluctuation in water level then it is possible to work stone during dry periods and allow the quarry to flood through the wet season. This is the 'minimal' impact method of sub-water table working, although, the creation of flooded workings can significantly modify groundwater behaviour in their vicinity. More commonly, where stone is won from beneath the water table it is by dewatering the aquifer. Water is pumped out from a central low sump in the quarry floor, from abstraction boreholes around or within the quarry, from galleries, or from trenches.

Pumping from a quarry will reduce the hydraulic head and thus drawdown water levels in the rock draining to the quarry. In those limestones which are relatively homogenous something approaching a classic cone of depression may develop. However, the anisotropic nature of most limestones means that an uneven 'zone' of depression is more likely, with preferential development along the zones of highest permeability. The form of the zone will also be affected by any permeability barriers, such as clay or vein mineral filled fractures. Water pumped out of the aquifer is usually discharged into a surface stream, and hence is likely to be lost from the local groundwater system thereby reducing spring flow(s), drying up water supplies and possibly changing the overall direction of underground flow (Fig. 6.1, Harrison et al. 1992).

In addition to diffuse flow seepage into sub-water table workings, phreatic conduits may be directly intersected with potentially severe problems. The area at risk will be a function of conduit geometry and the transmissivity of inter-conduit blocks.

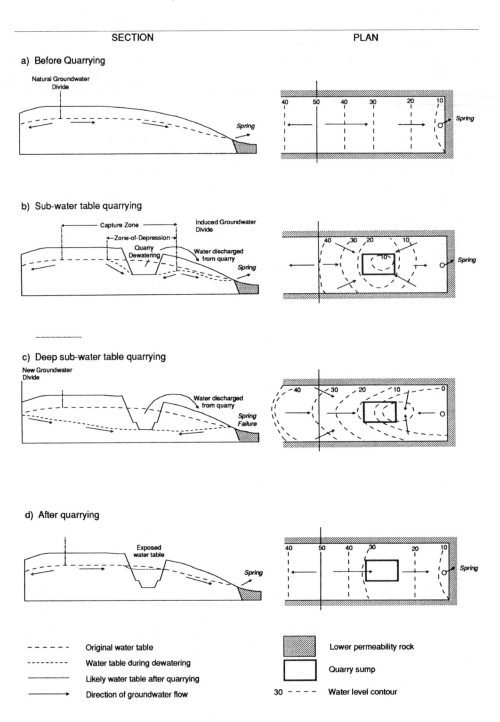

Figure 6.1. Schematic diagram showing the potential effects of sub-water table quarrying on the water table (after Harrison et al. 1992).

For example, in the Mendip Hills, where conduits are developed down dip, they tend to be looping in long section, the gradient of their rising and falling limbs being determined by the local dip. The height difference between the peak of one limb, and trough of the next may be considered as the amplitude of the conduit. Where conduits have a high amplitude, with peaks at the water table, then the area at risk is reduced to that between the quarry and the first section of the conduit which reaches the water table. If the limestone is of a high transmissivity then leakage can take place from one section of a conduit to another, thereby increasing the area at risk to derogation. However, if the conduit flows along the strike of the limestone, then conduit development may be more regular in long section, showing no distinct rising and falling limbs. In this case conduit gradient would be shallow, and the area at risk to derogation would depend on the level at which the conduit was intersected. Water loss from conduits can occur even where they are not directly intersected. For example, Edwards et al. (1991) demonstrated that quarry dewatering near to phreatic conduits can induce groundwater flow out of such systems and into the surrounding diffuse flow zone. In this case, repeat dye traces from a series of sinking streams which were undertaken at various stages of sub-water table working indicated an increase in loss of water from phreatic conduits as the zone of depression increased around the quarry sump.

Sub-water table quarrying may also impact upon surface water courses fed by karst springs. Firstly, there is the potential for contamination from increased suspended sediment loads and from pollutants such as fuel oils. Although regulatory authorities may specify limits on the amount of sediment that discharge water can contain, these limits may be exceeded during periods of heavy rainfall. A second possible impact is flooding downstream of the dewatering discharge point. As the requirements for dewatering are often greater in winter when surface watercourses also tend to have a greater discharge, the potential for flooding increases. The increases in flood flows will be accompanied by decreases in base flows and some streams may cease to flow during drier parts of the year.

In some cases it may be possible to mitigate against the impacts of sub-water table quarrying by well designed water management plans. For example, water abstracted from mineral workings may be recharged to the aquifer. However, this relies on the presence of suitable recharge sites sufficiently removed from the abstraction point to avoid re-circulation and the sites also have to be well integrated with the aquifer to accept the often large volumes of water pumped. In Britain, mitigation of derogation is often carried out in preference to such technical solutions. For example, supplies lost or reduced by dewatering may be compensated for, or replaced by water from another source such as the mains supply. Streams fed by springs which have dried up can be augmented by water discharged from the quarry. If this water is stored, either in a small surface reservoir, or by constructing a large sump in the base of the quarry, then it may be possible to maintain surface water courses during the summer months.

However, both options are problematical: surface reservoirs are expensive and rely on suitable nearby sites, and quarry sumps limit lateral development and become inoperable as the quarry penetrates the saturated zone to greater depths. In some countries, such as Germany, legal regulations are such that water abstraction takes priority over limestone quarrying, and workings must remain at least 2 m above the known water level in the vicinity of the quarry (Michel 1988).

In addition to direct hydrogeological impacts, large scale dewatering of limestone aquifers to permit mining operations has resulted in extensive surface lowering and sinkhole collapse. The dewatering of open pits for stone extraction is a more recent phenomenon and has involved individual sites rather than large regions.

6.2.4 *Predicting the impacts of sub-water table quarrying*

Quarrying in the unsaturated zone generally results in relatively local impacts, such as increased run-off, reduced water quality, and re-routing of recharge water through the aquifer. These may be considered as water management issues which do not impinge on the large scale groundwater resource. In contrast, sub-water table quarrying may have large scale, long-term impacts which must be predicted when quarry workings are proposed. Prior to assessing these impacts it is necessary to characterise the aquifer, especially to determine the degree of karstification. This is especially important as prediction of impacts is more problematic the more karstified an aquifer is. Karst aquifers can be defined in terms of their three end member attributes of recharge, flow, and storage (Smart & Hobbs 1986). Recharge is ranged between concentrated and dispersed end members, the former being characterised by large inputs at discrete points, the latter by smaller inflows at a much larger number of points. Flow varies between conduit and diffuse end members. Conduit flow occurs in large open channels with relatively high velocities giving rise to turbulent flow. Diffuse flow is confined to tight fractures and pores with small openings; velocities are low and flow is laminar obeying Darcy's Law. Storage varies between high and low end members and includes water stored in the soil, subcutaneous zone, conduits, and the saturated zone and is expressed in terms of the storage volume and annual recharge. The end members of the recharge, flow, and storage types can be plotted on three orthogonal axes to form a cube, the position of an aquifer within which can aid selection of the most suitable methods for predicting impacts. Four groups (and two sub-groups) have been defined based on the type of recharge, flow and storage present (Fig. 6.2).

Group 1. Represents aquifers with high storage, conduit flow and variable recharge. The prediction of the impact of quarry dewatering is the most difficult in these aquifers as it is highly dependant upon the likelihood of the workings intersecting an active conduit. As the storage in these aquifers is significant they can present a substantial water resource, hence the potential impact can also be significant. The group can be split into two sub-groups: 1a) with concentrated recharge, and 1b) with dispersed recharge. Where concentrated recharge is dominant the potential impact is much greater and more difficult to predict. Although the resurgence(s) for such concentrated inputs can often be determined, the exact course of the conduit is not normally known along its full length. Intersecting one of these conduits may dewater a large area down gradient from the quarry and could present significant problems in disposal of the discharge. Such impacts are also possible in aquifers with dispersed recharge although the conduits are usually less well developed in the upper reaches of spring catchments.

Group 2. Represents aquifers with low storage, conduit flow and variable recharge. Prediction of impacts within this group is no simpler than for group 1, but with low storage the number of water supplies and size of springs supported by the

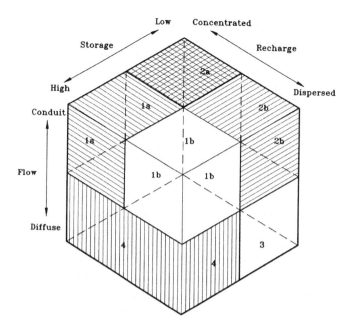

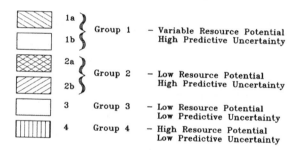

Figure 6.2. Carbonate aquifer classification – Assessment of the potential impacts from sub-water table quarrying (modified after Smart & Hobbs 1986).

aquifer is likely to be much smaller. As with group 1, this group can be sub-divided according to recharge type, the potential impact due to the possibility of the intersection of conduits fed by sinking streams being higher where concentrated recharge is present.

Group 3. Represents aquifers with dispersed recharge, diffuse flow and low storage. These are 'minor or non-aquifers' and present no problem from a hydrogeological viewpoint for sub-water table quarrying. They are not a significant water resource, but if the potential impact must be determined then formulae applicable to homogeneous aquifers can be used to give a best estimate. In such cases, the limitations of these formulae to fissure flow systems must be emphasised when quoting predicted impacts. The importance of local geology and topography should always be considered.

Group 4. Represents aquifers which have diffuse flow, high storage and variable recharge. Because these aquifers have high storage they provide a useful resource which may be derogated by sub-water table working. They may also support moderately large springs, which in turn may support stream/river base flow. The potential impact can, as with group 3, be assessed 'using standard formulae' providing that their limitations are understood. Within the central area of the cube an aquifer may fall into more than one group. In cases such as these a judgement will have to be made as to which are the most suitable methods to determine the potential impact of sub-water table quarrying. There are a number of methods of determining where an aquifer falls in the proposed classification (Fig. 6.2). The first is by field observation and/or desk study utilising geological and topographic maps, borehole information and abstraction data, as well as caving reports, dye trace studies etc. Spring hydrographs can also be examined to determine if discharge is flashy or not. Examples of the effect of varying recharge, flow, and storage on the spring hydrograph are given in Smart and Hobbs (1986). A third method is to continuously monitor the electrical conductivity of spring discharge water, the coefficient of variation of which (for a representative time period) can indicate the degree of karstification. In a similar manner, turbidity pulses may also be used (Hobbs 1988). Finally, direct measurements of aquifer permeability in a statistically significant number of, and suitably located, boreholes can be used (Smart et al. 1991). In most cases, only limited information will be available and a subjective estimate of the position of the aquifer within the classification will have to be made.

Once the nature of a karst aquifer has been defined, the most suitable method of prediction can be assessed, or if prediction is not possible the risk can be determined. Methods include the use of numerical formulae and computer models for more homogeneous aquifers, to plotting the position of sinking streams, cave passages and springs in karstified aquifers. Given the technological advances in computer systems, it is now possible to successfully model very complex aquifers, even those with a moderate degree of heterogeneity.

6.2.5 *Potential impacts upon cessation of mineral working*

Once mineral working has ceased at a site it may either be left dormant, or be redeveloped. If a sub-water table working is left dormant then it will fill with water at a rate depending upon local rainfall, evaporation, and groundwater flow. During the early part of this period any derogated springs will be unlikely to recommence flowing as water will be moving to storage in the quarry void. Given that this will have a storativity of 100% as opposed to 1 to 3% for karstified limestone, groundwater levels may take some time to recover. Furthermore, because the void will have an infinite transmissivity the quarry will, to all intents and purposes, have a flat water table surface. Thus, groundwater levels close to the quarry will never recover to their previous levels and some springs may never return.

Disused limestone quarries are often used as waste disposal sites, and unless considerable care is taken there may be rapid contaminant migration and minimal dilution of leachate prior to arriving at a spring (compare Sections 5.4 and 5.7.4). Edwards & Smart (1989) quote one instance in the UK of landfilling in a disused Carboniferous limestone quarry where no liner system was employed. Leachate from the landfill moved rapidly to a spring over 2 km distant resulting in gross contamination. Subsequently the landfill was capped with a low permeability layer in an attempt to reduce leachate generation. Similarly, Bodhankar & Chatterjee (1993) document waterborne disease and serious illness resulting from the contamination of public water supplies by pollutants from an unlined limestone quarry used as a disposal site for urban waste. For this reason the development of disused quarries in karstified limestone as landfill sites is not generally recommended. Where they are utilised their location is of some importance (Sendlein & Palmquist 1977) as are points at which to monitor for potential leachate migration (Quinlan et al. 1986). It has been suggested that deeper quarries may be more suitable for landfill because fissure size and frequency tends to decrease with depth. However, it is important not to assume that this is always the case since water tracing and related studies carried out independently by the authors have revealed high groundwater flow velocities at depths of up to 50 m beneath the floors of already deep quarries in areas with no known conduit development.

Although the main focus of the use of worked out quarries for landfills is associated with licensed landfilling, illegal 'fly' tipping can also be problematic, especially in small isolated workings. At such locations there is no regulation and no liner present to limit any leachate migration, therefore the potential for impact upon local springs/groundwater abstractions may be significant. In recent years, planning authorities in the UK have placed much greater emphasis on the end use of worked out quarries. Consequently, many planning applications now propose schemes for the restoration of mineral workings to include nature walks around flooded workings which can also be used for sport fishing, sub-aqua diving and boating.

Acknowledgements. Thanks are due to Paul Hardwick who provided helpful comments on an earlier draft and assisted one of the authors (JG) with much of the work on which this paper was based.

6.3 REHABILITATION OF THE LUNE RIVER QUARRY, TASMANIAN WILDERNESS WORLD HERITAGE AREA, AUSTRALIA
(David Gillieson & Ian Houshold)

6.3.1 *Destruction by quarrying*

Limestone has been quarried in Australia since 1798, but until very recently little or no treatment of the abandoned workings has occurred. In Tasmania limestone from the Lune River quarry has been used as a pH control in the electrolytic refining of zinc. Most resource conflict over limestone revolves around visual and water pollution, as well as loss of recreational amenity and conservation values. Most limestones in Australia are flat lying, have considerable overburden and often have the karst watertable close to the surface. Therefore limestone bodies with high relief are ideal for mining and are often the most cavernous. Most of these are found in the Eastern Highlands extending south into the island State of Tasmania.

Quarry rehabilitation is a rapidly growing field in Australia and at present three karst sites are being treated: Mount Etna in central Queensland, Wombeyan in southern New South Wales, and Lune River in the Tasmanian Wilderness World Heritage Area. The latter project is being funded by the Federal government under the World Heritage Properties Conservation Act 1983. The quarry overlies the Exit Cave system, an extensive and complex cave whose geomorphic and faunal values caused it to be inscribed on the World Heritage listing in 1989. The Exit Cave system has 26 km of mapped passages and has formed along three main genetic axes (Goede 1969). The cave is home to a rich terrestrial and aquatic fauna adapted to the continual dark and cold. In several places the walls are covered with awesome displays of glow worms which simulate the night sky. The cave fauna consists of more than 30 species, including 11 troglobites (Eberhard 1992, 1995).

A number of impacts have been observed in Exit Cave and its tributary caves as a result of the quarry operations (Houshold 1997):
- Removal of cave passages and their contents by quarrying;
- Destruction of palaeokarstic fills by quarrying;
- Increased sedimentation of fine clays in Little Grunt Cave (underlying the quarry) and the hydrologically connected Eastern Passage of Exit Cave;
- Recurrent turbidity in Eastern Passage and Exit Creek;
- Changes in pH, conductivity and sulphate ion concentrations (to 150 $mg \cdot l^{-1}$) in passages draining the quarry;
- Re-solution of speleothems by acidified drainage waters due to oxidation of sulphides from paleokarst fills;
- Reduced densities of indicator species of hydrobiid molluscs (*Fluvidona* sp. nov.) in passages draining the quarry (Eberhard 1995).

Because the continued operation of the quarry was producing these major geomorphic, water quality and biological impacts on the cave, the quarry was closed in August 1992 under World Heritage legislation. Following the preparation of a rehabilitation plan, a joint Commonwealth – Tasmanian team started active rehabilitation in April 1993 (DASETT – TasPWH 1993). The challenge was then to rehabilitate the quarry and the affected parts of the cave without further impacting the karst values and ecosystems. Reshaping of the quarry by restoration blasting techniques was deemed inappropriate because of the sensitivity of both the geomorphology and the biology of the cave.

6.3.2 *The rehabilitation plan*

The primary objective of the rehabilitation has been to protect the World Heritage values of Exit Cave system by returning the ecosystem processes within the quarry area to as close as possible to their original state. As far as can be ascertained, this is the first attempt to return underlying karst processes to something approaching a natural state. The main issues for rehabilitation are the integrity of the underground drainage, its water quality and the cave invertebrate populations. A secondary objective has been to maintain a high degree of interconnected secondary porosity in the quarry for effective recharge and to simulate as much as possible the original polygonal karst drainage and its forest cover.

The key concept in the rehabilitation is the simulation of the high secondary porosity of a polygonal karst network at a range of spatial scales from that of the whole Exit Cave system down to the diffuse infiltration points within an area of 100 m^2.

This hydrological approach is somewhat different to conventional quarry rehabilitation methods, which lay emphasis on the control of surface drainage, erosion control and aesthetics. Although these factors are important at the Lune River Quarry, they are only a small part of the rehabilitation process. The maintenance of underground drainage, water quality and cave invertebrate populations are of paramount importance for the World Heritage values of Exit Cave.

To achieve these objectives the quarry was subdivided into a number of small closed drainage basins (0.1 to 0.2 ha), each of which has a karst sink or infiltration zone (Fig. 6.3). Drainage control has been achieved by several methods. Impermeable bunds of clay and rubble have been constructed along the outer edge of the benches, about 500-1000 mm high and 1000-1500 mm wide. Similar bund walls have been placed to subdivide the benches into 0.1-0.2 ha internal drainage basins which simulate size of the depressions of polygonal karst (Fig. 6.4) on adjoining Marble Hill. Each sink is protected by a filter structure (Fig. 6.5) and areas under clay fans have additional structures to limit the movement of clay after rain. This simulation of natural karst drainage should be adequate to restore the dispersed nature of water flow into the Exit Cave system.

On most benches open cavities were present up to 80 m deep. These demonstrate the reality of open, direct hydrologic connections into the cave system over most of the quarry. Where benches sloped steeply outwards, diffuse infiltration areas of mulch, gravel and straw bales have been constructed to slow water draining over the edge of a face into a gully.

On most benches a depth of between 200 and 300 mm of sandy topsoil has been

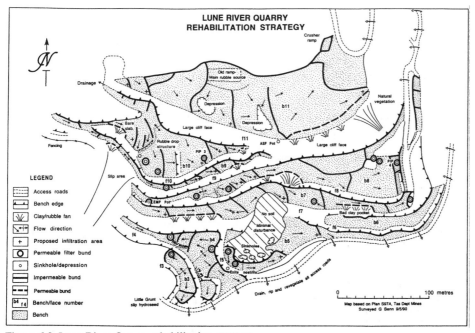

Figure 6.3. Lune River Quarry rehabilitation strategy.

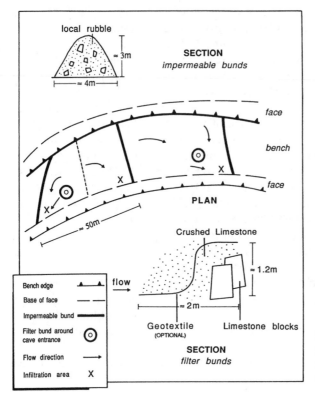

Figure 6.4. Layout and construction details of small catchment areas or bunds, Lune River Quarry rehabilitation.

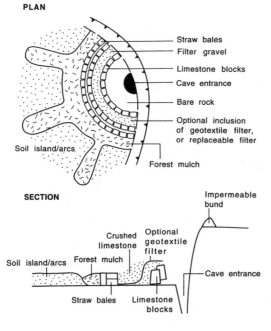

Figure 6.5. Layout and construction details of treatment of stream sinks and infiltration areas Lune River Quarry rehabilitation.

spread and mounded to give numerous small hollows and mounds for detention of rainwater and the areas revegetated with locally occurring species (Houshold 1997).

6.3.3 *Monitoring water quality*

Only limited dye tracing experiments, all under low flow conditions,have been carried out so far in the vicinity of the quarry (Kiernan 1993). Flow directions at high flow are unknown and re-activation of fossil passages is quite possible, involving drainage capture in the upper parts of the quarry. Water quality monitoring in Exit Cave has been carried out in order to evaluate the effectiveness of the rehabilitation programme (Houshold 1992, 1997). This has shown that hydrograph peaks in the transfer of solutes and fine sediments into the cave have flattened, that more controlled and diffuse infiltration of water in the quarry is occurring, and that the supply of sulphide-rich fine clays has been reduced.

6.3.4 *Karst rehabilitation objectives*

Some objectives for environmental rehabilitation learnt from the Lune River project can be summarised below:

– Restore the hydrology of the site by simulating as much as possible the drainage characteristics of the unimpacted karst. Reducing peak runoff by the creation of small internal drainage basins which simulate dolines in polygonal karst is an effective way of restoring near-natural infiltration rates and their spatially diverse patterning to allow soil and subcutaneous zone recharge.

– Control sediment movement at the source by the use of control structures and filters, and construct adequate filters at stream sinks to prevent entry of sediment into the karst hydrological system.

– Control active soil erosion and sediment entry to the karst system by stabilising the soil surface (using hydromulching on steep areas) at the sediment sources and encouraging cryptogam growth which is the soil's first defence against erosion.

– Establish a stable vegetation cover, preferably of perennial plants. A diverse vegetative cover with viable seed and the right structure to maintain geomorphic and biotic processes is not only aesthetically pleasing but in the long term will effectively moderate karst processes. The quality of substrates used in rehabilitation is crucial.

– Get the soil biology working again. Allowing the colonisation of the site by soil biota, especially the colonial insects necessary for litter breakdown, will enhance the recovery of nutrient cycling and produce a good soil structure for plant seedling establishment and water infiltration.

– Monitor progress above and below ground. The success or failure of the rehabilitation can only be assessed with meaningful data, preferably collected on an event basis so as to allow calculation of loads of solutes and sediments entering the karst drainage system.

– The complex nature of drainage and filter structures means that daily supervision is mandatory during machine work phases. Plant operators may be unfamiliar with karst rehabilitation principles, and environmentally costly mistakes may be made inadvertently.

6.4   THE INFLUENCE OF MINING ON KARST IN THE MANSFELD AREA, GERMANY (Wolfgang F. Grimmelmann)

6.4.1   *Geologic setting*

The Mansfeld Syncline (Fig. 6.6) is situated between the Harz Mountains and the city of Halle, about 150 km SW of Berlin (Brendel 1976, Reuter & Tolmacev 1990). It comprises an area of about 400 km^2.Geologic units influenced by mining in this area are listed in Table 6.1 and the total thickness of soluble rocks within the sequence listed above is shown in Table 6.2. The Pre-Tertiary geology is shown on Figure 6.7.

The development of the present landscape began at the end of the Cretaceous when the Mansfeld Syncline was formed. Solution-induced depositional basins contain Paleocene and Eocene beds. Jankowski (1964) found that during these periods the dissolution of halite led to the formation of large basins containing a polycyclic sequence of sediments, whereas the dissolution of gypsum resulted in smaller basins with monocyclic sedimentation. Karstification has continued up to the present. These are caves in the gypsum and anhydrite with ceiling heights of 35 m and diameters of more than 30 m. Caves in the limestone, dolomite and marlstone are much smaller.

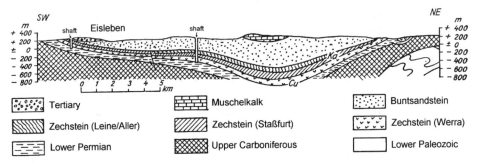

Figure 6.6. Geologic cross-section through the Mansfeld Syncline (Ka = potassium chloride, Cu = ore bearing layer) (after Kölbel 1971).

Table 6.1. Geologic units in the Mansfeld Syncline.

Geologic unit	Lithology	Maximum thickness (m)
Quaternary	Sand, gravel, silt, clay	150
Tertiary	Silt, clay, lignite, sand	185
Middle Triassic (Muschelkalk)	Limestone, marlstone	120
Lower Triassic (Buntsandstein)	Sandstone, siltstone, mudstone, gypsum, marlstone, limestone, dolomite	790
Upper Permian (Zechstein)	Rock salt, anhydrite, gypsum, mudstone, potassium salt, limestone, marlstone, sandstone	720
Lower Permian (Rotliegendes)	Conglomerate, sandstone, mudstone	300

Table 6.2. Thickness of soluble rocks in the Mansfeld Syncline.

Rock	Maximum total thickness (m)
Rock salt and potassium salt	500
Anhydrite and gypsum	300
Limestone, dolomite, marlstone	150

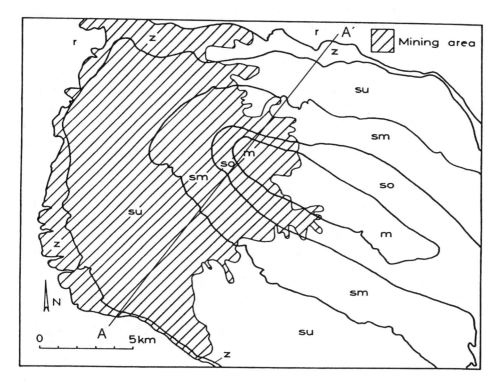

Figure 6.7. Pre-Tertiary geology and mining in the Mansfeld area. r = Lower Permian; z = Upper Permian; su, sm, so = Lower Triassic; m = Middle Triassic, A-A´= cross section Figure 6.6.

There are solution voids in the halite at depths of more than 250 m. Their maximum volume is not known as they have been avoided by mining. The variety of karst landscape features in the Mansfeld Syncline ranges from dolines and solution-induced depressions to stream sinks, dry valleys and karst springs.

6.4.2 *The effect of mining*

Bituminous marlstone was the object of mining in the Mansfeld area, The ore-bearing bed is situated near the base of the Upper Permian (Figs 6.6 and 6.7). It has an average thickness of only 0.3 m and contains a variety of metals, the economically most important are copper and silver. Mining began around the year 1200 AD with shallow shafts above the groundwater level. The outcrop of the ore-bearing bed is between 140 and 280 m a.s.l. in most of the area. Following the dip of the seam mining moved to greater depths, where the groundwater level had to be lowered. This posed technical problems, due to the high rate of inflow through karst conduits. The first drainage galleries were constructed at levels between 100 and 160 m a.s.l. before the year 1600. Drainage through galleries reached the lowest level (72 m a.s.l.) in 1879 with the completion of the main gallery which had a length of 31 km. Water from lower levels was also pumped into this gallery. According to Jung & Spilker (1972) the total mining area was about 150 km^2, with a total volume of 44 M m^3. Mining ended in 1969 at about 1000 m below surface.

Accelerated solution of rocks and increased subsidence were the main effects of mining. Solution grew most dramatically in areas where mining increased the contact between groundwater and salt. According to Jung & Liebisch (1966) the volume of water drained and pumped from mines and drainage galleries in the period 1880-1964 was about 1330 M m^3. As the average concentration of NaCl was 0.17 kg·l^{-1} during that time, the total volume of rock salt dissolved within 84 years was about 98 M m^3. The solution of gypsum and anhydrite was also increased by mining, not only because of the improved hydraulic contact and the accelerated groundwater flow, but also due to the higher chloride content of the groundwater, which raised the solution rate of the sulphate. Based on data from Kiel (1958) it is estimated that the volume of gypsum dissolved during the period 1880-1964 was about 2 M m^3. The increase of the dissolution of limestone, dolomite and marlstone caused by mining is thought to be small, compared to the solution of rock salt and gypsum. It is estimated, that the total volume of carbonates dissolved during the period 1880-1964 was less than 100,000 m^3.

Subsidence due to mining began above the groundwater level in places where unsupported workings collapsed below caves in gypsum and limestone. In areas where the groundwater level had to be lowered, gypsum and limestone caves collapsed due to loss of buoyant support. In areas where considerable thicknesses of rock salt had been preserved, solution through mining caused subsidence as well. According to Lorenz (1962) subsidence triggered the inflow of about 123 M m^3 of groundwater and surface water into a mine in 1892, and caused additional solution and subsidence. A number of similar events happened after that and influenced the decision to abandon mining in the Mansfeld district completely. Figure 6.8 shows areas where considerable subsidence continued up to 1960. The coincidence of subsidence and the area of mining activity (Fig. 6.7) is obvious. The extent of the area where the groundwater level was at least temporarily influence by mining is not exactly known. It is larger than the mining area.

Subsidence in the form of dolines is mainly caused by the collapse of caves in gypsum and limestone. Uniform subsidence is mostly due to relatively even solution of rock salt. Non-uniform subsidence is mainly caused by local and heavy solution of rock salt, but partly also by the solution of gypsum, anhydrite, limestone, dolomite and marlstone. Subsidence which was neither caused nor increased by mining is found in the minor part of the Mansfeld Syncline only. Reliable data on the degree of subsidence have been collected only from about 1880 AD onwards. Since that time hundreds of new dolines have formed, most of them with diameters of less than 10 m, the largest having diameters over 200 m and depths of more than 40 m. A number of dolines and areas of increased subsidence have been attributed to sudden inrushes

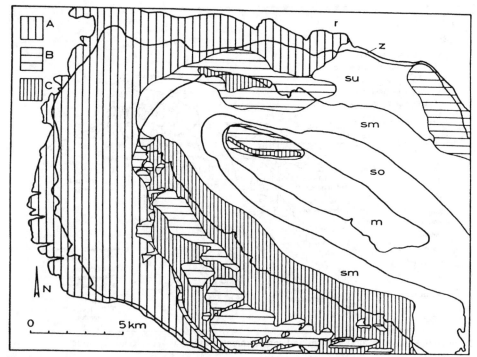

Figure 6.8. Subsidence in the Mansfeld area (after Kammholz 1964). A = subsidence mainly in the form of dolines; B = relatively uniform subsidence; C = non-uniform subsidence.

of water into mines. After such events the rate of subsidence exceeded 3 m per year in some places. Brendel (1976) estimated, that subsidence in the period 1890-1970 was equal to a volume of 75 M m^3. This result compared to the estimated of the volume of workings and the solution of rock salt and sulphates (although the periods compared overlap), permits the assumption, that in 1970 there was still much more void volume below surface than at the time when mining began. The extent of subsidence after 1970 confirmed this assumption.

6.5 THE INFLUENCE OF GOLD-MINING ACTIVITIES ON THE WATER QUALITY OF DOLOMITIC AQUIFERS (David B. Bredenkamp)

6.5.1 *Hydrogeology*

In the Republic of South Africa gold-mining in the Witwatersrand formations underlying the dolomite in the East and Far West Rand caused: 1) the formation of sinkholes and surface subsidence which endanger property and human lives, 2) dewatering of major aquifers to allow gold-mining to continue to greater depths, 3) decline of spring flow and failure of boreholes, and 4) pollution of dolomitic aquifers.

Carbonate rocks classified as Proterozoic Chuniespoort dolomite (Fig. 6.9) is overlain by a thick succession of Pretoria Group (shales and quartzite) and is under-

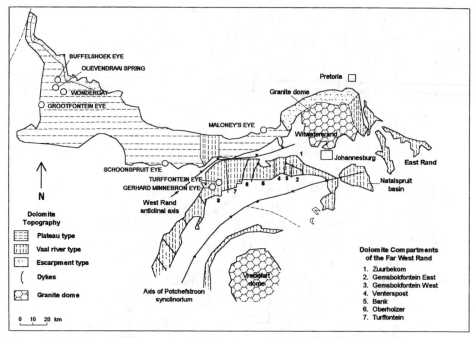

Figure 6.9. Some structural features and compartments of the Far West Rand dolomite.

lain by a band of Black Reef Quartzite Formation which overlies the gold-bearing Witwatersrand Reefs. The impact of gold-mining in the East and Far West Rand is of particular significance. The dolomite consists of four Formations: two layers of chert-poor dolomite, and two chert-rich formations. Intrusive dykes subdivide the dolomite into compartments as is shown for the Far West Rand.

The rate of sinkhole formation has increased in areas where mines dewater the dolomite. Brink (1979) estimated that sinkholes triggered by mining activities had caused the death of 38 people up to 1979. Human activities which have increased the likelihood of sinkhole formations include local disruption or concentration of surface runoff, increased infiltration due to leakages from water reticulation and sewage mains, and a dramatic lowering of piezometric levels due to dewatering in areas of high risk, as in the West Rand. In addition the formation of sinkholes could be caused by abutment for the roof over a void, provided by dolomite pinnacles or the sides of grikes, arching in the residuum, development of a void below the arch, a disturbing mechanism to trigger roof collapse, e.g. the introduction of water, or erosion and removal of binding material (collapse is often triggered by seismic tremors caused by mining activity) and reactivation of palaeo-sinkholes.

6.5.2 *Impact of mining on water quality East Rand*

Statistical analysis of available groundwater quality data for the East Rand (Walton & Levin 1993) rather surprisingly showed that the impact of pollution is still fairly localized (Fig. 6.10). An extensive study carried out in the East Rand area (Scott

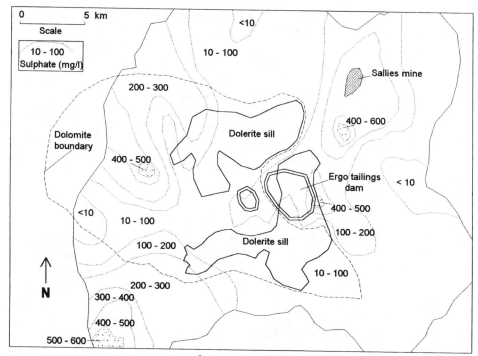

Figure 6.10. Sulphate distribution (mg·l^{-1}) in the East Rand gold-mining area indicating local areas of mine pollution (Walton & Levin 1993).

1994) assessed the interaction between recharge mainly from the overlying dolomitic aquifers, and pumpage from the underlying gold-bearing reefs to predict the extent and impact of the pollution likely to occur when mining becomes uneconomic and dewatering ceases.

The Ergo tailings dams for the gold extraction from old mine dumps show little sulphate pollution. The degradation of water quality in mining areas results mainly from the reaction of water with sulphide minerals, which produces acid rock drainage. Gold-bearing reefs can contain up to 3% (by weight) of pyrite (FeS_2). The surface rock piles, sand and slimes dumps, as well as back-fillings and spoil heaps in stopes and haulages, all contain pyrite. When exposed to air and water it oxidises to sulphuric acid, which upon further reaction generates sulphates, dissolution of heavy metals and an increase in dissolved salts. Concentrations of manganese and sulphate are reliable indicators of mining pollution in acid-mine drainage (Scott 1994).

Two mined-out basins have been formed under the dolomite of the East Rand, a highly industrialized and densely populated area whose streams carry very varied effluent in addition to mining waste water. The dolomite plays an important role by acting as a buffer to bring the acidic mine waters closer to desired pH levels.

Ingress of mainly dolomitic water to the stoping areas occurs via natural faults but also along secondary fractures resulting from the collapse of tunnels or the gradual closing of stopes. The mine excavations represent zones of high permeability which, during active mining operations, collect groundwater. Once the mines become defunct these highly permeable zones act as conduits which spread the polluted groundwater. When the piezometric levels have fully recovered, some of the low-lying shafts will become points of effluence. A model to simulate the water quality of the final outflow from the East Rand basins is not possible because of too many uncertainties and a lack of data. The best predictions thus far have been based on samples taken from water dammed underground, and of the rising waters in areas where inflow has become dominant. Ways of alleviating the pollution problem would be to rework and remove some of the mine waste heaps, end the disposal of water of bad quality, clean the polluted water discharged by industries, and, by ef-fective management, control the input of pollutants into the streams.

Far West Rand Dolomite

The Venterspost compartment was the first to deliberately dewater the dolomite (Fig. 6.9). Abstraction from the Gemsbokfontein compartment declined from an initial rate of 170 M l per day in September 1984 to about 68 M l per day by 1994. This mine returns part of the underground pumpage to an adjacent aquifer. The initial rate of replenishment was about 2 M l per day, but later had to be increased to 12 M l per day to limit the groundwater to within 7 m of drawdown. The sulphate concentration of this water is about 300 mg·l^{-1} compared to about 50 mg·l^{-1} in the receiving com-partment. Use of the high sulphate water to trace the plume of contamination proved unsuccessful, and it appears that the recharge water returns to the mining stopes along narrow, but highly permeable conduits.

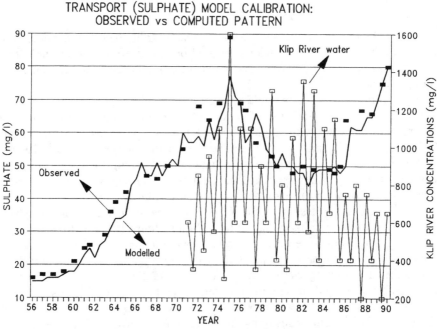

Figure 6.11. Comparison between measured and simulated values of sulphate concentrations in the Zuurbekom compartment in relation to that of the Klip River.

Pollution of sources of groundwater

The Zuurbekom compartment which has been pumped since about 1899 features a polluted dolomitic aquifer. The bulk of the sulphate contamination is derived via the Klip River from mining upstream. The sulphate in the Klip River, measured from 1970, shows high variations in concentration (500 to over 1000 $mg \cdot l^{-1}$), compared to natural concentrations of 50 to 70 $mg \cdot l^{-1}$ (Fig. 6.11). A good simulation of the interactive processes and of the pollution spread was obtained by Simonic (1993) using a standard USGS MOC model. The model incorporated mixing processes, the effects of abstraction and solute transport, the role of leaking dykes, and the probable buffering capacity of the aquifer material.

Manganese pollution of drinking water supplied from the Vaal river was linked to a slimes dam situated on top of the dolomite polluting the base flow contribution from the dolomitic aquifer (Bredenkamp & Verhoef 1981). Polluted waters were found in several boreholes as well as in a small spring emerging on the river bank. Different techniques, including measurements of ^{18}O concentrations, proved that the pollution had been caused by the mine.

6.5.3 *Impact of exploitation on use of available water*

Dewatering by mines has produced valuable information enabling reliable estimates of aquifer transmissivity, recharge and storativity to be obtained (Bredenkamp et al. 1994). A revised interpretation of pumping tests proved that the dolomite conforms to fractured aquifer flow.

Reduction of spring flow in dolomitic areas
Exploitation of the dolomitic aquifers will diminish runoff from the catchment, mainly by reducing the spring flow or base flow. Groundwater levels as well as the flow of springs are linearly related to the cumulative departures from the average rainfall (Fig. 6.12). The best correlations were obtained by summation of the rainfall differences over m-months (short-term memory) relative to the average rainfall over n-months (long-term memory).

The flows of major dolomitic springs also correspond to the moving average rainfall over a period of about 72 to 120 months. Natural fluctuations in water quality should therefore be inversely related to the average rainfall over n-months, and to the cumulative rainfall departures. However the real-time water quality response is also affected by delayed propagation in the unsaturated zone.

Natural water quality fluctuations in dolomitic springs
Water from a dolomitic spring represents an areally and temporally integrated sample of the quality over the recharge catchment. It serves as a reference from which to establish the degree of contamination in a catchment relative to natural conditions. Base-line values of natural concentrations can be derived by examining water quality measurements of dolomitic springs in different climatic regions. Inter-comparison of water quality of different dolomitic springs reveals the natural hydrological equilibrium of the aquifers, and can confirm whether admixture of non-dolomitic recharge or contamination occurs (see Fig. 6.13). The Turffontein and Gerhard Minnebron eyes represent typical cases of water polluted by mining upstream. The sulphate con-

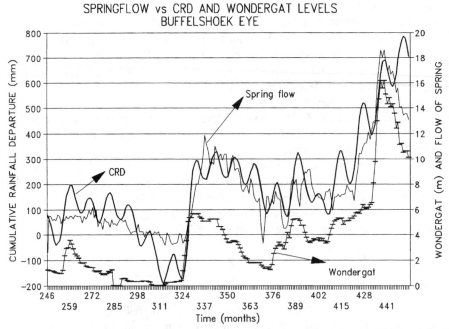

Figure 6.12. Correspondence between both the cumulative rainfall departures (CRD) and the Wondergat levels, and the flow of dolomitic springs, shown for Buffelshoek eye.

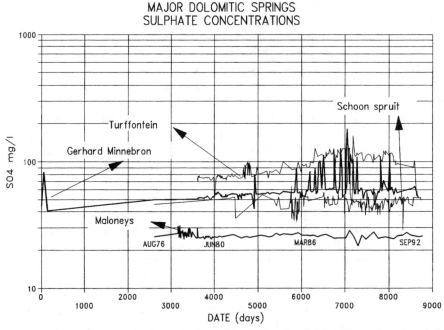

Figure 6.13. Sulphate concentrations in the polluted West Rand dolomitic springs, in relation to those of other large springs shown in Figure 6.9.

centrations of these springs show a gradual logarithmic increase of constituents, unlike those of uncontaminated dolomitic springs, e.g. Maloneys eye. The low sulphate concentration of the Maloneys eye and the apparent high age of the water can only be explained by natural admixture of quartzitic water.

6.5.4 *Future studies*

The study of dolomitic aquifers is entering an exciting phase following the firm establishment of methods for estimating most of the critical parameters. Significant progress has been made in understanding the interactions of abstraction and deterioration in quality. This knowledge can lead to better solutions and timely remedial action, as well as more effective management of the effects of mining and aquifer exploitation.

6.6 COLLAPSE AND PIPING INDUCED BY HUMAN ACTIVITY IN THE OLKUSZ LEAD-ZINC EXPLOITATIVE DISTRICT OF THE SILESIAN UPLAND, POLAND (Andrzej Tyc)

6.6.1 *Hydrogeological conditions*

The Olkusz lead-zinc exploitative district is situated in the eastern part of the Silesian Upland. It is very well known as a typical example of lead-zinc sulphide deposits of the 'Mississippi Valley-type' associated with palaeokarst features in Middle Triassic limestones and dolomites (Muschelkalk). Most of the carbonate rocks are covered by impermeable clays of Upper Triassic and Lower Jurassic age as well as being buried by Pleistocene fluvioglacial sediments, up to 70 m thick. The geology of the area influenced by mining is shown on Figure 6.14.

Under natural conditions a confined, partly artesian aquifer exists (Motyka & Wilk 1984, Motyka 1988). Exploitation of lead-zinc deposits and of groundwater for water supply caused development of a hydraulic cone of depression of 350 km^2 area in response to a pumping rate of 380 m$^3 \cdot$min^{-1} (6.3 m$^3 \cdot$s^{-1}). The base of the induced drawdown lies below most of the open and filled karst features encountered by drilling and in mine galleries. Contemporary artificial drainage is connected with a pre-Quaternary system of karst water circulation (Motyka 1988; Tyc 1989, 1990).

As a result of mining within the Triassic carbonate aquifer new hydraulic components have appeared; for example, artificial water galleries which take over the role of master conduits in the functioning of the karst circulation. Rapid evolution of open phreatic tubes into vadose channels takes place. Examples of such processes are common in the Olkusz lead-zinc mines (Tyc 1989, 1994).

6.6.2 *The influence of mining on collapse and piping processes*

Beside the transformation of the hydrogeological system within the Triassic carbonate aquifer, exhumation of infilled endokarst and palaeokarst has also taken place. The emptying of palaeo-drainage systems as a result of groundwater withdrawal is a factor in the development over 100 induced collapses and over 50 pipes in Olkusz

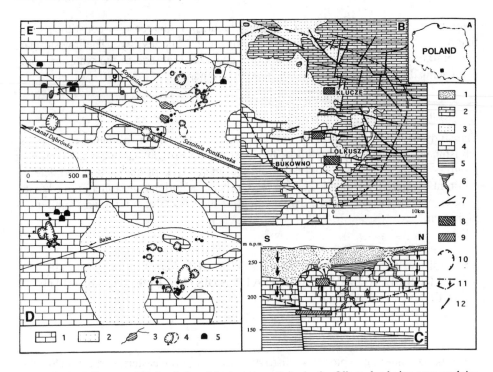

Figure 6.14. Collapse and piping induced by human activity in the Olkusz lead-zinc ores exploita-tion district of the Silesian Upland (Poland). Explanations: A = Location of the Olkusz lead-zinc ores exploitation district; B = Geological situation (without Quaternary deposits); C = Section showing development of collapse and piping processes: 1 = Quaternary deposits (sands, clays), 2 = carbonate rocks of (Jurassic), 3 = impermeable rocks (Upper Triassic to Lower Jurassic), 4 = carbonate rocks (Triassic), 5 = impermeable rocks (Palaeozoic), 6 = fillings of karst forms (palaeokarst), 7 = main faults, 8 = location of areas of collapse and piping occurrence, 9 = mine galleries, 10 = extent of the cone of hydraulic depression in Triassic rocks, 11 = decrease of under-ground water level due to lead.zinc ores and water exploitation, 12 = direction of sediment migra-tion within paleokarst due to underground water withdrawal. Situation and geology of areas of collapse and piping occurrence; D = 'Olkusz' mine; E = 'Pomorzany' mine: 1 = carbonate rocks (Triassic), 2 = impermeable rocks (Upper Triassic), 3 = hydrological network, 4 = collapses and subsidence, 5 = caves and large cavities discovered in mine galleries.

district. The spatial distribution and geology of the phenomena is shown on Figure 6.14. Forms develop in three areas: 61 induced collapses 2-90 m wide and 1-21 m deep are located in the southern part of the Olkusz district (D on Fig. 6.14), 50 in-duced collapses of 3-120 m width and 1-40 m depth occur in the central part (E on Fig. 6.14) and more than 50 smaller forms, mostly pipes, in the Biala Przemsza val-ley. An example of large-scale collapse in the central part of the area is shown on Photo 6.5.

The origin of the landforms induced by human activity in the Olkusz lead-zinc exploitative district is complex. Partly they are caused by underground exploitation – the post-exploitation collapse of galleries (most of the collapses in area D on Fig. 6.14 are of this type). Observations of collapse events in mine galleries as well as

Photo 6.5. Large-scale collapse in the central part of the Olkusz district (Photo: Tyc).

connections between induced forms and the palaeo-relief of the area (mostly of karstic origin) show that they are complex karst forms induced by human activity.
Change in hydrogeological conditions and the opening of the artificial galleries into
large infilled caves and cavities has caused collapse of consolidated internal karst
sediments and the washing out of unconsolidated karst infillings.

6.7 IMPACTS ON SALT KARST DUE TO SOLUTION MINING AND TO THE PETROLEUM INDUSTRY: KANSAS AND TEXAS, USA
(Kenneth S. Johnson)

6.7.1 *General remarks*

Salt is highly soluble; it is the most soluble of the common rocks that are widespread
in the world. Ground water in contact with salt will dissolve some of it provided the
water is not already saturated with NaCl. For extensive dissolution to occur, it is necessary for the brine, thus formed, to be removed from the salt deposit; otherwise the
brine becomes saturated, and the process of dissolution stops.

Four basic requirements are necessary for salt dissolution and karst development
to occur (Johnson 1981): 1) a deposit of salt against which, or through which, water
can flow, 2) a supply of water that is unsaturated with NaCl, 3) an outlet, whereby
the resulting brine can escape, and 4) energy (such as a hydrostatic head or density
gradient), to cause the flow of water through the system. When all four of these requirements are met, salt dissolution and karst development can be quite rapid, in
terms of geologic time. Human activities that are most likely to aid in meeting these
four requirements are those that involve drilling boreholes into or through subsurface
salt deposits, and then intentionally or inadvertently allowing unsaturated water to
enter into the borehole. If uncontrolled dissolution of salt occurs, then a cavity can
increase in width to the point where the roof will no longer be supported. Collapse of
the roof, followed by a series of successive roof failures, can cause the cavity to migrate upward. If the original cavity is large enough and shallow enough, the distur-

bance can reach the land surface, resulting either in land subsidence or catastrophic collapse. Several recent reports that document some of these adverse impacts in the United States are by Walters (1978), Dunrud & Nevins (1981), Baumgardner et al. (1982), Ege (1984), Coates et al. (1985) and Johnson (1987).The two activities that most commonly entail drilling boreholes into or through salt deposits are: 1) solution mining, and 2) petroleum activity.

6.7.2 *Impacts from solution mining of salt: Kansas and Texas, USA*

Solution mining is the process of extracting soluble minerals, such as salt or potash, by: a) introducing a dissolving fluid (i.e. water) into the subsurface, b) dissolving the minerals and forming a brine, c) recovering the brine, and d) extracting the mineral from the brine (usually by evaporation). Solution mining typically entails creation of one or several large underground cavities that are filled with brine (Marsden & Lucas 1973). In fact, some solution-mining operations are carried out specifically to create such a cavity for the underground storage of various liquids and gases; the liquids and gases are then injected into the cavity to displace the brine, and they can be recovered later by reinjecting the brine.

Solution-mining cavities are created in bedded salts, salt domes, and salt anticlines. The cavities typically are 10-100 m in diameter and are 10-300 m high, both dimensions being based largely on the thickness of the salt and the depth to the top of the cavity. Cavities can also be joined together hydraulically, thus allowing fresh water to be pumped down one borehole and brine to be extracted from the other; such cavities may end up quite long and relatively narrow. It is possible for cavities to become larger or shallower than planned, as a result of uncontrolled dissolution or unanticipated geologic or engineering/construction problems, and a number of these cavities have produced surface subsidence or collapse structures. Dunrud & Nevins (1981) reported 10 areas of solution mining and collapse within the United States alone, and additional sites are known from many other parts of the world. Most solution-mining collapses result from cavities formed 50-100 years ago, before modern-day engineering safeguards were developed; proper, modern design has virtually eliminated this problem in new facilities. Two well-documented collapse structures in the United States are Cargill Sink, in Kansas, and Grand Saline Sink, in Texas.

Cargill Sink, Kansas

Walters (1978) investigated a major sink caused by solution mining for salt in a brine field at Hutchinson, in Reno County, Kansas (Fig. 6.15). The collapse occurred on October 21, 1974, on the property of Cargill, Inc., and reached a diameter of about 60 m within 4 hours. Settlement continued until the afternoon of October 23, 1974, when the crater stabilized with a diameter of about 90 m and with a maximum depth of about 15 m. The volume of the crater was calculated to be about 70,000 m³.

Salt has been solution mined on these properties since 1888 (Walters 1978). The Permian Hutchinson salt here is about 105 m thick and occurs at a depth of about 130 m below the ground surface (Fig. 6.15). The salt is overlain by red and gray Permian shales, and these, in turn, are overlain by about 20 m of water-saturated, loose Quaternary sand. The locations of many of the earlier brine wells are not known and the dissolution methods often were uncontrolled; therefore, the location

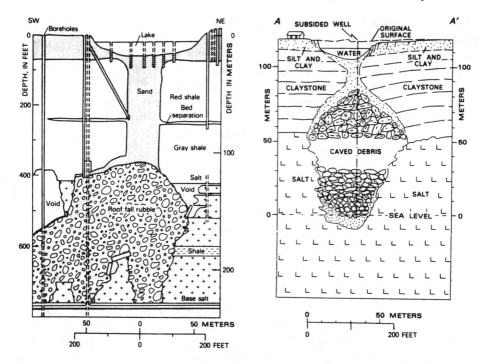

Figure 6.15. On the left is cross section through Cargill Sink, Kansas (modified from Walters 1978); the cavity shape is hypothetical. On the right is cross section through Grand Saline Sink in Grand Saline salt dome, Texas (modified from Dunrud & Nevins 1981).

and extent of many of the solution cavities on this property are unknown. The sink developed within the area of an active brine field that included both operating and abandoned wells. Embraced within the sinkhole was a well that had been drilled in 1908 and finally was plugged and abandoned in 1929.

Post-subsidence test drilling of the Cargill Sink area showed that a northeast-trending cavern had developed in the Hutchinson salt beneath the sink (Walters 1978). Elongation of the cavern parallels a line of brine-producing wells that were hydraulically connected. The span of the cavern roof is more than 400 m in its long dimension, but less than 90 m in its short dimension. Apparently the roof span exceeded the capacity of overlying shales to support the overburden. Therefore, failure of the roof caused collapse of successive overlying rock units until the uppermost rock layer finally collapsed into the water-filled void. At this point, the water-saturated Quaternary sands flowed into the cavity, creating the surface sink. The sand flowed down and now fills a chimney that is about 30 m in diameter and located below the center of the sinkhole (Fig. 6.15).

Grand Saline Sink, Texas
A large sinkhole developed suddenly in the city of Grand Saline, Texas, on April 27, 1976 (Dunrud & Nevins 1981). The sink occurred at the site of a well that had produced salt by solution-mining methods from 1924 to 1949 (Fig. 6.15). The borehole

penetrated the Grand Saline salt dome at a depth of about 60 m. Salt in the dome is overlain by about 50 m of Eocene claystones, and these, in turn, are overlain by about 10 m of unconsolidated Quaternary silt and clay.

Failure occurred in two stages (Dunrud & Nevins 1981). First, a hole 4-6 m in diameter and more than 15 m deep opened up. Second, the hole widened rapidly as slabs around the rim toppled into the sink, and it reached a diameter of about 15 m a few hours after the initial collapse. A total of about 8500 m^3 of silt and clay ultimately was displaced into the underground cavity. Ege (1984) points out that a similar collapse took place just east of this sink in 1948.

6.7.3 *Impacts from petroleum industry activity*

Petroleum activity is less likely to cause collapse features above salt deposits than is solution mining. Petroleum-industry activities that may lead to adverse impacts on salt karst include the drilling of exploration, production, or salt-water-disposal boreholes into, or through, subsurface salt units. Unintentional dissolution of the salt can create a cavity that is as large and as shallow as those created in solution-mining activities. If the cavity becomes too large for the roof to be self-supporting, successive roof failures may cause the collapse to migrate upward and perhaps reach the land surface. The few collapses related to petroleum activity involve boreholes drilled long ago, before development of proper engineering safeguards pertaining to drilling-mud design, casing placement, and salt-tolerant cements. Two well-documented collapses in the United States are Wink Sink, in Texas, and Panning Sink, in Kansas.

Wink Sink, Texas
The Wink Sink, located near the town of Wink in Winkler County, Texas, formed on June 3, 1980, and within 24 hours had expanded to a maximum width of 110 m (Baumgardner et al. 1982). Two days later, the maximum depth of the sinkhole was 34 m and the volume was estimated at about 159,000 m^3. The collapse occurred near the middle of the Hendrick field, a giant oil field that has been operating since 1926; one abandoned oil well (the Hendrick well 10-A, drilled in 1928) was incorporated within the sink itself, and a second oil well was plugged and abandoned later because of its proximity to the sinkhole. It appears that the Wink Sink resulted from an underground dissolution cavity that migrated upward by successive roof failures, thereby producing a collapse chimney filled with brecciated rock (Fig. 6.16) (Baumgardner et al. 1982, Johnson 1987). The dissolution cavity had developed in salt beds of the Permian Salado Formation, which is about 260 m thick and is about 400-660 m beneath the Wink Sink. Natural dissolution of salt beds in the Salado Formation in Winkler County and other areas of West Texas and New Mexico is well known, but the dissolution and collapse associated with the Wink Sink apparently resulted from, or at least was accelerated by, oil-field activity in the immediate vicinity of the sink.

The Hendrick well 10-A, an abandoned oil well, was located at the site of the sinkhole, and it appears likely that it was a pathway for water to come in contact with the Salado salt (Johnson 1987) (Fig. 6.16). In all likelihood, the well was drilled using a fresh-water drilling fluid that enlarged or washed out the borehole within the salt sequence. Poor cement jobs, and possible fractures in the cement lining, may

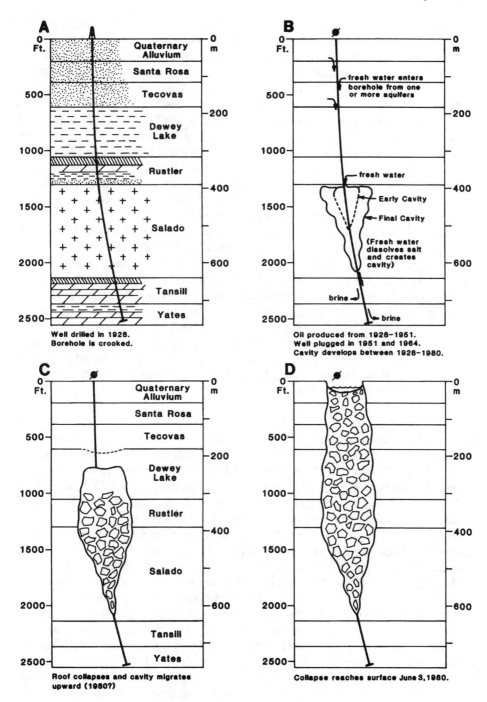

Figure 6.16. Cross section through Hendrick well 10-A showing the possible relationship of the well to salt dissolution and the development of the Wink Sink, Texas (Johnson 1987).

have opened pathways for water movement up or down the borehole outside of the casing. Because of undoubted borehole enlargement during drilling in the Salado salts, the small amount of cement reportedly used to set the casing in the hole was enough to cement only the lower part of the hole; thus leaving most of the salt section uncemented behind the casing (Johnson 1987). Casing in the well probably was perforated by corrosion due to production of great quantities of oil-field brine; parallel the casing corrosion that was observed in the nearby Hendrick well 3-A, which had a similar history. Use of explosives to realign well 10-A while drilling in the underlying Tansill Formation not only fractured the rock and increased its permeability locally, but also may have fractured the cement lining farther up the borehole. In addition, final removal of casing from the well in 1964 left an unlined borehole in the interval from the base of the Santa Rosa aquifer to the top of the Rustler Formation for a period of 16 years, until formation of the Wink Sink.

All of the above-mentioned activities, although consistent with standard industry practices during the life of Hendrick well 10-A, would have aided in conducting fresh water from shallow aquifers down the borehole to the salt beds (Johnson 1987) (Fig. 6.16B). Outlets for the high-salinity brine, formed by dissolution of salt in the borehole, included the porous and permeable strata underlying the Salado Formation, as well as possible preexisting dissolution channels within the Salado. Thus, a dissolution cavity may well have been formed around well 10-A, probably in the upper part of the salt sequence (Fig. 6.16B), and this cavity eventually would have become sufficiently large to permit collapse of the roof (Fig. 6.16C). By successive roof failures, the cavity then migrated upward until it finally reached the land surface and created the Wink Sink (Fig. 6.16D).

Panning Sink, Kansas
Rapid subsidence and collapse occurred on April 24, 1959, around a salt-water-disposal well (Panning well 11-A) in the Chase-Silica oil field of Barton County, central Kansas (Walters 1978). Within a 12-hour period, at which time subsidence ceased, the sinkhole expanded to about 90 m in diameter and had a water level 15-18 m below the land surface. Four days later, the size of the sink had not increased but the water level had risen to 3.5 m below the surface.

Walters (1978) postulated the following sequence of events leading to development of the Panning Sink. The Panning well 11-A, drilled in 1938 as an oil well, penetrated 91 m of Permian Hutchinson salt at a depth of 298 to 389 m (Fig. 6.17). Fresh-water drilling fluids dissolved the salt in the borehole to a diameter of 1.4 m, and this washed-out zone was not cemented behind the 15.2 cm-diameter casing. In 1946, the depleted oil well was converted to a salt-water-disposal well. Only the lower part of the salt section was then cemented behind the casing. Brine was disposed of by gravity flow through 12.7 cm tubing set within the casing. Disposal was into the Arbuckle Group dolomite at a depth of 995 m. In 1949, the tubing was removed and water was injected directly down the casing. Disposal water contained about 28,000 parts per million sodium chloride, and from 1946 through 1958 the well received about 1.8 million m^3 of this brine, an average of about 0.3 $m^3 \cdot min^{-1}$.

Corrosion of the 15.2-cm casing caused leaks, which allowed much of the unsaturated brine to circulate across the salt face and then flow downward into the underlying Arbuckle dolomite (Walters 1978). A large cavern, more than 90 m in diame-

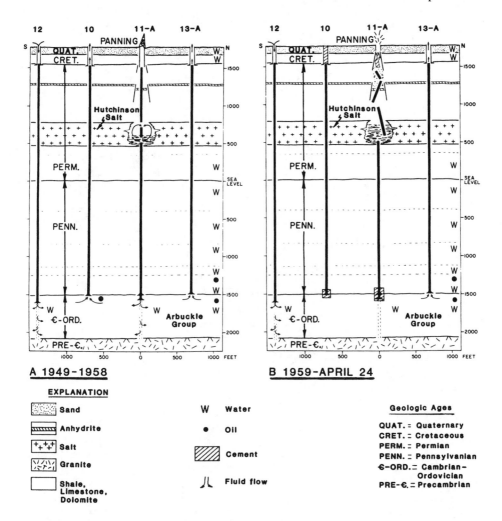

Figure 6.17. North-south cross section through the Panning disposal well 11-A showing the results of salt dissolution and the development of the Panning Sink, Kansas (modified from Walters 1978).

ter, was created by dissolution of the salt (Fig. 6.17). Successive roof falls into the cavern caused the water-filled void to migrate upward gradually until surface subsidence, ponding of water, and tilting of the derrick forced abandonment of the well in January 1959. Final collapse did not occur, however, until nearly 4 months later (April 24), when the uppermost mass of bedrock that spanned the cavity finally collapsed into the water-filled void. With this failure of Cretaceous bedrock (shales and sandstones) at a depth of 30 m, the overlying Quaternary sands and gravels, saturated with fresh water, were free to flow into the cavity and create the surface sink in a matter of hours.

CHAPTER 7

Karst water exploitation

ANTONIO PULIDO BOSCH (co-ordinator and main author)
Department of Geodynamics, Faculty of Sciences, Granada, Spain

7.1 INTRODUCTION

7.1.1 *Overview of exploitation methods*

The fact that karstic materials cover some 12% of the earth's continental surface (Ford & Williams 1989) offers an idea of the economic importance of karst. In areas such as Southeast Asia or the Mediterranean, karstic aquifers constitute the primary source of water. An estimated 25% of the world's population is supplied largely or entirely by karst (Williams 1993). In Spain e.g. it has been calculated that carbonate karstic materials cover roughly 100,000 km^2 with an annual mean recharge of 2000 M m^3 (Pulido Bosch 1995). With reserves estimated at 20,000 M m^3, karst represents the greatest source of water supply in the entire Iberian Peninsula. The domestic, agricultural and industrial supply is estimated at some 300 M m^3 per year.

The water resources related to karstic aquifers have been exploited since time immemorial (Mijatovic 1975, Burger & Dubertret 1984). The most ancestral method of exploitation has consisted simply of taking the water directly from the spring (Fig. 7.1a); this procedure presents grave drawbacks in dry periods, when the flow does not meet demand, or stops altogether. In these periods the lack of automated systems for pumping obliged the users to excavate galleries in the springs to collect the water from the interior of the aquifer (Fig. 7.1b). In other karstic regions with a saturated level near the surface and natural points of access (*cenotes* in Mexico and *casimbas* in Cuba, and many other examples in China), the water was originally extracted directly by manual methods (Fig. 7.1c).

The simple drawing of spring water continues to this day, but only when flow exceeds demand. However, when the supply falls below demand, the need arises to implement regulatory systems, of which there are many types. The construction of a dam in the spring (Fig. 7.1d) offers various advantages, such as the use for storage of the zone in which the piezometric level fluctuates – normally a quite karstified volume of rock; sluices or floodgates allow the release of the water as needed. The scheme can be complemented with horizontal wells below the dam, with a valve (Fig. 7.1e).

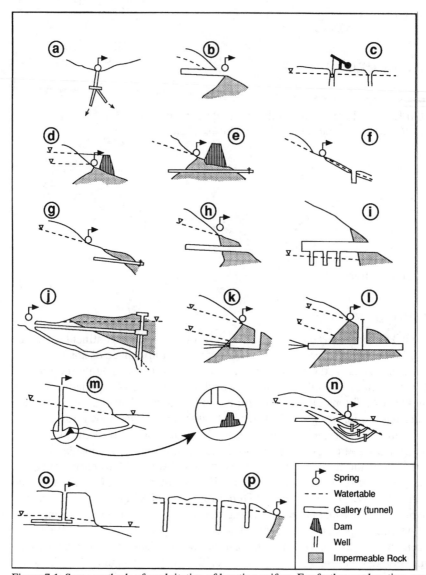

Figure 7.1. Some methods of exploitation of karstic aquifers. For further explanation, see text.

In other cases, it is simpler to pump water from wells situated near, above or be-low the water source, which can be vertical (Fig. 7.1f) or horizontal (Fig. 7.1g). Given that the discharge area is highly karstified, the specific capacity can be very high. More sophisticated is the collection method using galleries situated below the altitude of the spring, excavated into the first few metres of the impervious materials (Photo 7.3). The placing of sluices in the contact zone between the impervious and aquifer materials allows the release of water as desired (Fig. 7.1h).

Another procedure is the combined use of gallery and vertical wells drilled through impervious materials at the margins or into the aquifer materials themselves

(Fig. 7.1i), or a variant such as the collection method in the spring of Lez near Montpellier, France (Fig. 7.1j, Photo 7.2). In some cases, a well is opened of large diameter in the bordering impervious rocks, and at a certain depth a gallery branches off, in the interior of which horizontal drilling is carried out or vertical wells are dug (Fig. 7.1k); this scheme can be completed by a gravity drain at the same level as the gallery (Fig. 7.1l).

There are also more unusual collection methods, such as attempts to make use of submarine springs (Photo 7.4). If the principal conduit discharging into the sea can be located a dam is built in the conduit, and then a well is dug inland and equipped with a pump (Fig. 7.1m). The dam allows freshwater to flow to the sea, but theoretically it prevents sea water from entering. However, apart from being costly, the system does not guarantee the maintenance of the original water quality, given that the small conduits and the rock matrix itself play important parts in the storage and transmission of the water, as these can allow seawater to penetrate inland.

The best-known example is the attempt to exploit the undersea springs of Port-Miou (Cassis, near Marseille, Potié & Tardieu 1977). In this case the installation described was complemented by a barrier at the gallery's outlet to the sea. The intention of this was to prevent sea water from penetrating directly into the large conduit. With this construction, the saline content of the pumped water was considerably reduced. Other means of collecting water in coastal areas consist of the closing of conduits connected with the sea to guarantee good water quality (Golubinka, described in Pavlin & Fritz 1978; Fig. 7.1n), or in the case of a well-gallery combination just above sea level (Mijatovic 1984a, b and c), as a means of avoiding salinization (Fig. 7.1o).

Perhaps the most effective present-day method is exploitation by wells (Fig. 7.1p) drilled by cable tool, rotary or rotary-percussion with a hammer at the bottom (Photo 7.1). In addition to allowing source location near the site of the water demand, the pumping systems currently on the market are capable of extracting large flows from substantial depths and with consistently smaller diameters. This procedure permits true regulation of the system, so that pumping can be increased during times of maximum demand and decreased for supplies to recuperate during periods of low demand or rain.

There are many other methods which are used, or which were used in the past. For example, the springs of Ras-el-Ain, Syria, situated 5 km from Tiro and 700 m from the sea, and in which the water rises from fractures in the marls, were exploited by the Phoenicians, who constructed towers at the discharge points at higher altitudes. The water would rise in the towers some 6 to 8 m and thereby be distributed through a system of aqueducts over a large coastal strip (Mijatovic 1975).

7.1.2 *Relationships between aquifers and exploitation methods*

As seen above, the methods for exploiting karstic aquifers are extremely diverse, and thus in each case it is necessary to know which method might be the most appropriate. On the other hand, karstic aquifers also cover a wide variety of types, from underground rivers (with storage limited to the main conduit, with high transit velocities and with minimal residence time) to aquifers with great inertia (with large reserves, relatively homogeneous karstification and even matrix porosity). In all

Photo 7.1. Water exploitation from a confined karst aquifer using pumping wells, Umm Er Rad-huma aquifer, Wadi As Sah'ba, Eastern Province, Saudi Arabia. The mainly fossil water resources in arid areas are extremely sensitive with regard to quality changes. Due to lack of recharge in arid areas ground water mining frequently occurs (photo: Hötzl 1978).

Photo 7.2. Capture of the Lez spring, Montpellier, South France. The natural outflow of the Vau-clause-like spring is seasonally interrupted by large abstractions by the nearby new pumping station for the water supply of Montpellier (photo: Hötzl 1992).

Photo 7.3. Gallery capture of the Innsbruck water supply in strongly karstified limestones of the Northern Limestone Alps, Austria. Galleries located at the base of the aquifer can cause extensive depletion of karst aquifers (photo: Krauthausen 1970).

Photo 7.4. Capture of the partially submarine karst spring of Kiveri, Peloponnesus , Greece. The capture makes use of the less dense water inside the wall, so that water can flow by the natural gradient in the collector system without being directly affected by the sea water. Due to mixing processes within the aquifer, the spring water already shows increased salt content (photo: Reichert 1987).

cases, karstic aquifers exhibit a marked heterogeneity and preferential directions for flow, resulting from the karstification.

The character of the functioning of the system and its degree of karstification, when its recharge comes only from infiltration of rainwater falling over the outcrop, can be deduced from the simple examination of the hydrograph over which rainfall is superimposed. That is, when the response of the spring to rainfall is immediate and the flow abruptly diminishes when the rain stops, the system is highly karstified, has little regulatory power, and possibly scant reserves. In this sense, one simple parameter which can allow one system to be compared with another is the Q_{max}/Q_{min} relationship, that is, the higher the relationship, the more karstified the system. Given that the dimensions of the carbonate rock-mass plays a role in the response of the springs, it is necessary to use more advanced treatments, such as the analysis of recession curves (Drogue 1972, Mangin 1975) or, better still, the hydrograph corresponding to several cycles using correlation spectral analyses (Mangin 1981a and b, 1984, Padilla et al. 1994) or the deconvolution (de Marsily 1978).

In massifs with great inertia, considerable reserves and high degree of memory, exploitation by means of wells, following geological and hydrogeological criteria, can give very favourable results. This is the case with many of the aquifers in the Mediterranean area (Spain, Morocco, Algeria, Tunisia and Libya), intensely exploited for the last 30 years. On the contrary, in aquifers, characterised by the absence of reserves, rapid circulation and restricted to practically one very transmissive conduit, or very few conduits, the exploitation is limited to finding and tapping the conduit and/or establishing some type of construction in the stream itself.

7.1.3 *Impacts of water exploitation*

The waters of karstic aquifers are valuable economic resources. Exploitation can constitute a positive socio-economic force, visible in countless examples around the world. However, the exploitation of karst water is often accompanied by a negative impact which eventually affects the environment and, ultimately, human interests. The effects of exploitation will be described according to whether these are direct or indirect. Regarding the former, without attempting to be exhaustive, mention can be made of the depletion of piezometric levels, compartmentalisation of the aquifer, increased exploitation costs, deterioration of water quality, abandonment of wells, alteration of river regimes, changes in wetlands, and legal problems. Indirect effects include soil salinization, progressive desertification, induced collapses, changes in physical properties of aquifers, and induced pollution due to overpumping. Useage and/or regulation of surface waters (dams, canals.etc.) must impact on groundwater.

Four short contributions accompany this chapter, all concerned with impacts due to exploitation. An example of over exploitation is presented by Pulido Bosch et al. (Section 7.2), showing the economic, technical and water-quality problems induced in the Sierra del Cid aquifer (Alicante, Spain). Bono (Section 7.3) gives a good example of collapse induced by overexploitation: the sinkhole of Dogatella (Italy). Biondic et al. (Section 7.4) present an interesting example of the influence of seasonal overpumping in a sea-water intrusion in Bakar Bay (Croatia). Finally, Kovalevsky (Section 7.5) gives some ideas about the influence of intensive withdrawal on karstification and other processes; hydrodynamic, hydrochemical and hydrobiological changes.

7.1.3.1 *Direct impacts of water exploitation*

Depletion of piezometric levels. The exploitation of groundwater by any of the systems which include pumping bear the intrinsic consequence of a lowering of the piezometric level, which can either occur in the immediate zone only or affect a broad area. The depletion can be occasional, recovering after the pumping is stopped, or it can be continual. The magnitude of the drop depends essentially on the local hydraulic parameters, on the flow pumped, the total pumped, and on the recharge regime of the system. When pumping is continuous, with average flows exceeding recharge and for a time frame sufficiently long as to avoid local imbalances, the result is over-exploitation, or mining exploitation of the aquifer (Pulido Bosch et al. 1989, Candela et al. 1991, Simmers et al. 1992), which in some cases can practically empty the formation (Pulido Bosch et al. 1995).

Compartmentalisation of aquifers. In areas of complex tectonic structure, where the morphology of the impervious substratum can be quite irregular, with some sectors raised and others depressed, the lowering of the piezometric level can cause thresholds to be whereby different sectors have completely different evolutions. A classical Spanish example is the Quibas system (Rodriguez-Estrella 1986), which was initially considered a single system, but after the beginning of intense exploitation, seven sub-units were well defined by their piezometric levels and by the physico-chemical characteristics of the water. Some of the sub-units maintain springflow, whilst others have a piezometric level of more than 100 m below the initial level.

Increased exploitation cost. Due to the piezometric fall in proportion to the exploitation, the height to which it is necessary to pump the water increases and therefore energy consumption also increases. In addition, the depth of the well may prove inadequate and thus require deepening. Another cost increase may result from the pumping equipment. One example of this, perhaps exceptional, is the aquifer of Crevillente, where the piezometric level fell some 95 m within four years (1979 to 1983), and the cost of extracting well water went from 2.9 pts·m^{-3} to 10.9 pts·m^{-3} (1 US\$ = 150 pts during this period; Pulido Bosch 1985). To this increase, the expenses of transport and distribution to consumption points must be added, as well as the depreciation of the installations. At Crevillente, these expenses amounted to 7.2 and 21 pts·m^{-3}, respectively.

Deterioration of water quality. Exploitation can mobilise low-quality water, bringing about a rather pronounced mixing process. The low-quality water may have a natural origin, or reflect pollution. One case in particular would occur where the substratum or some of the edges of the aquifer consist of evaporite material of Keuper facies, for example. There are instances of increases in total mineralization, causing the water to change from calcium bicarbonate to sodium chloride after intense exploitation. In some cases it is possible that water quality has a vertical zoning (gravitational), with waters becoming steadily more saline with depth.

Seasonal overpumpage. Much more frequent is the case of coastal aquifers in which the exploitation and subsequent fall of the piezometric level either encourages a wedge of saline water or changes its morphology to the up-coning type, as in the Campo de Dalías (SE Spain), where conductivity values exceed $10,000 \, \mu S \cdot cm^{-1}$ in the vortex of the cone (Pulido Bosch 1993). Seasonal overpumpage can also induce sea water intrusion, even if the annual recharge amount is higher than the discharge (Calvache & Pulido Bosch 1994).

Abandonment of wells. In certain cases, exploitation of karstic aquifers can force the abandonment of wells because of water quality or quantity. The case of the coastal aquifer mentioned above involved the abandonment of a series of wells and the drilling of others in less deteriorated areas. When the piezometric level falls excessively, some wells may reduce their yields, and hence be abandoned, if there is not a feasible alternative (for example deepening, if the diameter is not adequate).

There are situations where the need for abandonment is far more evident, as when the system or the productive zone is emptied locally, leaving the water at a level in the aquifer which is little karstified (LeGrand & Stringfield 1971, Milanovic 1981).

Alteration of river regimes. Although karstic areas are usually characterised by having few water courses of perennial superficial flow, some in fact do exist. In these cases, there is a river-aquifer relationship which may be affected in terms of quantity as a consequence of the extractions from the aquifer. When pumping reaches substantial volumes the river can become dry during extended periods, with subsequent ecological impacts. In Spain, there are numerous examples of this, although perhaps the best known is the Guadiana River (Llamas et al. 1992, Martínez-Alfaro et al. 1992). The regime of this river has changed around the source area, as a consequence of the exploitation of related aquifers (Parques de las Lagunas de Ruidera and Tablas de Daimiel).

Changes in wetlands. As with rivers associated with karstic aquifers, wet zones fed by these aquifers can suffer the consequences of groundwater exploitation. Such wet zones can be the surroundings of the springs themselves, or be places where the piezometric level is above ground level. The lagoons of Ruidera in the province of Ciudad Real (Spain) provide a clear example of the great impact on an aquifer of $2700 \, km^2$ which receives an average recharge of $12.6 \, M \, m^3 \cdot a^{-1}$ and in which $3.5 \, M \, m^3 \cdot a^{-1}$ is pumped (DGOH et al. 1994). Despite the fact that the extractions are low in relation to the total recharge, various lagoons have dried up and the rest have shrunk notably.

Legal problems due to third party problems. The first legal problems arising from the exploitation of karstic aquifers involve the traditional users of the natural springs of the system. The springs reduce in flow in the drier seasons, coinciding with the periods of greatest demand. The rights acquired over many years of use can paralyse any exploitation which is not initiated by these same users.

In many countries, there have been numerous examples of attempts to regulate karstic aquifers, for which costly studies have been carried out with many wells for tests and exploitation being drilled. Examples in Spain include Deifontes (province of Granada), where five wells have a pumping capacity of $2.35 \, m^3 \cdot a^{-1}$, and Pego (province of Alicante), where the capacity is similar and there are two spectacular extraction installations which could not be put into operation because of users resistance.

Much more common are the problems of different wells mutually reducing yield. This can lead to lengthy and costly judicial proceedings to protect users' acquired rights. See also Chapter 8 of this book.

7.1.3.2 Indirect impacts of water exploitation

Soil salinization. Irrigation with karstic waters in which the saline content has increased due to exploitation can lead to the salinization of irrigated soils. There are some examples in which, in addition to the high salt content in the waters, aridity causes the evaporation of an important fraction of the water applied, resulting in further saline deposits in the soil.

In other cases the damage to the soil and/or vegetation increases due to the presence of some phyto-toxic element. An example of this situation is the Andarax (Almería, SE Spain), where the waters of the deep aquifer (dolomitic limestone, marmoreal limestone and marble) contain considerable quantities of boron (Pulido Bosch et al. 1992). In this area, several thousand hectares have had to be abandoned in the last two years. The origin of the salinization is not only attributed to the deep aquifer, but also to the detrital aquifer, which has in fact contributed a greater quantity of salinity.

Progressive desertification. Desertification is a consequence of the salinization of the soil and the overexploitation of the water. Lands which, after having been converted to irrigation, have to be abandoned because of soil salinization and/or an insufficient quantity of irrigation water, prove especially vulnerable to erosion. The agricultural practices in semiarid regions such as SE Spain, part of Morocco, Algeria and Tunisia protect against erosion, especially in tree cultivation (citrus, for example) and in terracing. The abandonment of these lands after the death of the trees favours the appearance of rills, gullies and rampant piping processes (García-Ruiz et al. 1986).

Induced collapse. There are abundant examples throughout the world illustrating karstic collapse, many of which reflect a clear relationship with aquifer exploitation (LaMoreaux 1991). Water acts as a stabilising element which supports part of the load; a drop in the piezometric level reduces the resistance of the subsoil, risking collapse. In other cases, the exploitation causes the reactivation of conduits and karstic hollows filled with decalcified sediments. Rapid removal of these materials which contributed to the pre-existing equilibrium, can provoke the rock cover to collapse for lack of support (Garay 1986).

The countries registering the most collapses are perhaps China and the United States (Volker & Henry 1988). The average dimensions of the sinkholes are less than 20 m in diameter and less than 10 m deep. The form can also be quite variable, from circular to elliptical, times elongated. In the south of China, the collapses induced by pumping number about 3000 since 1960. Common results of collapse are the destruction of buildings and damage to roads, railroads, pipes and transmission lines.

The damage caused by the collapse in Winter Park (Florida) in May 1981 involved costs of probably more than 4 million dollars (Dougherty & Perlow 1987).

When gypsum is involved, the processes can be far swifter due to the higher degree of solubility (Johnson 1986). Although not related to pumping, but rather to ground-water, the instances of collapse and subsidence in saline diapirs and other evaporite accumulations are also abundant; the potassium mines of Cardona (Catalonia, Spain) had to be abandoned, and the traditional mining region of Wilyzca (Poland) is cur-rently in experiencing serious trouble with regard to ground stability. (See also Chapter 6).

Changes in the physical properties of aquifers. The entire karstic aquifer is dynamic, from the moment when the karstification processes begin and for as long as favour-able conditions persist. These processes are exceedingly rapid on a geological scale (Bakalowicz 1979). With the exploitation of groundwater, alterations are induced in the hydraulic heads, producing changes in groundwater flow paths, which become concentrated near the areas of greatest extraction. This can be accompanied by an in-crease in the karstification potential, and thus an increase in the volume of voids. The karstification is accelerated if, as a result of the exploitation, there is mixing of water which generates sub-saturated water with respect to calcite, for example karstifica-tion by water mixing (Bögli 1980).

A particular case would be coastal aquifers (Back et al. 1986, Back 1992) in which the transition between freshwater and salt water is the site of a series of reac-tions of dissolution, precipitation and ion exchange, which translates as a change in the volume of voids. The predominant final result is a notable increase in the karsti-fication. Nevertheless, it should be noted that although the processes are rapid in re-lation to geological time, they are not appreciable on the scale of a human lifetime.

Induced pollution from long distances due to overpumping. Because pumping can produce a cone in a transitory regime (the usual regime in nature), the radius of in-fluence is increased in proportion to time. Consequently, pollution processes can be-come operative in sectors of the aquifer which would not be affected in a natural re-gime. The mobilisation of contaminants can also occur from toxic substances injected deeply in the carbonate aquifers, a common practice in Florida, for example. The decompression in the upper aquifers as a consequence of exploitation can induce an increase in pollution through the fractures or imperfections in injection wells.

7.1.4 *Final considerations*

Together with the undeniable socio-economic benefits offered by water exploitation in karstic aquifers, there are a series of possible direct or indirect negative effects which need to be taken into account in order to plan a rational water-use programme. Given the broad range of possible scenarios both in the karstic aquifer and in the sur-roundings, there is no universally applicable formula, but rather each case deserves individual analysis to assure the adoption of an appropriate plan of action.

7.2 OVEREXPLOITATION OF KARSTIC AQUIFERS; THE EXAMPLE OF SIERRA DEL CID (ALICANTE, SPAIN)
(Antonio Pulido Bosch, Jose Migual Andreu & Antonio Estevez)

7.2.1 *Geological and hydrogeologic situation*

The aquifer of the Sierra del Cid is situated in the central part of the province of Alicante (SE Spain). It is a deep system where exploitation began in the 1960s. Between 1970 and 1975 the number of wells grew to 44, with pumping mainly in the SW portion of the aquifer (Serreta Larga; Fig. 7.2). The aquifer occupies some 50 km^2 and embraces both the upland of Sierra del Cid and Serreta Larga. The aquifer is composed of carbonate rock, represented principally by a Cretaceous limestone formation more than 200 m in thickness.

From the geological viewpoint, the system belongs to the Betic Cordillera. The series of the Sierra del Cid unit begins with a member of sandy marls and limestone marls, reaching 300 m in thickness and dating from the Albian. This is overlain by gravely limestones from Lower Cenomanian. On top of this formation lie limestones separated by thin marly layers and finally limestones from the Upper Cenomanian. Over the former group another lithological unit is concordantly superimposed, this one belonging to the Turonian-Senonian and made up of limestones, marly limestones and marls. Bioclastic calcarenites are unconformably superimposed on Cretaceous rock with alternations of sandy marls as well as some conglomerate layers, all dated to Eocene and Miocene. Finally, the uppermost strata of the series also unconformably, are Plio-Quaternary detrital materials. The mountainous relief of the Sierra del Cid is largely surrounded by clay and marl, including evaporites, all of which belong to the Triassic.

Of the rocks described, Cenomanian limestone constitutes the principal water-bearing material. The more carbonate rich layers from the Senonian also form part of the system. In addition, the bioclastic limestones of Miocene age are permeable, but due to the tectonic complexity and the lithological variation in the area, the hydraulic connection with the rock comprising the aquifer system is not entirely in evidence. The impervious substratum of the principal aquifer is composed of marl and limestone marl of the Albian and this also constitutes the eastern and northern edges. The western edge is occupied by impervious Triassic materials.

The principal recharge source is direct infiltration of rainfall over the permeable surfaces. Average annual rainfall of the zone is around 400 mm, and the average annual temperature is 14.5°C. Infiltration is estimated at more than 2 hm$^3 \cdot$a^{-1}. Outflow of the system at present takes place exclusively through some wells still functioning, though not continuously. With respect to the hydraulic characteristics of the aquifer, we lack data from pumping tests. Nevertheless, it is known that there are wells capable of extracting 100 l$\cdot$s^{-1} without causing appreciable piezometric depletion, suggesting that the transmissivity values are rather high.

7.2.2 *Exploitation and piezometric evolution*

The available piezometric record is more or less continuous from 1975. The piezometric depletion prior to 1975 is given by the IGME-Iryda (1978); levels plunged

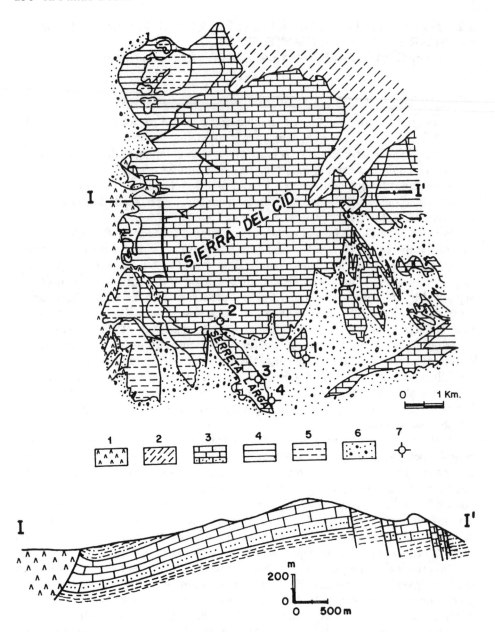

Figure 7.2. Hydrogeological scheme and geological cross-section of the Sierra del Cid. 1 = Triassic gypsum and clay (Keuper), 2 = Albian marls, 3 = bioclastic limestone and limestone from the Cenomanian, 4 = Senonian limestone marl and marl, 5 = Tertiary marl and calcarenite, 6 = Quaternary detritic materials, 7 = well; I-I': location of the geological cross-section.

–60 m in the period 1969 to 1974, equivalent to a depletion in piezometric level of –12 m·a⁻¹. This situation indicates heavy pumping, which in 1974 reached 2.4 M m³. The only record of levels prior to this date are three measurements, given by Piezometer No. 3 from the southern sector of the Serreta Larga from 1968 to 1976. The piezometric level was at –79 m in June of 1968, –83 m in November 1972 and –129 m in January 1976. The effects must have become noticeable beginning in 1974, the most spectacular case was that shown by Piezometer No. 2, located to the NE of Serreta Larga, in which the water level was at –63 m in October 1974 but –226 m in 1976, that is, a depletion of more than 160 m in less than two years.

Beginning in 1974 (Fig. 7.3), the piezometric level followed a strong downward trend, recording a great depletion in 1976. In January of 1977 the level was 280 m in depth but over 330 m in January 1978. In this year the level fell about 50 m. These data from this piezometer show that the level depletion was 174 m from 1975 to 1983, although the losses through wells in the Serreta Larga which during the same period lowered the level to 190 m (Pulido Bosch 1986), when added to the 60 m before 1975, register an emptying of the aquifer of around 250 m.

Pumping diminished progressively over the years; thus, in 1977 pumping was about 1.1 M m³ but only 0.7 M m³ in 1981. In the case of the Serreta Larga sector, the maximum depletions were recorded during 1979 (40 m; IGME 1983, 1986). In the time evolution of the piezometric lowering, there are at least three factors: the volume pumped, the rainfall amount, and the geometry of the system. In relation to this last mentioned factor, it must be noted that, in keeping with the structure of the zone, the surface area saturated reduces as depth increases, and therefore with the same amount of pumping, the depletion would increase with the progressive emptying of the system. The reduction in pumping is proportional to abandonment of wells. As a meaningful datum, it should be pointed out that already by 1977, wells which continued to be active represented only 45% of total. Wells were abandoned for various reasons: some went dry; others, due to structural deficiencies, could not accommodate the equipment necessary to pump at such depths, whilst others showed excessive deterioration in the chemical quality of water (Pulido Bosch 1986). Since 1984 the trend of the system has been the opposite, as only a few wells have remained active. Recovery has been a natural process due to rainwater infiltration exceeding the amount of water pumped. Thus the more or less continuous piezometric rise in level has reached around 270 m, representing a gain of nearly 115 m, equivalent to a mean of 19 m·a⁻¹.

Finally, at the end of 1991, when the most exhaustive monitoring of the aquifer system began, the rise in the piezometric level slowed to something over 5 m between November 1991 and 1992. During this period only 2 wells continued to function, and therefore the recovery continued throughout the hydrological cycle, almost without lessening during the dry period. In the middle of 1993 the increased pumping of 4 wells caused a drop of around 3 m, between May and November, but afterwards the recovery began again and has continued to the beginning of 1994. Pumpings increase during 1994 – total amount of more than 0.2 M m³·a⁻¹ – producing a slight depletion in the piezometric level (Fig. 7.3).

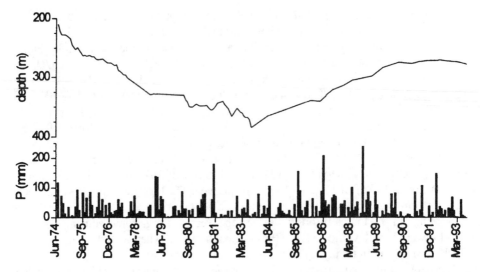

Figure 7.3. Piezometric evolution in well No. 1 and monthly rainfall measured in a nearby rain gauge (for location, see Fig. 7.2).

7.2.3 *Physico-chemical characteristics of the water*

In general, physico-chemical characteristics of the water indicate a predominance of chlorides over the rest of the ions, at concentrations exceeding 1 g·l^{-1}. Sodium constitutes the principal cation, and thus, despite the water origin being a karstic aquifer, the facies is sodium chloride.

The origin of the salinity may be the dissolution of the hyper soluble Triassic components of Keuper. Although the analytical record is not extensive and the determinations do not correspond to the same points of the aquifer, a comparison of the data corresponding to the years in which the levels were lowest does not differ notably from the last few years, chlorides maintaining predominance over the entire period, as shown in the columnar diagram (Fig. 7.4). Consequently, it appears that the spatial component directly controls the water quality; the location of the wells in the zones near Triassic evaporites apparently determines to a large degree the saline content of the water. Nevertheless, a vertical hydro-geochemical zonation cannot, a priori, be excluded, with the water increasing in saline content according to depth, as proposed for nearby areas (Pulido Bosch 1985, 1991).

7.2.4 *Discussion*

The above describes a spectacular example of groundwater over-exploitation followed by a recovery in piezometric levels due essentially to a considerable reduction in pumping. The carbonate nature of the aquifer demands special attention in view of the water storage far below the level of the springs.

The explanation for the processes described is not obvious; we cannot propose the massive karstification by water mixing as might be the case for coastal carbonate

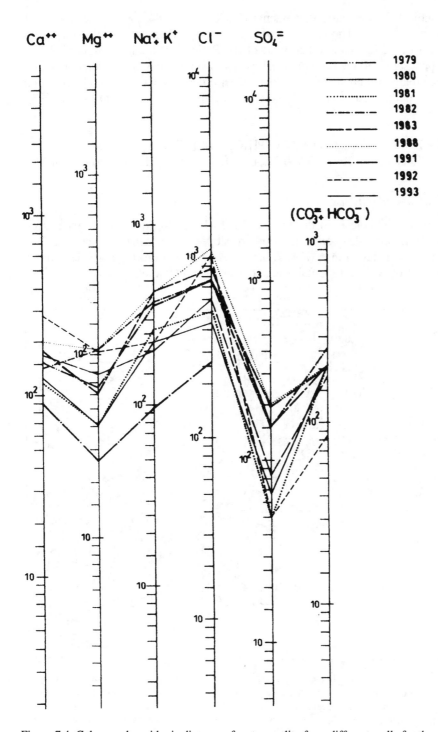

Figure 7.4. Columnar logarithmic diagram of water quality from different wells for the years indicated.

aquifers subjected to eustatic movements (Back et al. 1984), given that the sea is many kilometres away. The palaeo-karstification has had an active influence, as has been demonstrated for many other areas of the Betic Cordilleras (Vera et al. 1987), although persistence to the present is less evident from the cavities generated in these processes. Finally, it may be relevant to consider processes derived from liberation of deep CO_2 in mountain areas related to recent orogeny (Cerón & Pulido Bosch 1993).

7.3   A CASE STUDY OF CATASTROPHIC SUBSIDENCE:
   THE SINKHOLE OF DOGANELLA – CENTRAL ITALY (Paolo Bono)

7.3.1   *Location and land use*

The Doganella Sinkhole is in the Pontina Plain (Lazio Region) which extends between the Tyrrhenian Sea coast, the southern slopes of the Albano volcano and the Lepini Mesozoic karstic range (Fig. 7.5). The Pontina plain to the northeast links to the foothills of the Lepini Mountains at between 10 and 60 m elevation.

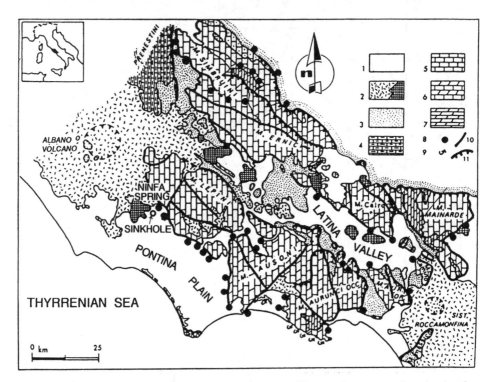

Figure 7.5. Hydrogeological scheme of the region. 1 = Pliocene-Quaternary deposits (marine, brackish and continental facies), 2 = Quaternary volcanic products and travertine (crossed lines), 3 = Upper Miocene marly-arenaceous deposits, 4 = Mesozoic p.p.to Cenozoic p.p. sequence of 'Sabina' facies (marls, cherty limestones), 5-6 = Mesozoic sequence of Lazio-Abruzzo facies (limestones and dolomites), 7 = Infra-liasic dolomitic outcrops, 8 = spring, 9 = submarine spring, 10 = normal fault, 11 = inverse fault and overthrust (modified after Boni et al. 1980).

The sinkhole of Doganella village formed on August 22, 1989. The collapse is in a flat area 1.5 km southwest of Ninfa karstic spring and 3.5 km northwest of three natural sinkholes bordering the Lepini karstic range (Bono et al. 1993).

The cavity is in an agricultural area allocated to seasonal crops. Some hundred metres from the collapse, Actinidia (kiwi) plantations have become particularly common since 1980, after vine cultivation became uneconomical. Cereal crops (wheat; maize), lucerne and seasonal vegetables represent the most common cultivation in the Doganella area. Irrigation of crops is extensively done from wells with a mean depth of 50 m (min. 20 m; max. 100 m). Wells are equipped either with submersible pumps or with large capacity centrifugal pumps many of them capable of supplying yields between 10 and 30 $l \cdot s^{-1}$. Well productivity however is generally higher.

The day before the collapse formed, the farmer did not notice any soil depression or fractures during grass mowing operations. On the August 22, 1989 in the early afternoon hours, a cavity was identified by the farmer about 1 m in diameter with 'deep walls shaped like an overturned funnel'. During the next few days after its discovery the hole enlarged rapidly becoming a source of concern for the farmers of that area. The increased enlargement of the sinkhole produced the collapse of the nearby rural road over a distance of 20 m (Fig. 7.6).

7.3.2 *Sinkhole site: geological, climatic and hydrological conditions*

The area surrounding the Doganella collapse is flat with a ground elevation of 30 m. Below 1 m of soil, are reworked volcanic deposits and other unconsolidated materials, as observed in the vertical walls of the sinkhole. The hole is elliptical. A survey was carried out on February 15, 1992 which defines an area of 520 m^2 with major and minor axes of 30 m and 21 m, respectively. The cavity bottom is irregular with the maximum depth 33 m from ground level. The sinkhole is water filled to a depth of 4 m from the surface and is hydraulically connected to the water table aquifer of the volcanics. Lowering of the water level of about 3 m during the period 1990-1993 has been observed.

Low piezometric levels usually occur in the summer season (driest period); however, negative fluctuations have been repeatedly observed in the winter-spring period following the resumption of irrigation by well pumpage. The volume of the collapsed material (22,000 m^3) provides an estimate of the hypothetical karst cavity buried by volcanics.

Drill logs of several boreholes in the Pontina Plain provide identification of three depositional phases of travertine correlated to the active stages of the Albano volcano. In particular the travertine near the sinkhole site is buried by 30-40 m of mostly loose Quaternary deposits.

A decrease in precipitation in the Pontina plain occurred from 1983 to 1991, identifying the last drought period of the time series (Fig. 7.7). The minimum year precipitation was recorded in 1989 (712 mm; effective precipitation: 144 $mm \cdot a^{-1}$).

During January-August 1989 there were severe climatic conditions with only 66 mm of effective precipitation against 261 mm for the same period in 1988 (effective precipitation: 312 $mm \cdot a^{-1}$). Piezometric surveys carried out in the Pontina plain in 1972 and 1992, demonstrate drought periods as evidenced by the rainfall time series from 1921 to 1991 (Fig. 7.7). The 1967-1976 drought period was more severe

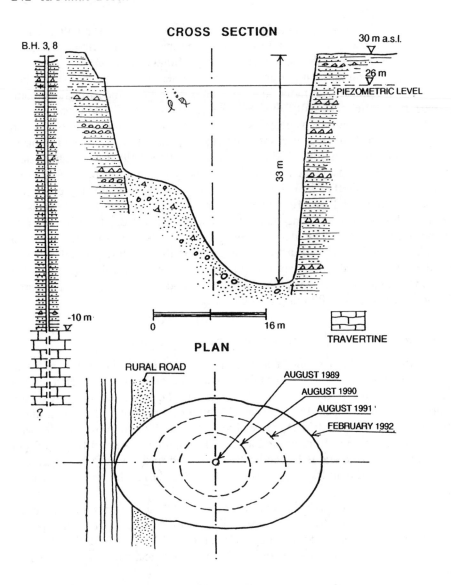

CROSS SECTION

B.H. 3, 8

30 m a.s.l.

26 m

PIEZOMETRIC LEVEL

33 m

-10 m

0   16 m

TRAVERTINE

PLAN

RURAL ROAD

AUGUST 1989

AUGUST 1990

AUGUST 1991

FEBRUARY 1992

?

Figure 7.6. Doganella sinkhole. Topographic survey (February 1992). Surface area
520 m × 2930 m × 21 m, maximum depth 33 m. The stratigraphy is from boreholes 3 and 8 located
to the south of the collapse.

than the latest (1983-1991), with the minimum rainfall of 600 mm in 1970. During
the 20 year period, there was a general decline of piezometric levels in the Pontina
plain in the range 0.25 m·a^{-1} to 1 m·a^{-1}. The former value is applicable to the area of
Doganella. Similar water level declines can be related to two water filled natural
sinkholes in the area southeast of Doganella, allowing an estimate during the period
1977-1992 of a mean negative variation of 0.1 m·a^{-1}.

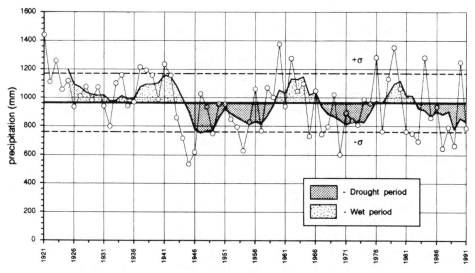

Figure 7.7. Latina rain gauge station (12 m a.s.l.). Precipitation time series (period 1921-1991) and 5-year moving mean curve.

The hydrometeorological conditions between 1983 and 1991, are also correlated with the general discharge decrease of Lepini karstic springs (mean discharge: 15 $m^3 \cdot s^{-1}$). In particular Ninfa Spring (discharge: mean 2 $m^3 \cdot s^{-1}$; min 0.6 $m^3 \cdot s^{-1}$; max 4.1 $m^3 \cdot s^{-1}$) is the closest to Doganella sinkhole and has the highest elevation (30 m) of the regional aquifer. Therefore, it shows a more intensive decline in discharge starting from 1984, with a minimum value in August-September 1989. According to information acquired from personnel of Ninfa Natural Park, the level of the lake in the springs area suddenly dropped 0.4 m during the days immediately before the Doganella sinkhole formed. Although climatic-hydrologic conditions similar to those of 1983-1991 period were recorded also in 1970 and again more severely in the 1945-1960 period, it is not known if there were additional sinkholes in the Pontina plain before the latest drought (1983-1991). Both decreasing rainfall and warmer air conditions during most of 1987-1989, caused severe soil aridity. In the winter and spring seasons there also was generally less precipitation and, consequently, farmers were compelled to practice field irrigation by well pumpage even during wet periods. Although piezometric data for wells for the period that preceded the formation of the Doganella sinkhole are not available (except information on decreasing discharge values of Ninfa spring), it is likely that during August 1989 the greatest depletion of piezometric levels was due to simultaneous pumping of a large number of wells in that area.

7.3.3 *Conclusions*

In the Doganella area the travertine formation is buried to depths of 40 m by mostly volcanic deposits. Water in the travertine is under a geostatic pressure of about 7.2 bar while the pressure in the confined aquifer is ordinarily 2.6 bar. The interstitial pressure in August 1989 was considerably decreased as a consequence of prolonged overpumping of the aquifer. Because of these hydraulic conditions the increased effective pressure limited the resistance of the inferred karst cavity cap. The cavity should be water-filled as a chamber of large volume.

A shallow earth tremor (3-4 km deep) of 3.2 magnitude was registered by the Seismic National Unit (Istituto Nazionale di Geofisica 1990) on August 8, 1989 at 00.41 h local time, with the epicentre located 20 km west of Ninfa spring. The acceleration on the Doganella area was approximately 0.02 g.

This earthquake was the highest in intensity and the nearest to Ninfa area during 1989. Thus it cannot be excluded that the seismic shock could have triggered the collapse of the karst cavity at a time of severe depletion of groundwater resources.

This condition is further implied by the sudden decrease of Ninfa lake level (about 0.4 m) in the days immediately before the sinkhole formed as this could probably be related to the seismic event.

The sinkhole formed by the progressive cavitation of unconsolidated volcanic and sedimentary deposits as a result of loss of support due to the collapse of the karst cover. The cavity grew deeper and migrated towards the top of the unconsolidated deposits until the roof (soil cover) collapsed on August 22, 1989.

Further catastrophic subsidence cases like Doganella sinkhole are predictable in the Pontina plain during periods of hydrologic stress, drought conditions and overpumping similar to those that occurred in the 1983-1989 period. It is also predictable that, even with less severe climatic conditions, if the demand of groundwater resources should increase, catastrophic subsidence could result. Although seismic records on the Pontina plain account for shocks of low magnitude, such events could trigger further collapses in geological and hydrogeological settings like Doganella.

7.4   IMPACT OF THE SEA ON THE PERILO ABSTRACTION SITE IN BAKAR BAY –CROATIA
(Bozidar Biondic, Franjo Dukaric & Ranko Biondic)

7.4.1 *Location and problem*

The catchment areas of the main springs and rivers in the Croatian coastal area are located in the carbonate karst massif of Dinarides which generally lies parallel to the coast. At the locations where the sea intrudes deeply into the mainland and in border areas of large structural formations, karst aquifers are laterally open to the sea enabling direct sea intrusion deep into the mainland. This applies to islands, too, where mainly local drainage basins are under significant influence from the sea. A very unstable equilibrium of salt and fresh water is usually disrupted by over-exploitation during dry summer periods. This creates enormous problems in unconnected littoral water supply systems, so that the water supply to particular cities or even regions is often cut off. This was the case in the Zadar-Biograd region, too.

Croatian experts have considerable experience in exploration works and attempts to remedy the intrusion of the sea into coastal karst aquifer, as have our neighbours in the Mediterranean who deal with similar problems. Numerous springs and spring zones have been explored in Croatia, from Istria to Dubrovnik and even up to Boka Kotorska.

Impelled by the difficulty of finding solutions to these problems and by numerous unsuccessful activities performed world-wide, the International Association of Hydrogeologists (IAH) organised a number of Salt Water Intrusion Meetings (SWIM). In 1991, the Association published a book of selected papers which has considerably changed the approach to the exploration of coastal areas with pronounced intrusion of salt water. The experience of our Italian neighbours (Cotecchia, Fidelibus and Tulipano 1991) is of special interest for us. They have worked in the karst region of the peninsula of Salento on the opposite side of Adriatic Sea (Section 7.5). The published papers and personal contacts have had extreme significance for the planning and realisation of exploration works in temporary saline abstraction structures located in the Bakar Bay near Rijeka. Our intention is to show, using the example of exploration works performed in the abstraction structure Perilo in Bakar, the way in which we in Croatia strive to solve this complex problem.

Croatian researchers theoretically stick to the postulates of Gyben-Herzberg Law based on the difference in the density of salt and fresh water, which were improved by the researchers in our country during the past thirty years. According to Breznik (1973), the Dinaric Karst aquifers are mainly heterogeneous and there are two possible ways of contamination by salt sea water. The first one is directly through submarine springs and the second one, which could be of special interest regarding our conditions, is within the karst massif itself. There are very few papers up to now in which theoretical postulates are drawn on the basis of observation, especially the observation of deep-seated parts of karst aquifer. The results of exploration works in the abstraction structure Perilo in the Bakar Bay could be interesting since these works have been specially designed for the exploration of deep-seated parts of the discharge zone, i.e. potential mixing zone of fresh and salt water under pumping conditions.

7.4.2 *Hydrogeological description*

The Perilo abstraction structure captures the waters of the coastal springs at Jaz and Perilo in the town of Bakar (Fig. 7.8). The structure is located in the upstream part of the town. It consists of a 50.35 m deep vertical pit (–1.5 m b.s.l.), and a 55 m long horizontal drainage gallery. The horizontal gallery cuts through the highly permeable limestone within impervious flysch sediments. By numerous exploratory boreholes and groundwater tracings through these boreholes, it has been established that the groundwater flows through this gallery to the springs in Bakar. Measured water discharge quantities in dry summer periods point to a possible water intake of up to $150 \ l \cdot s^{-1}$. However, test pumping of the structure carried out in 1979 has established a quantity of $240 \ l \cdot s^{-1}$ without the increase in salinity exceeding potable water standards. This is the basic quantity which has been exploited from this structure to date.

The abstraction structure was completed at the end of summer 1985, and prepared for exploitation. In the summer of 1985, the structure suffered significant salinity ($2500 \ mg \cdot l^{-1}$ Cl) and was out of operation until the end of a very long dry period. The abstraction structure has been saline again several times since, so that hydrogeological investigations started in 1992 in order to give protection against sea intrusion.

Figure 7.8. Map of Croatia showing the area investigated.

The above investigations were directed towards defining the geometry and structure of the aquifer on the basis of detailed hydrogeological and geophysical investigations. Exploratory drilling has been aimed at checking the results obtained, on the definition of hydrochemical and hydrodynamic elements of the aquifer during possible salinity in dry summer periods.

When improving the hydrogeological map, the emphasis has been put on structural and tectonic elements and on the elaboration of aquifer microtectonic parameters. The result has considerably changed the view of the form of the immediate discharge zone of the Bakar springs. It has been concluded that the carbonate rock on which the town of Bakar has been constructed is a gravitational tilt block, i.e. the eroded remainder of this block which covers one shell of well permeable limestone within flysch, which is the main groundwater drain towards the springs of Jaz and Perilo situated on the sea coast (Fig. 7.9). The Perilo abstraction gallery has been built in this limestone imbricate structure. Therefore, this tilted gravitational block hides the actual geological relations underground.

Impervious flysch sediments mainly surround the limestone formations, so that the discharge is concentrated on a very narrow coastal area. The hydrogeological situation is such that, in order to determine boundary conditions of the planned remedial works, it would be of extreme importance to define the location of flysch within the zone of limestone aquifer shell intersection.

Geophysical investigations in the scope of this project gave good results as regards the definition of geometry and indication of system functioning. Already in the first phase of exploration works, very low geoelectrical resistance (0.8 to 3.5 Ω) pointed to the presence of the sea in the aquifer even in the periods with entirely fresh spring water.

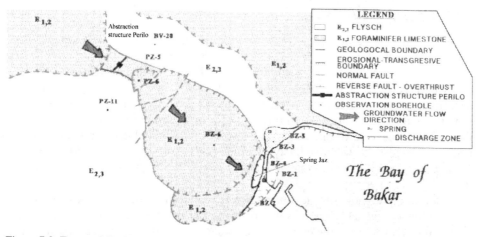

Figure 7.9. Zones of discharge at the town of Bakar.

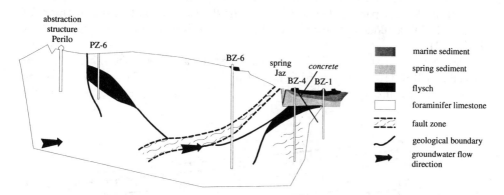

Figure 7.10. Longitudinal hydrogeological cross-section from Perilo capture towards Jaz Springs, Bakar, Croatia (not to scale).

Exploratory drilling is aimed at the solution of two main problems. The first one relates to the disposition of rocks of various hydrogeological characteristics in the spring zone – boundary conditions of remedial works, and the second one to the observation of changes in the aquifer during dry summer periods.

The longitudinal hydrogeological cross-section shows clearly the relation of permeable and impervious sediments in the aquifer. Hydrogeological investigations and exploratory boreholes have determined the disposition of sediments in the cross-section (Fig. 7.10). As regards the assessment of impermeability, it should be emphasised that limestones are water permeable, flysch entirely impervious and the marine sediments have such permeability as not to allow concentrated discharge from karst aquifer so that there are no submarine springs. At the same time, it is not an absolute insulator between fresh and salt water, as has been proved by observations conducted during dry periods.

7.4.3 Results of aquifer observations

Several exploratory boreholes have been drilled in the zone of possible sea water intrusion for detailed monitoring of aquifer behaviour. However, one of the boreholes located on the sea coast most accurately illustrates the behaviour of aquifer during dry periods (Fig. 7.11). In 1993, the dry period began in the early spring, so that high summer temperatures occurred when water outflows of the springs were already very low. Precipitation which occurred on August 8, increased the inflows and freshened the entire system. However, the flows decreased rapidly and dry conditions continued until August 25, when spillway quantities on springs rose due to precipitation. Despite this fact, the entire system salinized during the nights immediately prior to the rise of the water table.

Maximum pumping of the abstraction structure Perilo, i.e. maximum over-exploitation of available fresh water reserves during the night of August 27/28, caused the breakdown of the entire aquifer system. This over-exploitation caused the intrusion of a saline water cone from the mixing zone, deep underground towards the abstraction structure. The salinization encompassed practically the entire aquifer. However, the saline regime did not behave as expected. The saline level gradually grew from deepest parts of the aquifer towards the surface, and during the night of August 27-28, a sudden breakdown of the entire fresh water system occurred (Figs 7.11-7.13).

The abstraction structure Perilo salinized first, meaning that the mixing zone of fresh and salt water was drawn from deeper parts of aquifer to the surface, i.e. to the abstraction structure, due to increased pumping. Thus, a sub-pressure has been cre-

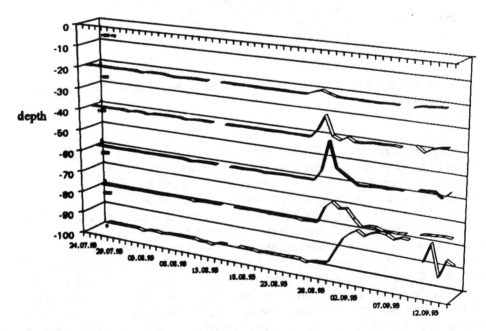

Figure 7.11. Groundwater salinity in the borehole BZ-4.

l/s

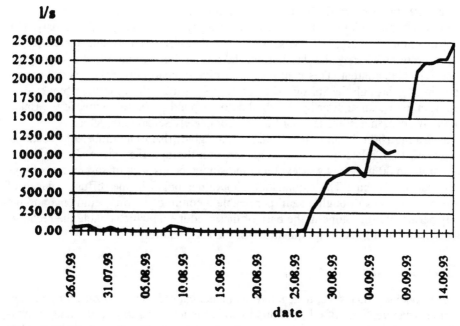

Figure 7.12. Discharge from the Jaz group of springs.

m3/day

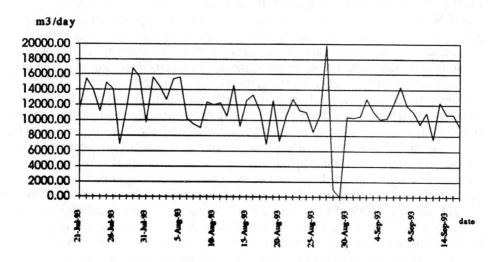

Figure 7.13. Pumped discharge from the Perilo abstraction structure.

ated in the entire aquifer. Such conditions caused a response in the entire system manifested by the intrusion of saline water through the discharge zone on the sea coast. Maximum sea intrusion (1,450 mg·l^{-1} Cl) was registered on the gauged bore-hole at a depth of 60 m below sea level on August 29, i.e. almost 2 days after the first observed salinity in the abstraction structure when the salinity in the abstraction structure Perilo had already begun to fall due to the larger quantity of fresh water coming from the background. The coastal spring site salinized simultaneously with the abstraction structure, but the salinity was of lesser intensity and there was no prior indication. Already the day before, the abstraction structure had a slightly increased chloride content. This is almost a positive indication that the sea did not intrude directly through the spring. The spring, rising from the limestone block separated from the Bakar aquifer by flysch, did not respond to the salinizing of the Bakar aquifer. We can, therefore, say that this is a completely separated system.

The sea intrusion through the discharge zone gives rise to a number of questions. First of all, they relate to the recorded intrusion through the spring site front at a depth of 40 to 80 m below sea level and the simultaneous lack of submarine springs in this part of the bay. The explanation should be sought in the diffuse balance of fresh and salt water through a semi-permeable medium of marine sediments and the highly permeable limestone in the base of these sediments which enable the recharge and discharge of limited water quantities of the karst aquifer.

7.4.4 *Conclusions*

We can conclude that several negative elements caused the salinization of the abstraction structure. First of all, it is due to an extremely long dry period and to over-exploitation of aquifer, whereby the equilibrium of fresh and salt water in the aquifer was disrupted. Furthermore it was established from water level data recorded in Bakar Bay that the maximum impact of the sea has been recorded during high tidal waves. When the first water wave reaches such a disturbed equilibrium it widens the mixing zone by its dynamics. Finally, maximum pumping raises this zone to the top of the aquifer. In this way, maximum salinization effects emerged at the very site of the abstraction structure. The response of the aquifer to this disrupted state is to recharge quantities of sea water until it establishes a new equilibrium.

During the last ten years, salinization of the Bakar aquifer occurred several times and it is not likely that anything will change without some intervention in the natural system. In our opinion, this intervention should provide for the following: control the rise of the mixing zone of fresh and salt water towards the abstraction structure Perilo and prevent the possibility of salt water intrusion in the aquifer. The control of rising could be accomplished by one borehole up to 250 m deep drilled very near to the Perilo abstraction structure. The sensors for the control of chloride content should be built-in at various depths and automatically connected to individual segments of pumping devices. Such a built-in control system would enable rational use of the abstraction structure during dry summer periods and avoid saline effects. The prevention of the possibility of salt water intrusion in the aquifer is a much more complex and expensive task. The areas in which the intrusion of saline water in deep karst underground has been recorded should be sealed by a grout curtain.

What recommendations could we give to research workers on the basis of our experience? First of all, to explore the deeper zones of an endangered karst aquifer, since we anticipate that in other regions, and especially in karst aquifers widely open to the sea, salt water fills deeper parts of the aquifer even when the springs are entirely fresh. The next element is the duration of the retention period of salt water in the karst aquifer and the changes in the position of mixing zones when the fresh water outflow is significantly reduced, i.e. detailed control of hydrodynamic and chemical changes of groundwater and the control of fresh water pumping on pumping sites. Finally, considerable attention should be paid to regional relations, since focusing on only the narrow discharge zone very often leads to a high probability of error occurring in the remedial project.

7.5 GROUNDWATER SALINIZATION IN THE APULIA REGION, SOUTHERN ITALY (Luigi Tulipano & Maria Dolores Fidelibus)

7.5.1 *General geological and hydrogeological features*

Murgia and Salento represent two important karstic aquifers in the Apulia region of southern Italy affected by sea water intrusion due to over-exploitation. The two aquifers are comprised of Mesozoic carbonate rocks including limestone, dolomitic limestone and dolomites. Great differences in the overall degree of permeability of the aquifers and in their spatial distribution, cause the hydrogeological features of Murgia to be markedly different from those of Salento (Grassi & Tulipano 1983). The Murgia aquifer (Fig. 7.14) is characterised by piezometric heads reaching to 200 m a.s.l. and by hydraulic gradients of 8%, whereas the Salento aquifer exhibits piezometric heads seldom exceeding 3 m a.s.l. and hydraulic gradients, on the average, are of the order of 0.2%. The Salento aquifer is very thin because the maximum hydraulic heads occurring allow the presence of salt water at a maximum depth of about 120 m below s.l., whereas the Murgia aquifer has a considerable thickness: wells drilled at more than 1000 m depth in the innermost areas of the territory, found very fresh waters. Figure 7.15 shows on a regional scale, the state of saline contamination in the Murgia and Salento aquifers: waters with saline contents characteristic of groundwaters not contaminated by the sea, that is about 0.5 $g \cdot l^{-1}$, are found in very broad areas of Murgia, whereas in the adjacent Salento only in much more limited areas.

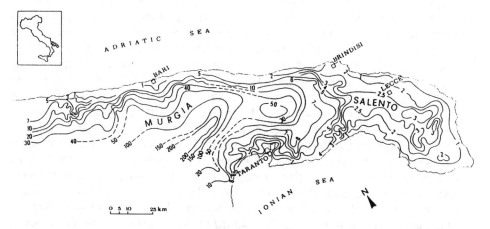

Figure 7.14. Trend of the piezometric surface of groundwater in the Murgia and Salento aquifers.

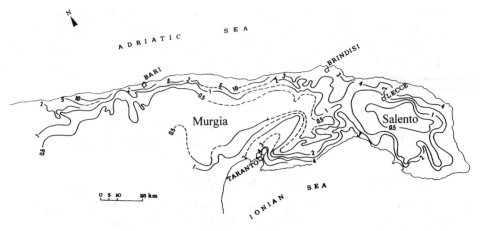

Figure 7.15. Saline content contour lines (g·l^{-1}) of Murgia and Salento groundwaters.

7.5.2 Fresh water – salt water equilibrium

The first signs of a decline in Apulian groundwater quality, due to an increase in the saline content, were noticed in the sixties, when a big effort was being made to improve agriculture, especially in the Salento area. The Water Research Institute of the National Research Council (C.N.R.-I.R.S.A.) embarked on a series of scientific research programmes, designed to clarify the matter of salinization due to over-exploitation. Hence, several observation-wells were drilled in the Salento area, thus making it possible to check on the position of the transition zone, to study its dynamic behaviour (Tadolini & Tulipano, 1979) and to verify the equilibrium conditions between fresh and salt waters (Cotecchia et al. 1986).

7.5.3 Groundwater temperature in relation to salt water intrusion

The input temperature of waters feeding the Murgia and Salento aquifers is equal to the average temperature of the rain water during the recharge period. Dependant on how long waters stay in the aquifer and on the depth which they reach, they will be subjected to the influence of geothermal flow tending to raise their temperature. Considering all the factors regulating the thermal regime of groundwater, it can be assumed that the longer the residence time in the aquifer, the larger will be the resulting increase in temperature of waters (Tulipano 1988). The underground saline waters of marine origin, present under the fresh groundwaters in the Salento, are to be regarded as practically still, tending to reach equilibrium with the geothermal flow: their temperature is consequently certainly higher than that of fresh groundwaters. Vertical temperature sections allow the reconstruction of horizontal sections at various depths. Figure 7.16 gives a map of groundwater temperature at sea level for the Salento aquifer and Figure 7.17 an example of a vertical section with extension of these values downward in the same aquifer. Hence they are following the trend of isothermal curves and therefore outline the salt water intrusion phenomenon in depth (Tulipano & Fidelibus 1988).

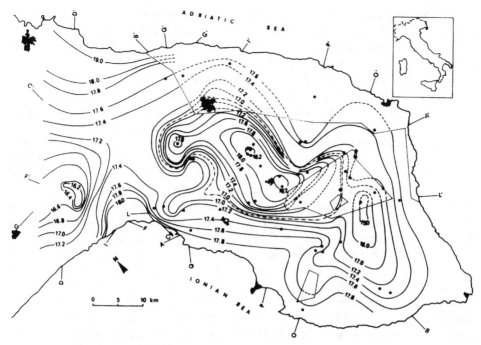

Figure 7.16. Temperature distribution (°C) at sea level for groundwaters of the Salento aquifer. Where the top of the aquifer is below sea level, isotherms are shown by hatched lines.

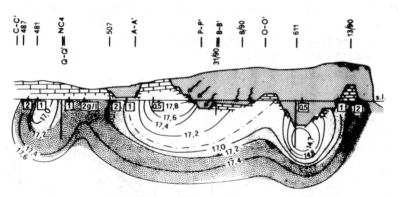

Figure 7.17. Example of vertical temperature section (L-L' in Fig. 7.16). In the section is shown the geometry of Mesozoic basement: post-Mesozoic covers are in grey. Numbers within squares represent the salt content of groundwaters (deduced from Fig. 7.16): dotted zones indicate salinized groundwater. Vertical scale is 50 times the horizontal one.

7.5.4 *Chemical parameters as tracers of relative age of intruding sea waters*

Considering the behaviour of major constituents during a salinization process, their ratios appear to correlate with the chloride concentrations expected for conservative mixing between fresh water and sea water, but their concentration gives rise to doubts as to the unique nature of the saline matrix in mixing or on the occurrence of modifying geochemical processes. The trend of the mixing that can be observed for groundwater sampled along each observation well, shows in fact, that different saline end members are present (Fidelibus & Tulipano, 1986).

While there is more calcium in the underground saline waters than in present-day sea water, magnesium, in relation to the total salt content, varies in the opposite direction, the maximum concentrations being found in present-day sea water and progressively lower values are to be found in the various underground saline waters sampled further inland. Consequently (Fig. 7.18) underground salt waters are identified by rMg/rCa values (r stands for meq/l) of around 3.5, while generally higher values of 4.5 are characteristic of present-day sea water, including those that have intruded the land-mass very recently and are subjected to cyclic flow. The alteration of the Mg/Ca ratio in the underground salt waters can be attributed to water-rock interaction which, in a carbonate aquifer is represented mainly by dolomitization, dissolution of carbonate matrix and precipitation. Progress in the dolomitization process takes account of the gradual decrease in the Mg/Ca ratio, which can be correlated with increasing residence times of underground salt waters in the aquifer.

Strontium concentrations are also modified during the changes in the Mg/Ca ratio, showing increasing values as Mg/Ca ratios decrease. Enrichment of Sr^{2+} in underground saline waters with respect to the mean concentration in sea waters (8 mg·l^{-1}) may exceed 100%. This is indicative of incongruous dissolution of the carbonate matrix: the Sr^{++} ions pass into solution together with Ca^{2+} during dissolution steps, but, when precipitation occurs, Ca^{2+} ions are preferentially involved, while Sr^{2+} ions remain in solution. This comparison confirms the hypothesis that different maturity degrees of underground saline waters are characterised by the Mg/Ca ratio, which corresponds in any case to an increase in residence time (Tulipano & Fidelibus 1995).

An important corroboration of the significance of this parameter comes from ^{14}C analyses of some underground saline water. Utilising the Mg/Ca ratio, in terms of

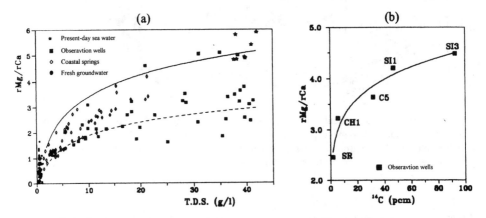

Figure 7.18. rMg/rCa ratio (r = meq/l) versus TDS (a) and versus ^{14}C contents (pcm) (b). Symbols refer to fresh waters (dots), coastal spring waters (rhombs), waters of different salinity from observation wells (squares) and present sea waters (asterisks), sampled in the Salento and Murgia aquifers.

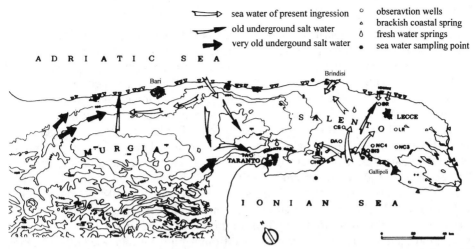

Figure 7.19. General outline of underground saline water circulation in the Murgia and Salento aquifer systems.

relative age, it is possible to follow the pathways of intruding sea water in the Murgia and Salento aquifers inland from the coastal fronts, where present intrusion occurs. Figure 7.19 represents a general outline of the slow circulation of underground saline waters. The reconstruction of the main pathways of saline waters was derived also from further information about the chemical evolution of the underground salt waters (Fidelibus & Tulipano 1990), obtained from the study of the brackish waters discharged by coastal springs.

7.6 THE IMPACT OF INTENSIVE WITHDRAWAL ON THE INTENSIFICATION OF KARST PROCESSES
(Vladimir S. Kovalevsky)

7.6.1 *General remarks*

Intensive water withdrawals from large centralised well fields and also mine water withdrawals inevitably cause the intensification of karst processes accompanied by formation of new sinkholes and collapses on the surface, deformations in buildings and communications routes and deterioration of the ecological situation (cf. Chapter 6). The considerable decline of karst water levels caused by the water withdrawal brings about ten-fold increases of gradients and velocity of groundwater flow, speeds up water exchange, assists the inflow of surface water that is aggressive and saturated with micro-organisms into the aquifers, so intensifying rock leaching. Water withdrawal is accompanied by removal of leaching products and loose sediments from infilled caverns of ancient covered karsts. All these can influence the course of karstification. As a result numerous sinkholes and collapses caused by water withdrawal have been observed in the United States, Russia, Lithuania and other countries.

7.6.2 *The example of the Moscow region*

Active exploitation of confined aquifers of Carboniferous limestones in the Moscow city region has reduced the piezometric heads to 10-70 m lower than the river and the originally confined groundwater levels. This causes the inflow of aggressive and polluted river water into the aquifers and changes balance, structure and hydrochemical conditions in the karstified medium. As a result mineralisation of exploited groundwater increases 1.5-2 times mainly due to magnesium and calcium ions with a decrease of sulphates and chlorides. The content of calcium increases in particular 15-20 times. Removal of the salts with the water pumped out, including suffosional removal of particles, has amounted to 100 $t \cdot a^{-1}$ per km^2. Concentrations of free carbonic acid in the groundwater amounted to 20 $mg \cdot l^{-1}$, analogous to the river water. The oxygen content increased 2-10 times and pH decreased to 4.9-5.5 from an original value of 7-8. The content of strontium, which indirectly characterises the leaching activity increased to 16-18 $mg \cdot l^{-1}$ from the background of 0.1-0.21 $mg \cdot l^{-1}$. The rate of the aquifer water exchange estimated from tritium has increased to 2-12 years, compared to more then 100 years in natural conditions at the beginning of the century. The content of helium in the aquifer water decreased abruptly 15-20 times due to its dilution with river water in the recharge zone. It was demonstrated earlier (Kovalevsky & Zlobina 1984, 1987, 1988) that a helium concentration survey can be used as an indirect but very sensitive method for revealing and mapping potential sites of reactivation of karstification.

Surface water reaching the aquifer caused the occurrence of up to 30 kinds hydrobionites amounting to 14,000 per ml, including 4 types of filamentous bacteria, 5 of diatoms, 7 of green algae, 3 of blue-green algae, 3 of protozoan, and also infusoria, shell amoebas, etc. The content of carbon dioxide amounted to 200 $mg \cdot l^{-1}$ in the zones of high micro-organism concentration. The weak acids are the product of their vital activity and it must be considered that this factor of accelerated karstification is one of the most important. The impact of various contaminants on karstification processes is not enough studied but is apparently significant. Thus, according to laboratory tests carbonate leaching increases 4-5 times in the presence of ash dumps washed with infiltrating water in the areas of thermoelectric plants. In this case water acidity and its under-saturation with respect to calcium is increasing. The increase of groundwater temperature in the areas of thermoelectric plants furthers the development of micro-organisms with the above mentioned consequences.

Thus, intensive water withdrawal causes not only hydrodynamic, but also hydrochemical, hydrobiological and temperature changes in aquifers and these changes may affect karst intensification both directly and indirectly. The most significant such changes in the territory of Moscow are readily apparent with areas of newly formed sinkholes, causing the failure of some buildings and pavements They can serve as an indicator for revealing zones of risky construction and inhabitation. This is particularly important in urbanised regions.

Karst processes in the well field terrain can be controlled by redistributing or decreasing water intake thus lessening gradients and velocity of groundwater movement.

Part 3: Implications and conclusions

CHAPTER 8

The management of karst environments

DAVID DREW
Department of Geography, Trinity College, Dublin, Ireland

HEINZ HÖTZL
Department of Applied Geology, University of Karlsruhe, Germany

8.1 THE DISTINCTIVE CHARACTER OF KARSTS

8.1.1 *The karstic hydrological system*

The main characteristics that set the hydrological systems of karst regions apart from those of other rock types are:
– Rainwater sinks underground instead of generating surface river systems
– Inputs (recharge) to the karstic aquifer may be concentrated via sinking streams as well as diffuse.
– The underground water moves and is stored in fissures enlarged by the dissolving action of the water.
– Rates of flow of the underground water are rapid in comparison with conventional groundwater flow velocities.
– Retention times for water in the aquifer are relatively short.
– The capacity for self purification of the water in the aquifer is limited.
– The karst aquifer evolves through time as solutional enlargement of fissures develops.
– Preferential pathways develop through the aquifer from recharge to discharge area with a hierarchical system of conduits draining the rock mass efficiently.
– Discharge from the aquifer is commonly in a concentrated form from springs.
Some of the attributes listed above apply to other hydrogeological situations but only in karst terrains are all of them important. The extent to which the criteria are present in any particular area will depend upon the degree of karstification of the soluble rocks which may be slight or intense, and the characteristics listed are likely to be fulfilled to the greatest extent in a karstic area with little cover of superficial deposits, well developed cave-conduits and a shallow groundwater flow system

8.1.2 *The karst ecosystem*

A further aspect that differentiates karst from areas underlain by non-karstifiable rocks is the complexity and intimacy of the links between the hydrological system

259

and the geomorphologic and ecological systems, to the extent that changes in any one aspect of the overall karst rapidly influences many other aspects of the system. In addition the sensitivity of karst terrains to imposed stress referred to elsewhere in this book means that a relatively slight stress at some point in the hydro-geo-eco karst system may cause a disproportionately large response.

These relationships require a comprehensive approach with regard to the protection of the natural karst water resources. It is not possible to deal just with the groundwater and its vulnerability without having in mind the whole karst ecosystem including the karst aquifer, its covering, the morphology, the natural vegetation and the different forms of land use by man. Therefore a sustainable activity demands an integrated management of the karst system, if it is to be successful.

Two flow diagrams (Figs 8.1 and 8.2) illustrate the interconnectedness of hydrological – geomorphologic – ecological conditions in karst terrains in relation to human activities.

The example selected involves a change of land use in a karst area, which might influence the distribution of precipitation water. In this respect there are two possibilities (a) increasing or (b) decreasing the surface runoff in a karst area. In the first flow diagram (Fig. 8.1.) the effects of increased surface runoff, e.g. due the conversion of an natural karst pasture into farmland, on water quantity and quality for each subsequent process are shown.

On the surface this might at first cause (Phase 1) the development of a drainage network due to the absence of or poor development of channels, and in the next step additional erosion. The absence of, or reduction of the quantity of infiltrating water will cause reduced groundwater recharge and thereby an increased residence time and reaction time for the available karst water. Due to the increased reaction time with the rocks higher mineralisation (water in chemical equilibrium with the rocks) will result by natural geochemical processes.

In Phase 2 the high sediment load of the flood water may lead to a plugging of swallow holes and filling up of the small openings. This can cause a significant drop in the ground water table elevation. In respect of the chemical composition of the underground water, an additional increase of total dissolved calcium carbonate will take place and due to changing saturation conditions precipitation of solutes will lead to sinter (speleothem) formation.

The increased surface runoff may find new surface flow routes and perhaps lead to new karstification (Phase 3) or perhaps to a reactivation of sealed palaeokarst. In conjunction with the lowered groundwater table this may cause the collapse of new sinkholes. With the increased sediment loading, attenuation of the dissolved components in the karst water will take place but also high turbidity due to the suspended sediment.

In Phase 4 the newly activated karst system will cause intensified infiltration and seepage of surface runoff underground. The generally contaminated surface water will result in an increased pollution of the karst ground water.

In a final stage (Phase 5) the new flow and discharge conditions may become stabilised. More and more a new karst flow system will be developed with conduit flow conditions prevailing. This means higher flow velocities, lower retention capacity and in terms of water quality, more polluted water with high seasonal and storm-generated variations.

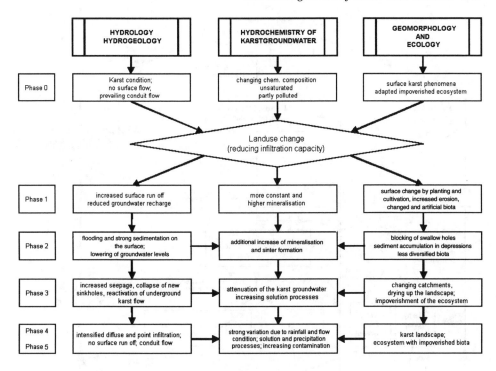

Figure 8.1. Possible consequences of land clearance in a karst area.

This scenario identifies the simple chain of cause and effect following a change of landuse. It does not take in account the additional processes which might be induced, for example by a change of the saturation conditions of the karst water – plugging of conduits by sinter formation or increased solution and widening of the water routes. A more complex chain of possible causes and effects initiated by land clearance to expose bare soil (a possibility in rural and urban environments) is presented in Figure 8.2.

Irrespective of rock type, certain changes in hydrological and sedimentary regimes will be consequent upon land clearance as the upper part of Figure 8.2 illustrates. In the karst terrain however, the intimacy and directness of the links between the surface and the subsurface hydrological systems may cause more immediate and dramatic changes. Subterranean flow routes may become blocked and inactivated whilst ancient, inactive routes may be re-excavated and become functional once more, often in an unpredictable manner. The examples given are hypothetical but are representative of many possible scenarios in the karstic environment.

To what extent have legislators and environmental planners and decision makers recognised the singularity of the karst environment?

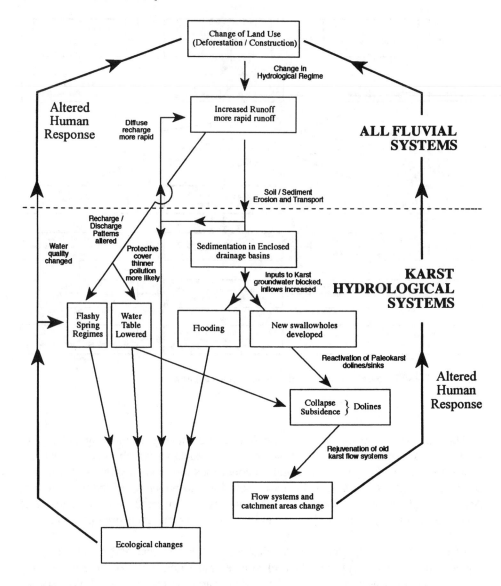

Figure 8.2. Schematic cycle of karst and water development due to human induced land use change (e.g. reclamation)

8.2   THE LEGISLATIVE RESPONSE TO THE KARST WATER ENVIRONMENT

8.2.1 *Introduction*

The concept of environmental management is now widely accepted as an integral part of the social and economic framework in most parts of the world. Environmental impact evaluations are commonly a part of the planning process and legislation to protect the environment is universal. This is true specifically regarding the protection of water resources including to a lesser extent, groundwater. For example in the USA there are numerous Federal statutes affecting water protection.

Similarly, other regulations affecting water protection exist within the European Union and via global agreements. All such regulations are of course applicable to karst areas and to karst groundwater but it is not common for water protection regulations or guidelines to be formulated in a manner which takes account of the unique characteristics of karst environments as detailed in Section 8.1. Isolated examples of karst specific legislation do exist. For example in Great Britain the Wildlife and Countryside Act (Limestone Pavements Orders) of 1981 protects areas of limestone pavements (lapies) from being quarried away for ornamental use whilst certain caves are protected as Sites of Special Scientific Interest. However, neither of these relate directly to water resources.

In recent years there has been the development of investigations into the appropriate management of karst environments and the formulation of codes of practice or even legislation that is karst specific. The three sections that follow summarise the main developments in this field in the USA, in Europe and in other areas of the world.

8.2.2 *Karst water protection in the USA*

Karst specific management codes and karst specific legislation has probably been developed in the USA to a greater extent than elsewhere. Such developments have occurred at Federal, State and local levels. The survey that follows is exemplary and does not purport to be in any way a comprehensive survey of a major and rapidly developing subject. Much of the material is derived from LaMoreaux (1995) and LaMoreaux et al. (1997) and from contributions to the fifth multidisciplinary conference on sinkholes *Karst Geohazards* (Beck 1995). In this publication Quinlan et al. (1995) remark on the significance of karst aquifers in certain areas:

'Groundwater is the source of 51% of drinking water in Tennessee, 31% of it in Kentucky, and 92% of it in Florida. At least 100 municipalities in Tennessee and 25 in Kentucky obtain some or all of their water from springs. No springs are used for municipal drinking water in Florida; all of it is pumped from [karstic] wells. ... protection for springs in Florida is important however, because they are a critical part of the aquatic habitat of numerous endangered species of animals and plants, as well as being important for recreational use.'

The Federal Environmental Protection Agency often delegates to State level and it is at State and more local levels that explicit recognition of karst is most apparent.

Devilbiss (1995) notes that at the level of local government, land use control is the most effective tool in managing karst environments. He describes non-regulatory, non-mandated practices that have arisen in Carroll County, Maryland to avoid problems in karst areas by means of informed, mutual agreement between planners and developers. According to Devilbiss the mitigation strategy may include:
– *Layout design* (i.e. use sinkhole-prone areas as open space or for reforestation, cluster to non-carbonate rock areas on site);

- *Stormwater management* alternatives (i.e. underground storage versus ponds, sheet flow versus concentrated containment);
- *Storm drainage design* (i.e. sheet flow versus ditchlines, piping or lining ditchlines to non-carbonate areas);
- *Road design* (i.e. realignments to non-carbonate rock areas, detailed testing to identify specific problem areas, roadway reductions).

A similar approach at local level adopted in north New Jersey is described by Schwartz & Drewes (1995). Aggressive land use planning policies are avoided, in part due to a lack of awareness of possible impacts of development on karst terrain. To help overcome this a North Jersey Resource Conservation and Development Council was formed incorporating a Limestone Resource Committee representing a wide range of interests and with the aim of supplying appropriate educational and support services as required.

'*Over the course of two years, the Committee prepared a model ordinance for land development activity in karst areas that addresses both groundwater protection and prevention of structural collapse... The Committee prepared the ordinance largely in response to significant development pressures in northern New Jersey, coupled with an absence of appropriate planning measures being taken at the local level*' (Schwartz & Drewes 1995).

At State level, Kentucky, with extensive karst areas, has developed karst specific legislation. For example (Harker & Ray 1995):

'*Kentucky Revised Statutes 433.871 through 433.885, enacted July 15, 1988, provide for direct protection of caves. These sections address dumping of dead animals, excavation permits for archaeological purposes, unlawful burning or removing of formations from caves exhibited to the public, and general protection of caves. ... The phrase 'which is large enough to permit a person to enter' in the definition of cave limits the protection afforded to karst systems. Although KRS 433.875 makes it unlawful' ... to store, dump, litter, dispose of or otherwise place any refuse, garbage, dead animals, sewage, toxic substances harmful to cave life or humans, or to store other such similar materials in any quantity in any cave,' the limited definition of cave reduces the scope of the law. Refuse, garbage, dead animals, and toxic substances can be detrimental to karst groundwater recharge features smaller than those that can be entered by humans. Thus, inadequate definitions of karst features may engender gaps in the regulatory protection of karst systems.*'

Specific measures designed to apply more rigorous protection to karst water systems are also in force:

The Kentucky Pollutant Discharge Elimination system requires permits for the discharge of pollutants from point sources into waters of the Commonwealth. Municipal sewage facilities located in karst regions often discharge to sinkholes or sinking streams. As many as 63 facilities including motels, mobile home parks, and schools discharged to sinkholes in 1988. Discharge limits are calculated to meet in-stream limits based upon effluent and stream flow. Since the characteristics of subsurface streams, including flow rate, are largely unknown or difficult and expensive to obtain, effluent limits for conventional pollutants are established based upon the assumption of no further dilution beyond the sinkhole. If the effluent discharges directly to a sinkhole, these limits must be met at the end of the pipe.

In Tennessee, also possessing very large areas of karst, landfill site regulations are karst specific (House 1995) The requirements that must be satisfied when landfill facilities are proposed within karst terrains are:

Rule 1200-1-7-04: Karst terrain – if a facility is proposed in an area of highly developed karst terrain (i.e. sink holes, caves, underground conduit flow drainage and solutionally enlarged fractures) the applicant must demonstrate to the satisfaction of the commissioner that relative to the proposed facility siting:
- There is no significant potential for surface collapse;
- The ground water flow system is not a conduit flow which would contribute significant potential for surface collapse or which would cause significant degradation to the ground water; and
- Location in the karst terrain will not cause any significant degradation to the local ground water resources.

Thus, in the USA karst oriented legislation (and indeed litigation) is emerging, although as many writers point out the definition of karst is often too restrictive and the people who apply or enforce the legislation are not necessarily wholly conversant with concepts of karst system functioning.

8.2.3 *Karst water protection in Europe*

The need to introduce ground water protection measures in Europe arose mainly from hygienic problems. Bacteriological pollution of water supplies at the end of the last century led to catastrophic epidemic diseases with many people dying in the expanding cities at the beginning of industrialisation (Section 5.7.2). Independent of which aquifer the groundwater was coming from the demand was great to keep the water clean. Therefore regulations were introduced by certain water suppliers for their catchment areas at first, then extending over wider areas as general rules for all exploited water resources were developed. There are now specific water laws in most countries.

Due to the nature of the main problem in the past, when even chlorination was not widespread, priority was given to the hygienic aspects of protection, which led to a bacteriologically defined subdivision of the area to be protected. Though the special vulnerability of karst aquifers with regard to bacteriological impacts was recognised in very early stages of the development (Hägler 1873), the regulations were, as mentioned rather general and referred to all types of aquifer without any specificity.

Depending on the importance of karst aquifers for water supply in different countries more and more additional provisions were added to define special protection measures for karst areas. However with the exception of a few States such details on karst have been added to the legal regulations only since the mid 1980's.

In connection with European Commission COST-Action 65 (EC-Cost 1995) an evaluation was performed on groundwater regulation in 16 States of Europe. The subsequent discussion will refer mainly to the report of this action supplemented by some newer developments. Table 8.1 gives an overview of source protection, whilst Table 8.2 refers to resource protection.

For source protection the concept of protection zones, minimising the necessary restrictions in the catchment areas by considering the increased self-purification capacity of the aquifer, is accepted in all countries. The regulations take note of two

Table 8.1. Source protection in different countries of Europe with special comments on karst areas (EC Cost 1995).

Country	Spring/well head protection areas	Inner protection zone(s)	Outer protection zones	Catchment area
Austria	Inner protection zone	Outer protection zone – 60 days	Prevention zone I – recharge area	Prevention zone II – monitoring area
Belgium (Walloon region)	I = Water supply zone Installations + 10 m radius	IIa = 24 hr >25 m – in karst include all preferential points of infiltration having connection with source. IIb = 50 days in confined In unconfined: >IIa + 100 m sandy >IIa + 500 m gravel >IIa + 1000 m karst	III = Observation zone or catchment area	
Croatia	Proposed: IA = Immediate spring IB = Immediate catchment area	Proposed: II = 24 hrs – zone of strict limitation III = 1-10 days – zone of limitation and control	Proposed: IV = 10-50 days – zone of limited protection	Specially protected zone – water reserve area
England and Wales	Wellhead	50 days	400 days	Catchment area
France	Well head area dependent on size of wellfield, also around sinkholes etc.	Inner protection – commonly 50 days		Outer protection (optional)
Germany	I = 10-20 m radius	II = 50 day travel time	IIIA = 2 km radius	III B
Hungary	Immediate 10-20 m radius	Inner protection >100 m (75 days for karst)		Catchment area
Ireland	1A = 0-10 m radius	1B = 10-300 m	1C = 300-1000 m	

Table 8.1. Continued.

Country	Spring/well head protection areas	Inner protection zone(s)	Outer protection zones	Catchment area
Italy	Each regional government responsible for own legislation			
Malta	Land expropriated and enclosed in immediate vicinity of well head	500 m radius or more where several zones meet		
Portugal		High permeability terrain (low filtration): 20-50 m Low permeability terrain (high filtration): 10-20 m areas with ≥ 50 m of impermeable cover: 5-10m radius	Low permeability terrain: 100-200 m High permeability terrain: 50-100 m Areas with ≥ 50 m of impermeable cover: 20m	
Slovakia	Wellhead = 10-50 m radius	Inner protection zone ≥ 50 days minimum 50 m	Outer protection zone	
Slovenia	4 protection areas – generally defined for 'larger' sources			
Spain	Immediate protection zone – 24 hours	Proposed: Second protection area 50 days (porous) 100 days (karstic)	Outer protection zone – 1 year	Remote protection
Switzerland	S1 = well/spring head and sink-holes (5-20 m)	S2 = 10 days, > 100 m – in karst include the contributing region not covered by sufficient impervious material – if necessary, S3 substitute for S2	S3 = at least as large as S2	Currently under discussion: Delimination of the sector from which ca. 90% of abstraction comes from
Turkey	I = -50 m -100 (karst)	II = 50 -250 m 100-500 m (karst)	III = Recharge area	Catchment area

Table 8.2. Groundwater resource protection in different countries of Europe (EC-Cost 1995).

Austria	Development of water management skeleton plan to coordinate all interests concerning water in any area.
Belgium	Regional government responsibility.
Croatia	Water protection regulations 1976; Chapter 4 – Karst Water Protection: – detailed regulation on a regional basis – groundwater protection an obligatory part of land-use planing
England and Wales	A = Major Aquifer: defined according to aquifer vulnerability B = Minor Aquifer: based on consideration of geological and soil parameter C = Non-Aquifer
France	Water planning and management schemes at the administrative basin scale (6 such main river basins in France) – then developed at sub-basin including aquifer or aquifer-system scale. 1992 law allows for delineation of a water planning management scheme for karstic aquifers with consequent severe controls on activities
Germany	Resource protection measures on a federal basis by a general law on the protection of waters
Hungary	Restrictions on contaminating sources. Yearly variable limitation of water exploitation
Ireland	Major aquifers (zone 2) Minor aquifers (zone 3) list of controls/prohibitions Poor aquifers (zone 4)
Italy	Each regional government responsible
Malta	Environmental impact assessment required for development proposals and if proposed development is in a source protection zone an assessment of impact on fresh water resource is required
Portugal	Resource protection zones based on hydrogeological investigations
Slovakia	Resource protection measures defined according to hydrogeological conditions
Slovenia	Water act 1974, 81, 91, Environmental protection law 1993 Land regulation act 1984 Cave protection act (in preparation1993)
Spain	Water law (1985, 1986): – definition of overexploited aquifer – protection against saline intrusion – prohibited activities and criteria for environmental impact assessments for all activities posing a risk to groundwater – waste disposal – wasteland reuse, spills and specific dangerous substances

Switzerland	A = Aquifers exploited for drinking water,or suitable for future exploitation
	 B = Less suitable aquifers
	     Areas adjacent to exploited or exploitable aquifers
	     Aquifers particularly well protected by overlying strata
	 C = All regions neither in A or B nor in source protection zones
	     S1, S2, S3, nor in 'groundwater protection areas'
	     Water Protection Law (1991):
	 The law requires a qualitative and quantitative protection against aquifer overexploitation and impairment

Turkey	General law on the protection of waters

main aspects, which have to be considered for the delineation of the different zones:
1. High velocity of karst water flow;
2. Special vulnerability of certain parts of the catchment.

Generally the catchment is subdivided into the spring/wellhead protection zone as well as into the inner and outer protection areas. The closer to the capture the more comprehensive are the restrictions for existing and new potentially polluting activities, which encompass the pollutant loading and the pollution control elements of risk. Some examples from different European countries are given in the following:

– In Austria guidelines for the use and protection of karst water resources were published in 1984 (ÖWWV 1984). The regulation differentiates between 'protection' and 'prevention zones'. The first includes the direct evaluated catchment, for which the regulations are compulsory, the second may include the whole karst area but the regulations are initially only recommendations for land use and land use planning. A new code of practice is given in ÖVGW (1995).

– In Belgium all preferential points of infiltration like swallow holes and other karst depressions, which have connections with the source or well, belong to zone II with highest protection priority, except for the source or well head area.

– A special protection law is planned for Croatia, a country, more than 50% of whose area belongs to the famous Dinaric Karst. To adapt the protection measures in the most effective way a concept with several strongly differentiated zones is planned; for example for the water captures of the city of Rijeka 6 zones are distinguished (Biondic & Dukaric 1993, EC-Cost 1995). The new approach bestows special care on the mountainous hinterland as the supposed recharge area. It is included as a zone of water management reserve, where in case of a change in landuse detailed hydrogeological reconnaissance studies are obligatory.

– A recently introduced French Law (3rd January 1992) includes specific consideration of karstic aquifers by allowing for a Water Planning and Management Scheme (Schema d'Ameagment et de Gestion des Eaux, SAGE). Within this scheme specific and potentially severe rules can be decreed for resource and source protection, even if protection zones have not yet been defined.

– In Germany a revised version of the old regulations, which defined the delineation of protection zone I (surrounding of well head), protection zone II (50 day travel time in the saturated zone), protection zone III (rest of the catchment area), were published recently (DVGW 1995). More attention is paid now to the behaviour of the different aquifer types. For karst and those fissured aquifers reacting hydrodynamically similarly to karst, special criteria for the delineation of zones II and zone III are given. They restrict on one hand the further subdivision of zone III, if there exists neither an impermeable cover system of a more than an 8 m thick clayey-silty layer nor a hydraulically completely separated perched aquifer system. Due to the fact that the 50 day line will lead to extremely extended zone II's in karst areas with restrictions that are practically not manageable (e.g. small cities, main traffic routes within the zone II), exceptions from the 50 day criteria are tolerated, but all relevant areas of high vulnerability have to belong to the zone II as it is shown in Figure 8.3. (GLA 1991, Hötzl 1996). Examples are: all slopes inclined towards the spring, all karst depressions if they drain extensive areas into groundwater, the surroundings of river seepage, deeply incised dry valleys and outcropping fault zones. Beside these

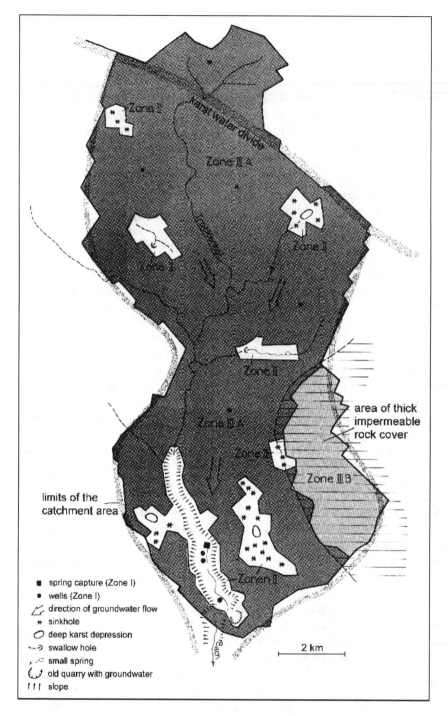

Figure 8.3. Example of protection zonation for a karst spring and well group on the basis of vulnerability mapping (GLA 1991). Subdivision of the protection zones I, II, IIIA and IIIB according to the German national regulations for drinking water supply (see Table 8.1).

Table 8.3. Assignment of levels of response (R) to the protection zones.

Vulnerability rating	Source protection			Resource protection*				
	Source site	Inner	Outer	Major aquifer		Minor aquifer	Poor aquifer	
				Karst	F and G**			
Extreme	R4	R4	R4	R4	R4	$R3^m$	$R2^m$	↓
High	R4	R4	R4	R4	$R3^m$	$R3^n$	$R2^b$	↓
Moderate	R4	R4	$R3^m$	$R3^m$	$R3^n$	$R2^b$	$R2^m$	↓
Low	R4	$R3^m$	$R3^m$	$R3^{mn}$	$R2^m$	$R2^a$	R1	↓
	→	→	→	→	→	→	→	

With R1 = acceptable; R2a, b, c, ... = acceptable in principle, subject to conditions in note a, b, c, etc. (the number and content of the notes will vary for each zone/sub-zone and for each activity); R3m, n, o, ... = not acceptable in principle; some exceptions may be allowed subject to the conditions in note m, n, o, etc.; R4 = not acceptable (EC-Cost 1995).

areas zone II has to include the whole upstream catchment to a distance from the spring of at least 1 km.
– In Ireland source protection areas are defined by an arbitrary fixed radius of 1 km, which is subdivided in three subzones, 0 to 10 m, 10 to 300 m and 300 to 1000 m. In a recent approach (Daly 1995) emphasises for the inner protection area the prevention of microbial pollution, for which a 100-day time of travel is proposed. It is pointed out that in karst areas where conduit flow is dominant the time of travel approach is not applicable, therefore in connection with the arbitrary fixed radius an increased distance is proposed for the sources. Karst aquifers are regarded as regionally important aquifers. In order to classify the risk level, vulnerability categories are used. They form the criteria for the definition of the levels of response to the risk of a particular potentially polluting activity (Table 8.3). This provides an improved classification for the protection of karst areas. The proposed Irish protection scheme was taken over also in the recommendations of the EC-COST Action 65 (1995).
– The Swiss water protection law dating from 1991 gives more emphasis to the protection of natural waters both from the quantitative as well as qualitative aspect (Hartmann & Michel 1992). Source protection is subdivided into protection zones and the criteria given in Table 8.4 are used for groundwater in general. For karst aquifers supplementary regulations are implemented. The zone S1 can include non-adjacent infiltration areas that are in a direct hydraulic connection with the tapped water. Zone S2 includes the karst area that is not protected by a sufficient impervious cover. If the zone is too large to allow for the execution of the relevant protective measures, the substitution of S3 for S2 is accepted. However, it is proposed that areas that present an enhanced danger to groundwater, e.g. in the vicinity of karst forms showing possible hydraulic connections, are restricted to S2 even if not directly adjacent.

8.2.4 *Karst water protection elsewhere in the world*

Elsewhere in the world and particularly in Africa and Asia, with the explosive rise of population, the growth of industry and/or intensive agriculture, pollution problems have become more and more widespread, restricting the direct use of surface and groundwater. Though the matter of environmental problems has been raised in these countries, detailed legal regulations for the protection of ground water have not, for the most part, come into force so far. Regarding the vulnerability of karst water; on one hand protection measures have been realised for special individual cases and also a growing degree of attention by scientists and water management authorities can be discerned. Examples are given for some such regions outside of Europe and North America.

China might well be the country with the largest use of karst ground water in the world. The carbonate rock area in China cover more than one third of its total territory and the karst groundwater comprises up about one quarter (800 km$^3 \cdot$a^{-1}) of the total groundwater resources (Yuan 1991). China enacted a law in 1984 'Preventing and Controlling Water Pollution' – a general measure; but Chapter 5, Item 32 of the law prohibits: *...any enterprises or institutions from using recharge wells, sinkholes, fissures or caves to discharge or pour, waste waters containing poisonous pollutants, polluted water containing disease, virus, or any other waste matters.* (Yuan Daoxian pers. comm.). Some examples of the application of protection zones in China were recently published by LiZhenshuan (1996) and Zhang (1996).

In parts of the world other than Europe and the USA comprehensive legislation aimed at protecting the environment is widespread, but karst-specific legislation or guidelines are much less common. New Zealand has a wide-ranging Resource Management Act which includes karsts but does not specifically identify them. However, the Department of Conservation has issued guidelines on the management of caves and karst resources (P. Williams, pers. comm.)

In Tasmania, Australia, the Forestry Practices Code by the Forestry Commission 1987 established specific guidelines for forestry operations in the karst regions of the island, differentiating between areas with very thin karstic soils and areas with thicker soils.

The pattern is similar elsewhere with broad measures to protect the environment being enacted at national or federal level with, in some instances, local legislation or guidelines formulated that are karst specific.

8.3   CONCLUSIONS

Gillieson (1996) remarks that 'Karst management must be holistic in its approach... Managers of karst areas should recognise that these landscapes are complex, three-dimensional natural systems... in karst, surface actions may be rapidly translated into impacts underground and elsewhere'

From the overview given in Section 8.2 it is apparent that an awareness of the distinctiveness of karst is developing. Often awareness is best developed at a local level in areas where the existence of karst and karst related phenomena are obvious and the need to manage it intelligently is manifest.

Karst phenomena such as certain caves or scenic areas are also afforded protection in many parts of the world, as sites of heritage or scientific importance rather than because they are an integral part of a karst system.

At national level, environmental guidelines are almost always couched in general terms and require to be interpreted in a karstic context at a more local level. Developments in the USA at State and district levels suggest that this process is underway. Pan-European projects such as the EC-Cost Action 65 and Action 620 (vulnerability mapping of karst areas) concerned with developing continent-wide recommendations on the protection of karst waters represent the opposite extreme but will ultimately have to be translated into karstic conditions at regional levels.

There are two aspects, one scientific and one educational that need to be further addressed:

1. The development of a more rigorous assessment procedure for individual carbonate rock terrains to better evaluate the degree of karstification and hence the vulnerability of the waters to particular impacts.

2. Increasing the awareness of the relevant personnel and institutions of the special characteristics of karst and the need for an informed approach to the management of karst areas.

CHAPTER 9

Conservation of karst terrains and karst waters: The future

DAVID DREW
Department of Geography, Trinity College, Dublin, Ireland

HEINZ HÖTZL
Department of Applied Geology, University of Karlsruhe, Germany

9.1 RISK ASSESSMENT IN KARST TERRAINS

Risk assessment provides a useful conceptual framework within which human activities may be evaluated in holistic terms of their impacts on karst systems. Using the definitions suggested by Warner (1992) impacts are modelled in terms of hazards; a hazard being described as: '*An event or continuing process, which if realised, will lead to circumstances having the potential to degrade, directly or indirectly, the quality of the environment*'.

A risk is the likelihood of a specified adverse consequence and a hazard presents a risk when it is likely to affect something of 'value' – karst water supplies or caves for example.

Expressed in these terms for a karst environment:

– *Hazard* includes the various human activities and combinations thereof described in Chapter 4-7 of this book. The intensity, extent and duration of an imposed stress need to be quantified relative to the system it impinges upon

– *Vulnerability* is the intrinsic property of the karst system and ultimately of karst water relative to the hazard(s). It is the main aspect of karst protection that has come to attract increasing attention in recent years. Semi-quantitative codification and measurement of the vulnerability of karst regions has been attempted by in the USA by Johnson & Quinlan (1995), by Hubbard (1993) and by Doerfliger et al. (1995) in Switzerland for example. Protection of karst water was the aim of these researches but they have the potential for a broader application in karsts.

– *Consequences* will depend on the 'value' of the karst resources in question and may be difficult to evaluate fully if complex chain reaction such as those described in Chapter 8 take place.

– *Risk* is: Probability of occurrence of adverse effects occurring × consequential damage.

– *Response* will be the engineering, political, legislative, management reaction that is appropriate to the karst situation

Table 9.1, based on Daly & Johnson (1996) and placed in the context of a karstic environment, summarises the procedure in relation to a hypothetical situation.

Table 9.1. Summary of risk assessment (adapted from Daly & Johnson 1996) as applied to a possible hazard in a karstic terrain).

Risk estimation

1	What can go wrong (Identifying the hazard)	e.g. Over abstraction from a karst aquifer
2	If something does go wrong how likely is it that there will be a significant negative impact? (Probability)	e.g. Depends on the amount of abstraction + nature of karst drainage system + stability of overlying rocks + nature of surface, etc.
3	What would happen if it did go wrong? (Consequences)	e.g. Possible loss of water supplies elsewhere + drying of wells and springs + diversion of karst water courses + surface subsidence + ecological changes (e.g. Figs 8.1 and 8.2, and flow diagrams in Chapter 1)

Risk evaluation

Is the risk acceptable and/or reducible?	$= (2) \times (3)$ above

Risk management

Risk management strategy (Informed decision making)	e.g. Modify abstractions; no houses near dolines

Determination of the impact of composite risks and quantification of the complete karst system are necessary for risk management to become a useful aid to the management of karst resources and this will be a difficult task. The development of a sufficiently flexible yet systematic framework for use karst terrains is as yet in its infancy.

9.2 RESTORATION OF KARSTIC GEO-ECOSYSTEMS

As with any specialised ecosystem adapted to specific environmental conditions, karstic ecosystems are very sensitive to any even marginal changes. The examples in the different chapters of the book have drawn attention to human activities, that intervene in this labile balance of the natural system and thereby cause problems and even damage to the karst groundwater. The often profound impacts which lead to a change in the available quantity and quality of karst groundwater are obvious and require restorative and precautionary measures in order to conserve the whole karstic ecosystem.

The discussion of the problems in the earlier chapters has already touched on the complex and mutual relationships within the karstic ecosystem. The direct connections between certain activities and the resulting changes in the karst water system were emphasised. It is necessary also to focus on the complex interrelationships within the whole system as described in Chapter 8, which mean that certain impacts, though originally not directly affecting karst water, might lead via a chain of subsequent processes to the eventual involvement and degradation of karst water. Therefore it is advantageous to include in considerations of restoration and conservation measures not only the karst groundwater, but the whole karstic ecosystem, including the indirectly connected non-karstified surroundings. The main goals of the restoration should be:

1. Removal of the impact source,
2. Restoration of the natural conditions within the karst system, and
3. Safeguarding the karst system for the future by additional precautionary measures.

With regard to impacts on the sensitive karst system the question arises as to whether restoration is possible at all after major impacts with significant damage. Fortunately some of the main disadvantages of a healthy natural karst groundwater system, the low storage and retention capacity together with the predominant conduit flow, are also the main factors in the rapid rehabilitation of the disturbed and damaged karst system. However, this positive diagnosis should not conceal the fact that it is necessary to distinguish between damages which are:

– Reparable just by removing the source of damage;
– Reparable only by comprehensive remediation measures within the karst system;
– Irreparable or reparable only over long periods of time.

The removal of the source of impacts is one of the preconditions of sustainable remediation measures. Especially in case of contaminants such as seeping waste water or landfill leachates the recovery of a karst water system and thus of the whole connected ecosystem can take place in a short time. The great amount of attenuation by karst water flow as well as the small amount of retardation and consequent small desorption later, can lead to self-purification within a short time. Examples were given in the previous chapters like that from the famous Skocjanske Jame (Slovenia), where the massive pollution of the Reka River disappeared with the abandonment of the industrial complex responsible (Section 5.7.4). Similar quick restoration may result from stopping over-exploitation of a karst aquifer or from the sealing of a traffic tunnel running through and partially draining a karst system (Section 5.7.7), so that the depleted karst water level can recover in a short time. But even if the original level can be restored again, this does not always mean that the original amount of water is available at each spring or well.

More difficulties are experienced in the remediation of a karst ecosystem when the contaminants have already penetrated into the karst system, like a oil spill, and form there a new source of contamination. In this case remediation has to be carried out within the system, which is generally more difficult. Especially in the case of a site with a thick vadose zone for instance, the oil can be distributed over large parts of the vertical section within narrow fissures and karstified joints. Under such conditions removal of the oil spill is nearly impossible. Extraction from remediation wells is not always very effective due to strong dispersion and/or the strong heterogeneity of the karst system, so that frequently the charcoal treatment at the natural outlets is the only realistic remediation work However, this treatment can last for several years until the concentrations reach the standard for drinking water.

Reparable impacts which also need comprehensive restoration measures are frequently connected with quarrying, mining and some kinds of agricultural activities. In many cases, as described in the previous Chapters they do not impact the karst water directly but via a chain reaction involving several parts of the karstic ecosystem the karst groundwater is degraded. Remediation has to restore the ecosystem as far as possible to its original natural condition, which can once again take a long time.

Examples of nearly irrevocable impacts or impacts which are reparable only within extremely long-time periods are urbanisation in general, deforestation with subsequent intense soil erosion and hazardous waste sites located in karst areas.

These restrictions on the direct restoration of karstic ecosystems impacted by a hazard require additional precautions to prevent in advance the contamination of the karst water. The karstic ecosystems including their groundwater are much too important to leave exposed to human activities which are associated with potential risks. Precautionary regulations are required, which together with responsible land-use management can be a guarantee of the conservation of our karst terrains as well as of the quantity and quality of the karst groundwater.

9.3   FUTURE ASPECTS OF KARST WATER MANAGEMENT

Uncertainty is the only certainty in terms of the future relationship between humans and the global environment and this is true irrespective of locale, of geology or of culture. There is little value in speculating idly about possible future scenarios in terms of human impact on the karst environment, but three seemingly probable future developments, each due to human actions and global in extent, are worth consideration in relation to their effects on karstic terrains and resources.

Global modification of climate (Section 5.7.12) may have a disproportionate impact upon karst terrains. If the cause of climatic change is anthropogenic then the likeliest single factor is increased atmospheric carbon dioxide levels, perhaps a doubling in the near future. This will increase the carbonic acid content of precipitation proportionately and so accelerate the 'natural' process of dissolution of limestone world-wide. However, this is a change, the consequences of which are unlikely to become manifest on a human time scale.

However, changes in precipitation, in temperature and in the resulting effective rainfall, may have significant, short term impacts on karst resources. In mature karsts the turnover time for recharge water is short, with only a small degree of storage within the carbonate aquifer. A diminution in precipitation may translate quickly into a lessening of available water resources due to the lack of temporal buffering in the karst groundwater system. Conversely, increased precipitation may cause flooding in either mature karst (rapid flow-through times) or immature karsts (limited storage capacity).

The complexity and sensitivity to stress of the karst system (Chapter 8) is such that it is necessary to look beyond changes in recharge amounts and to consider the effects of altered climate on ecology, on soils and on sediment transport systems in order to predict changes in karst groundwater. Many karst areas are balanced precariously between being fertile regions and functional deserts and climatic changes may unbalance some such areas. In a worst case scenario the effects of changed climates on karst systems could overwhelm any previous direct human impacts.

Possible changes in the workings of karst systems as a result of climatic changes are an indirect consequence of human actions and would operate as changes in the intensity and location of natural processes in karst terrains.

However, more direct human stresses on karst systems are probable and will simply be amplifications of the present day anthropogenic impacts that have been documented in this book.

For example, the great acceleration in the process of urbanisation and the development of mega-cities with all the associated negative impacts on sensitive ecosystems may be another virtually inevitable development which may damage many of the world's karst systems.

Many of the world's most densely populated and also less economically developed regions include extensive karst areas – large parts of southeast Asia, and meso-America for example. Direct pressures on karst water resources, for supply and for waste disposal may become the critical stress, requiring a very rigorous management system to sustain resources. This will be true whether populations continue to increase or whether increased prosperity curbs population growth but increases individual demands on water resources.

A related phenomenon is the world-wide increase in tourism. Tourism is an industry that seems certain to expand vastly in the future. The problems attendant on tourism have been discussed elsewhere in this book, but it is worthwhile repeating that many of today's remaining remote areas are karstic and are likely to be tomorrow's prime tourist venues. Indonesia, Vietnam, Laos, meso-America are examples.

The remediation of environmental contamination as well as the protection of our ecosystems are amongst the most important tasks for the preservation of our living conditions. Within this context there is a special and urgent requirement to conserve the functioning and efficiency of sensitive karstic ecosystems. With regard to the varied human impacts on karst groundwater described in this book the request for remediation and restoration is not sufficient by itself. Even with high economic and technological inputs some of the damage caused by the human activities is nearly irreparable at least within the short to medium term. The ongoing damage means that we must turn away from the illusory concepts of effective after-care to encompass the principles of the precautionary protection of karst groundwater and karstic ecosystems.

This implies the need for some legislative regulations as well as a sustainable planning and a proper management of our natural resources. In developing and using these resources, priority has to be given to satisfying basic needs and the safeguarding of ecosystems (Agenda 21, Rio Conference 1992). The acceptance and observance of such requirements by the public requires a certain degree of understanding of the natural processes and inter-relationships. Though karst is a widespread feature on our earth there is still great public ignorance as to its nature. It is one of the principal tasks of the individual hydrogeologists as well as of relevant organisations and societies to arouse public awareness of karst groundwater and karstic ecosystems.

References

Adamczyk, A.F., Motyka, J., Wilk, Z. & Witczak, S. 1988. Mining drainage of a karstic aquifer and the related problems. In *Karst Hydrogeology and Karst Environment Protection. Proc 21st IAH Congress Guilin* 1097-1104.

Adams, C.D. & Thurman, E.M. 1991. Formation and transport of deethylatrazine in the soil and vadose zone. *Journal of Environmental Quality* 20: 540-547.

Albritton, D. 1987. *Atmospheric Processes*. Chapter 4 in NAPAP (The National Acid Precipitation Assessment Program) Interim Assessment, Volume III: Atmospheric Processes and Deposition, pp. 4-1 to 4-81.

Aldwell, C.R., Burdon, D.J. & Sherwood, M. 1983. Impact of agriculture on groundwater in Ireland. *Environmental Geology* 5: 39-48.

Aldwell, C.R., Thorn, R.H. & Daly, D. 1988. Point source pollution in karst areas in Ireland. In Yuan, D. (ed.), *Karst Hydrogeology and Karst Environment* Protection *(Proceedings 21st IAH Congress, Guilin, China)* 2: 1046-105. Geological Publishing House, Beijing.

Alexander, M. 1985. Biodegradation of organic chemicals. *Environ. Sci.Technol.* 18/2: 106-111. Washington.

Alföldi, L. 1984. Connection between thermal springs in Budapest and mining activity. In Burger, A. & Dubertret, L. (eds), *Hydrology of Karstic Terrains: Case Studies, IAH* 1: 222-224.

Allison, G.B. & Hughes, M.W. 1972. Comparison of recharge to groundwater under pasture and forest using environmental tritium. *J. Hydrol.* 17: 81-95.

Anderle, N. 1950. Zur Schichtfolge und Tektonik des Dobratsch und seine Beziehung zur alpin-dinarischen Grenzzone. *Jb. Geol. B.-A.*, 94. Wien.

Anderson, M.G. & Burt, T.P. (eds) 1988. *Hydrological forecasting*, 729 p. Willy Publ.

Andrajchouk, V. & Klimchouk, A. 1993. Environmental change and human impact on karst in the western Ukraine. In Williams, P.W. (ed.), *Karst Terrains, Environmental Changes, Human Impact, CATENA supplement* 25: 147-160.

Andreo, B. & Carrasco, F. 1993a. Estudio hidrogeológico del entorno de la Cueva de Nerja. In Carrasco, F. (ed.), *Geología de la Cueva de Nerja.Trabajos sobre la Cueva de Nerja* 3: 163-187.

Andreo, B. & Carrasco, F. 1993b. Estudio geoquímico de las aguas de infiltración de la Cueva de Nerja. In Carrasco, F. (ed.), *Geología de la Cueva de Nerja. Trabajos sobre la Cueva de Nerja* 3: 299-328.

Andreo, B., Carrasco, F. & Sanz de Galdeano, C. 1993. Estudio geológico del entorno de la Cueva de Nerja. In Carrasco, F. (ed.), *Geología de la Cueva de Nerja. Trabajos sobre la Cueva de Nerja* 3: 25-50.

Appelo, C.A.J. & Postma, D. 1993. *Geochemistry, groundwater and pollution,* 536 p. Rotterdam: Balkema.

ASTM, American Society for Testing and Materials 1995. Standard Guide for Design of Ground-Water Monitoring Systems in Karst and Fractured-Rock Aquifers. ASTM Designation, D 5717-95, Annual Bool of ASTM.

Atkinson, T.C. 1977a. Carbon dioxide in the atmosphere of the unsaturated zone: An important control of groundwater hardness in limestones. *J. Hydrol.* 35: 111-123.

Atkinson, T.C. 1977b. Diffuse Flow and Conduit Flow in Limestone Terrain in the Mendip Hills, Somerset (Great Britain). *J.Hydrol.* 35: 93-110.

Atkinson, T.C., Bradshaw, R. & Smith, D.I. 1973. *Quarrying in Somerset.* Suppl 1, Hydrology and Rock Stability. Mendip Hills. A review of existing knowledge, Somerset County Council.

Audra, Ph. 1994. *Caving below the vineyards. The Entre-deux-Mers karst, Gironde, France. International Caver* 11: 3-10. Aven publications, Swindon.

Avidad, J. & García-Dueñas, V. 1980. Mapa geológico de España a escala 1:50.000, plan Magna, n° 1055 (Motril), I.G.M.E., 36 p.

Back, W. 1992. Coastal karst formed by ground water discharge, Yucatan, Mexico. In Back, W. & Paloc, H. (eds), *Hydrogeology of selected karst regions. IAH, Intern. Contr. to Hydrogeology* 13: 461-466. Heise, Hannover.

Back, W., Hanshaw, B.B., Herman, J.S. & Van Driel, J.N. 1986. Differential dissolution of a Pleistocene reef in the ground-water mixing zone in coastal Yucatan, Mexico. *Geology* 14(2): 192-197.

Back, W., Hanshaw, B.B. & Van Driel, J.N. 1984. Role of ground water in shaping the Eastern coastline of the Yucatan Peninsula, Mexico. In La Fleur, R.G. (ed.), *Groundwater as a geomorphic agent*: 281-293. London: Allen & Unwin, Inc..

Baehr, A.L., Hoag, G.E. & Marley, M.C. 1989. Removing volatile contaminants from the unsaturated zone by inducing advective airphase transport. *J. Contaminant Hydrology* 4: 1-26.

Bakalowicz, M. 1979. Contribution de la géochimie des eaux à la connaissance de l'aquifère karstique et de la karstification. Thèse Doct. Sci. Nat., Univ. P. Et M. Curie, Paris. 269 pp.

Bakalowicz, M. 1984. Water chemistry of some karst environments in Norway. *Norks. Geogr. Tidsskr.* 38: 209-214.

Bakalowicz, M., Plagnes, V. & Richard, J. 1996. Land management and sustainable development of karst groundwater the Larzac Plateau (France) as an example. In Rózkjowski, A., Kowalczyk, A., Motyka, J. & Rubin, K. (eds), *Karst-Fractured aquifers – Vulnerability and Sustainability*: 299-308. Univ. Slaskiego, Katavice.

Barany-Kevei, I. (ed.) 1995. Environmental effects on karst terrains (homage to Lasso Jacks). *Acta Geographica Szegediensis* 34: 213p. University of Szegediensis.

Barbee, G.C. 1994. Fate of chlorinated oliphatichydrocarbons in the vadose zone and ground water. *Groundwater Monitoring and Remediation, Nat. Groundwater Ass.* 9(1): 129-140.

Barchet, W.R. 1987. *Acidic Deposition and its Gaseous Precursors.* Chapter 5 in NAPAP (The National Acid Precipitation Assessment Program) Interim Assessment, Volume III: Atmospheric Processes and Deposition, 5-1 to 5-116.

Barker, J.F. & Nicholson, R.V. 1993. *Subsurface assessment for contaminated sites.* Handbook for the Canadian Council of Ministers of the Environment.

Barker, J.F., Cherry, J.A., Reindard, M., Pankow, J.F. & Zapico, M.M. 1989. Final report: The occurrence and mobility of hazardous organic chemicals in groundwater at several Ontario landfills: Research Advisory Committee Project No. 118 PL for Environment Ontario, 148 p.

Barker, J.F., Patrick G.C. & Major, D. 1987. Natural attenuation of aromatic hydrocarbons in a shallow sand aquifer. *Ground Water Monitoring Review* 7(1): 64-71.

Barner, W.L. 1997. Comparison of stormwater management techniques in a karst terrane in Springfield. In Beck & Stephenson (eds), *The Engineering Geology and Hydrogeology of Karst Terranes*: 253-258. Rotterdam: Balkema.

Barner, W. & Uhlmann, K. 1995. Contaminant transport mechanisms in karst terrains and implication on remediation. In Beck, B. (ed.), *Engineering and environmental problems in karst terrane. Proceedings of the 5th multidisciplinary conference on sinkholes and the environmental impacts of karsts:* 207-212. Rotterdam: Balkema.

Barrington, N. & Stanton, W.I. 1977. *Mendip, The Complete Caves and a View of the Hills.* 3rd. edition. Barton productions/Cheddar Valley Press, Cheddar, Somerset.

Baumgardner, R.W., Hoadley, A.D. & Goldstein, A.G. 1982. Formation of the Wink Sink, a Salt Dissolution and Collapse Feature, Winkler County, Texas. Texas Bur. Econ. Geology, Rept. Investig. 114.

Beck, B.F. (ed.) 1984. Sinkholes: their geology, engineering and environmental impact. *Proceedings of the 1st multidisciplinary conference on sinkholes,* 429p. Rotterdam: Balkema.

Beck, B.F. (ed.) 1989. *Proceedings of the 3rd multidisciplinary conference on sinkholes and the environmental impacts of karsts,* 384p. Rotterdam: Balkema.

Beck, B.F. (ed.) 1993. Applied karst geology. *Proceedings of the 4th multidisciplinary conference on sinkholes and the environmental impacts of karsts,* 295p. Rotterdam: Balkema.

Beck, B.F. (ed.) 1995. Karst Geohazards. Engineering and environmental problems in karst terrane. *Proceedings of the 5th multidisciplinary conference on sinkholes and the environmental impacts of karsts,* 581p. Rotterdam: Balkema.

Beck, B.F. & Stephenson, J.B. (eds) 1997. The engineering geology and hydrogeology of karst terranes. *Proceedings of the 6th multidisciplinary conference on sinkholes and the environmental impacts of karsts,* 516 p. Rotterdam: Balkema.

Beck B.F. & Wilson W.L. (eds) 1987. Karst Hydrogeology: Engineering and environmental opplications. *Proceedings of the 2nd multidisciplinary conference on sinkholes and the environmental impacts of karsts,* 467p. Rotterdam: Balkema.

Beck B.F. & Wilson W.L. (eds) 1988. Proceedings of the 2nd multidisciplinary conference on sinkholes and the environmental impacts of karsts. *Environmental Geology* 12(2).

Beck, B.F., Stephenson, J.B., Wanfang, Z., Smoot, J.L. & Turpin, A.M. 1996. Design and evaluation of a cost-effect method to improve the water quality of highway runoff prior to discharge into sinkholes. In *Proceedings of the 1996 Florida Environmental Expo, Tampa, Florida,* October 1-3: 155-164.

Behrens, H., Benischke, R., Bricelj, M., Harum, T., Käss, W., Kosi, G., Leditzky H.P., Leibundgut, Ch., Maloszewski, P., Maurin, V., Rajner, V., Rank, D., Reichert, B., Stadler, H., Stichler, W., Trimborn, W., Zojer, H. & Zupan, M. 1992. Investigations with natural and artificial tracers in the karst aquifer of the Lurbach system, Austria. *Steir. Beitr. z. Hydrogeologie* 43: 3-283. Graz.

Benavente, J. & Almécija, C. 1993. Estudio geomorfológico del entorno de la Cueva de Nerja. In Carrasco, F. (ed.), *Geología de la Cueva de Nerja Trabajos sobre la Cueva de Nerja* 3: 119-158.

Benischke, R. 1993. Zur Hydrogeologie des Höllengebirges. *ÖGG, Exkursionsführer* 14(2): 32-36. Wien

Benischke, R., Zojer, H., Fritz, P., Maloszewski, P. & Stichler, W. 1988. Environmental and artificial tracer studies in an Alpine karst massif (Austria). *Proc. IAH 21st Congress* 21(2): 938-947. Guilin.

Beogan, E. 1938. Il Timavo, Studio sullídrografia carsica subaerea e sotterranea. *Memorie dellÍstituto Italiano di Speleologia* II: 1-251. Trieste

Beriswill, J.A., Humphries, R.W., McClean, A.T. & Kath, R.L. 1995. Karst foundation grouting and seepage control at Haig Mill Dam. In Beck, B.F. (ed.), *Proceedings of the 5th multidisciplinary conference on sinkholes*: 363-370. Rotterdam: Balkema.

Berryhill, W.S. Jr. 1989. The impact of agricultural practices on water quality in karst regions. In Beck, B.F. (ed.), *Engineering and environmental impacts of sinkholes and karst; Proceedings of the third multidisciplinary conference*: 159-164. Rotterdam: Balkema.

Bertolani, M., Cigna, A., Maccio, S., Morbidelli, L. & Sighnolfi, G. 1991. The karst system 'Grotta Grande del Vento-Grotta del Fuime' and the conservation of its environment. *Proceedings of the International Conference on environmental changes in karst areas, Univ. Padova* 13: 289-298.

Biondic, B. 1985. Dependence of karst waterflow of Croatian Littoral upon its Geologic Structure. *V intern. Symp. on Groundwater, IAH, Taormina*

Biondic, B. & Dukaric, F. 1993. Water resources in the region of Rijeka. *Hrvatska vode* 1(3): 185-190. Zagreb.

Biondic, B. & Goatti V. 1986. Protection of Groundwater in Karst Areas of Croatian Littoral. *19th Congress IAH, Karlove Vary.*

Bock, P., Hötzl, H. & Nahold, M.E. 1990. Untergrundsanierung mittels Bodenluftabsaugung und In-Situ Strippen. *Schr. Angewandte Geologie* 9: 408 p. Universität Karlsruhe.

Böcker, T. 1984. Connection between lake spring of Heziv and mining activity. In Burger, A. & Dubertret, L. (eds), *Hydrology of Karstic Terrains: Case Studies, 1, International Association of Hydrogeologists:* 225-228.

Böcker, T. & Hegyi-Hovanyi, K. 1983. Multi-scale mathematical modelling of carbonate-type aquifers for the planning of dewatering systems in bauxite mining. In *Proc. 5th Int. Cong. of Int. Comm. for the Study of Bauxites & Aluminium, Zagreb:* 233-241.

Bodhankar, N. & Chatterjee, B. 1993. Pollution of limestone aquifer due to urban waste disposal around Raipur, Madhya Pradesh, India. *Env. Geol.* 23: 209-213.

Boegan, E. 1938. Il Timavo. Studio sull'idrografia carsica subaerea e sotterranea. *Memoria dell'Istituto Italiano di Speleologia* 2: 5-251. Trieste.

Bögli, A. 1980. *Karst hydrology and physical speleology:* 284 p. Springer, Berlin.

Bögli, A. & Harum, T. 1981. Hydrogeologische Untersuchungen im Karst des hinteren Muotatales (Schweiz). *Steir.Beitr.z.Hydrogeologie* 33: 125-264.

Böhler, U., Brauns, J., Hötzl, H. & Nahold, M. 1990. Drucklufteinblasung und Bodenluftabsaugung als kombiniertes Verfahren zur Sanierung kontaminierter Grundwasser Beobachtungen im Locker- und Felsgesteinen. In Arendt, F., Hinsenveld, M. & Brink, van den W.J. (eds), *Altlastensanierung '90, 3. Int. KfK-TNO-Kongress über Altlastensanierung, Karlsruhe, BRD* 2: 1157-1163. Kluwer Acad.Publ., Dordrecht, Boston, London.

Bolner, K. & Tardy, J. 1988a. Bacteriological and chemical investigations of dripping waters in the caves of *Budapest. Proc. Int. Symp. Physical, Chemical, and Hydrological Research on Karst, Kosice, Czech., 1988:* 102-108.

Bolner, K. & Tardy, J. 1988b. Chemical and bacteriological test series of stalactite waters in Budapest caves. *Proc. Int. Symp. Physical, Chemical, and Hydrological Research on Karst, Kosice, Czech., 1988:* 109-111.

Bolner, K., Tardy, J. & Nemedi, L. 1989. Evaluation of the environmental impacts in Budapest's caves on the basis of the study of the quality of dripping waters. *Proc. 10th. Int. Congr. Speleology (Budapest) 1989* II: 634-639.

Bonacci, O. 1987. *Karst Hydrology,* 184 p. Springer, Berlin.

Bonacci, O. 1985. Flooding of the poljes in karst. In *Proceedings 2nd International Conference on the Hydraulics of Floods and Flood Control, Cambridge, 1985:* 119-136. BHRA, The Fluid Engineering Centre, Cranfield,.

Bonaparte, R. & Berg, R.R. 1987. The use of geosynthetics to support roadways over sinkhole prone areas. In Beck B.F. & Wilson W.L. (eds), *Karst Hydrogeology: Engineering and environmental opplications. Proceedings of the 2nd multidisciplinary conference on sinkholes and the environmental impacts of karsts:* 437-446. Rotterdam: Balkema.

Boni, C., Bono, P. & Capelli, G. 1986. Schema idrogeologico dell'Italia centrale. *Mem. Soc. Geol. It., Rome* 35(1986): 991-1012 and 2 maps.

Boni, C., Bono, P., Calderoni, G., Lombardi, S. & Turi, B. 1980. Indagine idrogeologica e geochimica sui rapporti tra ciclo carsico e circuito idrotermale nella Pianura Pontina (Lazio meridionale). *Geol. Appl. Idrogeol.* 15: 203-247. Bari.

Boni, C., Pettita M., Preziosi, E. & Sereni M. 1993. Genesi e regime di portata delle acque continentali del Lazio, C.N.R., Roma.

Bono, P., Malatesta, A. & Zarlenga, F. 1993. Itinerario n. 3. In Cosentino, D., Parotto, M. & Praturlon, A. (eds), *Guide geologiche regionali Lazio. Soc. Geol. It.:* 117-130. BE-MA Milan.

Bosak, P. & Koroe, I. 1991. Mining versus karst hydrogeology: Vaclav graphite mine near Blizna, South Bohemia, Czechoslovakia. In *Proceedings of the International Conference on Environmental Changes in Karst Areas, Italy, September 15-27:* 177-183.

Boyries, P. 1987. Le karst de l'Entre-deux-Mers (Gironde). Université de Bordeaux III, Mémoire de Diplôme d'études approfondies, 168 p., Bordeaux.

Bredehoeft, I.D. 1982. *Ground-water model.* UNESCO Press 1: 235.

Bredenkamp, D.P., Van Rensburg, H.J. & Botha, L.J. 1994. Manual on quantitative estimation of groundwater recharge and aquifer storativity. Water Research Commission, Project K 8/142.

Bredenkamp, D.B. & Verhoef, L.H.W. 1981. Mangaanbesoedeling van die Vaalriver in die Klerksdorp, Orkney en Stilfonteingebied. Technical report GH 3184, Department of Water Affairs, Pretoria.

Brendel, K. 1976. Technogen beeinflußte natürliche Subrosions-Senkungen in der Mansfelder Mulde. *Z.geol.Wiss.* 4/8: 1115-1133.

Breznik, M. 1973. Nastanke zaslanjenih kraskih izvorov in njihova sanacija. *Geol. Razprave in Porocila* 16: 83-186. Ljubljana.

Brink, A. 1979. *Engineering geology of Southern Africa.* Building publications, 1, Silverton, Pretoria.

Brown, R.F. & Lambert, T.W. 1963. *Reconnaissance of ground water resources in the Mississippian Plateau region, Kentucky.* US Geol. Survey Water Supply Paper, 1603, 58p.

Budyko, M.I. 1988. The climate of the end of 20th century. *Meterologiya i Gidrologiya* 10: 5-25.

Burger, A. & Dubertret, L. (eds) 1975. *Hydrogéologie des terrains karstiques.* I.A.H. serie B, 3, 190 p., Paris.

Burger, A. & Dubertret, L. (eds) 1984. *Hydrogeology of karstic terrains: case studies.* IAH, Int. Contrib. of Hydrogeology 1. Heise, Hannover.

Burns, I.G. 1975. An equation to predict the leaching of surface-applied nitrate. *J. Agric. Sci.* 85: 443-454. Cambridge.

Caballero, E., Jiménez de Cisneros, C. & Reyes, E. 1996. A stable isotopc study of cave seepage waters. *Applied Geochesmistry* 11: 583-587.

Calvache, M.L. & Pulido-Bosch, A. 1994. Modelling the Effects of Salt-Water Intrusion Dynamics for a Coastal Karstified Block Connected to a Detrital Aquifer. *Ground Water* 32(5): 767-777.

Candela, L., Gómez, M.B., Puga, L., Rebollo, L. & Villarroya, F. 1991. *Aquifer overexploitation.* XXIII IAH Congress, 580 p., Canary Island.

Cañete, S. 1997. Concentraciones en Radón e intercambio de aire en la Cueva de Nerja. Univ. Málaga, Tesis de Licenciatura, 84 p., unpublished.

Carrasco, F. & Andreo, B. 1993. Características de las aguas de infiltración de la Cueva de Nerja (Málaga). *Geogaceta* 14: 9-12.

Carrasco, F., Andreo, B., Benavente, J. & Vadillo, I. 1995. Chemistry of the Nerja Cave system (Andalusia, Spain). *Cave and Karst Science* 21(2): 27-32.

Carrasco, F., Andreo, B., Liñán, C. & Vadillo, I. 1996. Consideraciones sobre el funcionamiento hidrogeológico del entorno de la Cueva de Nerja (provincia de Málaga). *Jornadas sobre Recursos Hídricos en regiones kársticas, Vitoria*: 249-263.

Celli, M. 1995. Aspects of human impact in the Monte Grappa Massif (Venetian Pre-Alps, Italy). *Acta Carsologica* 24: 147-155. Ljubljana.

Cerón, J.C. & Pulido-Bosch, A. 1993. Considérations géochimiques sur la contamination par le CO_2 des eaux thermominérales de l'aquifère surexploité de l'Alto Guadalentín (Murcie, Espagne). *C.R. Acad. Sci. Paris* 317(II): 1121-1127.

Chardon, M. 1995. L'Impact anthropique dans le Vercors. *Acta Carsologica* 24: 157-168. Ljubljana.

Chauve, P. & Mudry, J. 1980. *Premiere reconnaissance des somees de la Haute vallee du Doubs.* 26th Internat. Geol.Congress, Ann. Scientif. Univ. Besancon, Geol. 4(2): 81-90. Besancon.

Chiang, C.Y.J., Salanitro, J.P., Chai, E.Y., Colhart, J.D. & Klein, C.I. 1989. Aerobic biodegradation of benzene, toluene and xylene in an sandy aquifer – data analysis and computer modelling. *Ground Water* 27(6): 823-834.

Chieruzzi, G.D., Duck, J.J., Valesky, J.M. & Markwell, R. 1995. Diffuse flow and DNAPL recovery in the St. Genevieve and St. Louis limestone. In Beck, B. (ed.), *Engineering and environ-*

mental problems in karst terrane. Proceedings of the 5th multidisciplinary conference on sink-holes and the environmental impacts of karsts: 213-225. Rotterdam: Balkema.

Cigna, A. 1993. Environmental management of tourist caves: the examples of Grotta di Castellana and Grotta Grande del Vento, Italy. *Environmental Geol.* 21: 173-180.

Civita, M., Cucchi, F., Eusebio, A., Garavoglia S., Maranzana, F. & Vigna, B. 1995. The Timavo Hydrogeologic System: An Important Reservoir of supplementary water resources to be re-claimed and protected. *Acta carsologica* 24: 169-186. Ljubljana.

Coates, G.K., Lee, C.A., McClain, W.C. & Senseny, P.E. 1985. Closure and collapse of man-made cavities in salt. In Schreiber, B.C. & Harner, H.L. (eds), *Sixth International Symposium on Salt, The Salt Institute, Alexandria, Virginia* 2: 139-157.

Commonwealth Dept. of Arts, Sport, Environment and Territories (DASETT) & Tasmanian Department of Parks, Wildlife and Heritage (TasPWH) 1993. Rehabilitation Plan, The Lune River Quarry, Southern Tasmania, Joint report to World Heritage Planning Team, Department of Parks, Wildlife & Heritage, Tasmania, 26p.

Comune di Roma, I.W.S.A.-A.C.E.A. 1986. Il trionfo dell'acqua, 16 Congresso ed Esposizione Internazionale degli Acquedotti, Paleani Ed., Roma.

Congressional Research Service (CRS) 1984. Acid rain, a Survey of Data and Current Analyses. Report for the Subcommittee on Health and the Environment, Committee on Energy and Commerce, US House of Representatives, 98th Congress, 2d Session, US Government Printing Office, Washington, D.C., 954 p.

Coppa, G., Pediconi, L. & Bardi, G. 1986. *Water and aqueducts in Rome 1870-1984.* Sp. ed. 16th International Water Supply Congress and Exhibition, Ed. Quasar, Roma.

Cosentino, D., Parotto, M. & Praturlon, A. (eds) 1993. Guide geologiche regionali 'Lazio'. Soc. Geol. It., BE-MA Milan, 368 p. and 2 maps.

Cotecchia, V., Fidelibus, M.D. & Tulipano, L. 1986. Phenomenologies connected with the variation of equilibria between fresh and salt water in the coastal karst carbonate aquifer of the Salento Peninsula (Southern Italy). *Proc. 9th Salt Water Intrusion Meeting, Delft, The Netherlands*: 19-28.

Cotecchia, V., Fidelibus, M.D. & Tulipano, L. 1991. Phenomena related to the variation of equilibria between fresh and salt water in the coastal karst-carbonate aquifer of the Salento Peninsula. SWIM Papers IAH, 9-16, Hannover.

Cowling, E.B. 1982. Acid precipitation in historical perspective. *Environ. Sci. Technol.* 16(2): 110A-123A.

Coxon, C. & Thorn, R.H. 1989. Temporal variability of water quality and the implications for monitoring programmes in Irish limestone aquifers. In *Groundwater Management: Quantity and Quality (Proceedings of the Benidorm Symposium, October 1989).* IAHS Publication 188: 111-120.

Craig, D.H. (1987). Caves and other features of Permian karst in San Andreas dolomites, Yates field reservoir, West Texas. In James N.P. & Choquette P.W. (eds), *Paleokarst*: 342-363. New York: Springer-Verlag.

Crampon, N. & Bakalowicz, M. (eds) 1994. Basic and applied hydrogeological research in French karstic areas. In *EC-COST action 65 (1995). Hydrogeological aspects of groundwater protection in karstic areas. Final report. European Commission,* EUR 16547EN: 69-117. Luxembourg.

Crawford, N.C. 1982. Hydrogeologic problems resulting from developement upon karst terrain, Bowling Green, Kentucky. Guidebook for US Environmental Protection Agency Karst Hydrogeology Workshop, Nashville, Tennessee, 34 p.

Crawford, N.C. 1988. Karst ground water contamination from leaking underground storage tanks: Prevention, monitoring techniques, emergency response procedures and aquifer restoration. Environment problems in karst terranes and their solutions conference (2nd, Nashville, Tennessee). *Proceedings. National Water Well Association, Dublin, Ohio*: 213-226.

Crouch, D.P. 1993. *Water Management in Ancient Greek Cities.* 380 p., Oxford University Press, New York.

Daly, D. 1995. *Groundwater protection schemes in Ireland: A proposed approach.* Geol. Survey. of Ireland, Groundwater Section, 2/8/1995, 38 p. Dublin.

Daly, D. & Johnson, P. 1996. *Risk and risk management – a framework for groundwater management schemes.* Geological Survey of Ireland, Groundwater Newsletter 29: 3-4.

Daly, D., Thorn, R. & Henry, H. 1993. *Septic tank systems and groundwater in Ireland.* Geological Survey of Ireland Report Series, RS 93/1 (Groundwater).

Danchev, D., Velikov, B. & Damyanov, A. 1982. Effect of farming on underground water pollution in north-east Bulgaria. In *Impact of Agricultural Activities on Groundwater* (IAH International Symposium, Prague, Czechoslovakia), IAH Memoires, 26, 2, Novinar Publishing House, 71-83.

Dasinger, A. 1994. The challenges of characterizing and remediating DNAPLs. In *The Hazardous Waste Consultant.* Elsevier Science Inc., July/August, 29.

Davis, S.B. 1997. Interstate assessment of governmental regulations on landfills in karst areas. In Beck & Stephenson (eds), *The Engineering Geology and Hydrogeology of Karst Terranes*: 433-438. Rotterdam: Balkema.

Day, M.J. 1993. Human impacts on Caribbean and Central American karst. In Williams, P.W. (ed.), *Karst Terrains, Environmental Changes, Human Impact. CATENA supplement* 25: 109-125.

Day, M.J. & Rosen, C.J. 1989. Human impact on the Hummingbird Karst of Central Belize. In Gillieson, D.S. & Smith, D.I. (eds), *Resource Management in Limestone Landscapes: International Perspectives*: 201-214.

Deevey, E.S., Rice, D.S., Rice, P.M., Vaughan, H.H., Brenner, M. & Flannery, M.S. 1979. Mayan urbanism: impact on a tropical karst environment. *Science* 206: 298-306.

Denk, H.J. & Felsmann, M. 1987. Measurements and evaluation of a multicompartment model for estimating future activity profiles of radiocesium in undisturbed soil of pasture-land in North Rhine-Westphalia. In Feldt, W. (ed.), *The Radioecology of Natural and Artificial Radionuclides, Proc. XVth Reg. Conf. IRPA*: 182-187. Gotland, Sweden.

Devilbiss, T.S. 1988. An investigation into the movement of an agricultural pesticide within the groundwater system of a karst swallet. Eastern Kentucky University, M.S. Thesis, 99 p., Richmond Kentucky.

Devilbiss, T.S. 1995. A local government approach to mitigating impacts of karst. In Beck B.F. (ed.), *Karst Geohazards. Engineering and environmental problems in karst terrane. Proceedings of the 5th multidisciplinary conference on sinkholes and the environmental impacts of karsts*: 499-504. Rotterdam: Balkema.

Deyo, B.G., Robbings, G.A. & Binkhorst, G.K. 1993. Use of portable oxygen and carbon dioxide detectors to screen soil gas for subsurface gasoline contamination. *Ground Water* 31(4): 598-604.

DGOH, DGCA, ITGE 1994. *Libro blanco de las aguas subterráneas*, 135 p. MOPTMA, Madrid.

Dodge, E.D. 1984. Les characteristiques hydrogeologiques des aquiferes karstiques du Causse Comtal (Aveyron, France). *Hydrogeologie-Geologie de l'ingenier* 3: 241-252.

Doerfliger, N., Jeannin, P.Y. & Zwalen, F. 1995. Vulnerabilite des eaux dans les régions karstiques et délimitation des zones de protection. Rapport de la phase 2, Centre d'Hydrogeologie de l'Ùniversite de Neuchatel, 31.

Doerfliger, N. & Zwahlen, F. 1997. EPIK: A new method for outlining of protection areas in karstic environment. In Günay, G. & Johnson, A.I. (eds), *Karst water and environmental impacts. Proc. 5th International Symp. and Field Seminar, Antalya*: 117-124. Rotterdam: Balkema.

Domenico, P.A. & Schwartz, F.W. 1990. *Physical and chemical hydrogeology*, 824 p. New York: Wiley.

Dougherty, P.H. 1981. The impact of the agricultural land-use cycle on flood surges and runoff in a Kentucky karst region. In Beck, B.F. (ed.), *Proc. of the 8th International Congress of Speleology, Bowling Green, Kentucky*: 267-269.

Dougherty, P.H. 1983. Valley tides – a water balance analysis of land use response floods in a karst region, Sinking Valley, Kentucky. In Dougherty, P.H. (ed.), *Environmental Karst, GeoSpeleo Publications*: 25-36. Cincinnati, Ohio.

Dougherty, P.H. & Perlow, M. 1987. The Macungie sinkhole, Lehigh Valley, Pensilvania: Cause and repair. *Envir. Geol. Water Sci.* 12(2): 89-98.

Drew, D.P. 1983. Accelerated soil erosion in a karst area: the Burren, western Ireland. *Journal of Hydrology* 61: 113-124.

Drew, D.P. 1984. The effect of human activity on a lowland karst aquifer. In Burger, A. & Dubertret, L. (eds), *Hydrogeology of karstic terrains – case studies, International Contributions to Hydrogeology* 1: 195-201. Heise, Hannover.

Drew, D.P. 1996. Agriculturally induced changes in the Burren karst, western Ireland. *Environmental Geology* 28(2): 137-144.

Drew, D.P. & Coxon, C.E. 1988. The effects of land drainage on groundwater resources in karstic areas of Ireland. In Yuan, D. (ed.), *Karst Hydrogeology and Karst Environment Protection. Proc. 21st IAH Congress, Guilin, China* 1: 204-209. Geological Publishing House, Beijing.

Drew, D.P. & Magee, E. 1994. Environmental implications of land reclamation in the Burren, Co. Clare: a preliminary analysis. *Irish Geography* 27(2): 81-96.

Drogue, C. 1972. Analyse statistique des hydrogrammes de décrue des sources karstiques. *Journ. of Hydrology* 15: 49-68.

Dunrud, C.R. & Nevins, B.B. 1981. Solution mining and subsidence in evaporite rocks in the United States. US Geol. Survey, Misc. Inv. Series Map I-1298, 2 sheets.

Durán, J.J., Grün, R. & Ford, D. 1993. Dataciones geocronológicas (métodos ESR y series de Uranio) en la Cueva de Nerja. Implicaciones evolutivas, paleoclimáticas y neotectónicas. In Carrasco, F. (ed.), *Geología de la Cueva de Nerja. Trabajos sobre la Cueva de Nerja* 3: 233-248.

DVGW 1995. Richtlinien für Trinkwasserschutzgebiete, 1. Teil, Schutzgebiete für Grundwasser. – Dt.Ver.Gas- u. Wasserfach; Techn. Regeln, Arb.-Bl.W 101, 23 p., Eschborn.

Eberhard, S. 1990. Ida Bay karst study: the cave fauna at Ida Bay in Tasmania and the effect of quarry operations. Report to World Heritage Planning Team, Department of Parks, Wildlife & Heritage, Tasmania, 25p.

Eberhard, S. 1992. The effect of stream sedimentation on population densities of Hydrobiid molluscs in caves. Report to World Heritage Planning Team, Department of Parks, Wildlife & Heritage, Tasmania, 8p.

Eberhard, S. 1995. Impact of a limestone quarry on aquatic cave fauna at Ida Bay in Tasmania, *Proceedings, 11th Australasian Cave and Karst Managment Association Conference, Hobart, Tasmania*: 125-137.

EC-COST action 65 1995. Hydrogeological aspects of groundwater protection in karstic areas. Final report. European Commission, EUR 16547EN, 246 p., Office for Official Publications of the European Communities, Luxembourg,

Edwards, A.J. & Smart, P.L. 1989. Waste disposal on karstified Carboniferous limestone aquifers of England and Wales. In Beck, B.F. (ed.) 1989, *Engineering and Environmental Impacts of Sinkholes and Karst, Proc. of the 3nd Multidisciplinary Conference on Sinkholes and the Environmental impacts of karst*: 165-182. Rotterdam: Balkema.

Edwards, E.J., Hobbs, S.L. & Smart, P.L. 1991. Effects of quarry dewatering on a karstified limestone aquifer: A case study from the Mendip Hills, England. In *Proceedings of the Third Conference on Hydrogeology, Ecology, Monitoring, and Management of Ground Water in Karst Terranes, National Water Well Association*: 77-92.

Ege, J.R. 1984. Mechanisms of surface subsidence resulting from solution extraction of salt. In Holzer, T.L. (ed.), *Man-Induced Land Subsidence, Geol. Soc. Amer., Reviews in Engineering Geology* 6: 203–221.

Eiswirth, M. 1995. Charakterisierung und Simulation des Schadstofftransports aus Abwasserkanälen und Mülldeponien. *Schr. Angew. Geol. Karlsruhe* 39: 258 p.

Eiswirth, M. & Hötzl, H. 1997. Contaminant transport from leaky landfills in karst areas. *Proceedings of the 12th International Congress of Speleology and 6th Conference on limestone hydrology and fissured aquifers, La Chaud-de-Fonds (Switzerland)* 2: 213-216.

Ekmekci, M. 1993. Impact of quarries on karst groundwater systems. In Gunay, G. (ed.), *Hydrogeological Processes in Karst Terrains*, IAHS Publ. 207: 3-6.

Ellaway, E.M. 1991. A study of the hydrochemistry of a limestone area: Buchan, East Gippsland. University of Melbourne, Department of Geography and Environmental Studies, Unpublished PhD Thesis, 318p.

Ellaway, E.M. & Finlayson, B.L. 1984. A preliminary survey of water chemistry in the limestones of the Buchan area under low flow conditions. *Helictite* 22: 11-20.

Emmett, A.J. & Telfer, A.L. 1994. Influence of karst hydrology on water quality management in southeast South Australia. *Environmental Geology* 23: 149-155.

European Union-Council 1997. Vorschlag für eine Richtlinie des Rates über Abfalldeponien. EU-Kommissionsvorschlag 6692/97, ENV 412 PRO-COOP 136, Brüssel.

Ewers, R.O., Duda, A.J., Estes, E.K., Idstein, P.J. & Johnson, K.M. 1991. The transmission of light hydrocarbon contaminants in limestone (karst) Auqifers. In *Proceedings of the Third Conference on Hydrogeology, Ecology, Monitoring and Managment of Ground Water in Karst Terranes, US EPA and NGWA, Nashville, Tennessee*: 287-304.

Fabre, G. 1989. Les karsts du Languedoc méditerranéen . *Zeitschrift f. Geomorphologie* 75: 49-81.

Fairchild, D.M. (ed.) 1987. *Ground water quality and agricultural practices*. Lewis Publishers, Chelsea, Michigan.

Farmer, V.E. 1983. Behaviour of petroleum contamination in an underground environment. *Petroleum Ass. for Conservation of the Vanadian Environment, Seminar Proc.* II: 1-16.

Fernández, P.L., Gutiérrez, I., Quindós, L.S., Soto, J. & Villar, E. 1986. Natural ventilation of the paintings room in the Altamira Cave. *Nature* 321: 586-588.

Fetter, C.W. 1993. *Contaminant hydrogeology*, 458 p. New York: Macmillan.

Fidelibus, M.D. & Tulipano, L. 1986. Mixing phenomena owing to sea water intrusion for the interpretation of chemical and isotopic data of discharge waters in the Apulian coastal carbonate aquifer (Southern Italy). *Proc. 9th Salt Water Intrusion Meeting, Delft, The Netherlands*: 591-600.

Fidelibus, M.D. & Tulipano, L. 1990. Major and minor ions as natural tracers in mixing phenomena in coastal carbonate aquifers of Apulia. *Proc. 11th Salt Water Intrusion Meeting, Gdansk, Poland*: 283-293.

Field, M.S. 1989. The vulnerability of karst aquifers to chemical contamination. In Moore, J.E., Zaporozec, A.A., Csallany, S.C. & Varney, T.C. (eds), *Recent Advances in Ground-Water Hydrology (Tampa Florida, 1988), Am. Inst. Hydrol.*: 130-142. Minneapolis.

Field, M.S. 1993. Karst hydrology and chemical contamination. *Journal of Environmental Systems* 22: 1-26. Baywood, Amityville.

Finlayson, B. & Ellaway, M. 1987. Observations on the Buchan karst during high flow conditions. *Helictite* 25: 21-29.

Finlayson, B.L. 1985. Field calibration of a recording turbidity meter. *Catena* 12: 141-147.

Fischer, J.A., Greene, R.W. & Fischer, J.J. 1993. Roadway design in karst. In Beck B.F. (ed.), *Applied karst geology. Proceedings of the 4th multidisciplinary conference on sinkholes and the environmental impacts of karsts*: 219-224. Rotterdam: Balkema.

Flinspach, D. & Drescher, G. 1984. Die neuen Aufbereitungsanlagen im Egauwasserwerk. *Landeswasserversorgung, Schriftenreihe* 4: 5-14. Stuttgart.

Foose, R. 1967. Sinkhole formation by groundwater withdrawal: Far West Rand, South Africa. *Science* 157: 1045-1048.

Ford, D.C. (ed.) 1993. Environmental change in karst areas. *Environmental Geology* 21(3).

Ford, D.C. & Williams, P.W. 1989. *Karst geomorphology and hydrology*, 601p. London: Unwin-Hyman.

Forestry Commission, Tasmania 1987. *Forest practices Code*. Government Printer, Hobart, 46p.

Foster, S.S.D., Bridge, L.R., Geake, A.K., Lawrence, A.R. & Parker, J.M. 1986. The groundwater nitrate problem. A summary of research on the impact of agricultural land-use practices on groundwater quality between 1976 and 1985. British Geological Survey Hydrogeological Report 86/2, 95p., Natural Environment Research Council, Wallingford.

Fournier, E. 1919. Rapport sur les perturbations apportees dans le région du Bief Rouge par les travoux de ferement du mont d'Òr, 18 p., Jacques et Demontrand, Besançon.

Friederich, H. & Smart, P.L. 1982. The Classification of Autogenic Percolation waters in Karst Aquifers: A study in GB Cave, Mendip Hills, England. *Proc. Univ. Bristol Spel. Soc.* 16: 143-159.

Fritz, F., Bakun, S., Biondic, B., Polsak, A., Vulic, Z., Bozicevic, S. & Pavlin, B. 1984. Hydrogeologic Features of some northern Dalmatia littoral karst parts. In Mjatovic, B.F. (ed.), *Hydrogeology of the Dinaric Karst, IAH-Intern.Contr. to Hydrogeology* 4: 228-239. Heise, Hannover.

Furley, P.A. 1987. Impact of forest clearance on the soils of tropical cone karst. *Earth Surface Processes and Landform* 12(5): 523-529.

Gagen, P.J. & Gunn, J. 1987. Restoration blasting in limestone quarries. *Explosives Engineering* 1: 14-15.

Gams, I. 1993. Origin of the term 'karst', and the transformation of the Classical Karst (kras). *Environmental Geology* 21: 110-114.

Gams, I. (ed.) 1987. Karst and man. Univ. of Ljubljana. Study Group of Man´s impact in Karst. *Proc. Int. Symposium on human influence on Karst, Postojna, Yugoslavia.*

Gams, I. & Habic, P. 1987. *Man's impact in Dinaric Karst.* Guide book. International Geographical Union, Study Group Man's impact in karst: 1-205. Ljubljana.

Gams, I., Nicod, J., Julian, M., Anthony, E. & Sauro, U. 1993. Environmental change and human impacts on the Mediterranean karsts of France, Italy and the Dinaric region. In Williams, P.W. (ed.), *Karst Terrains, Environmental Changes, Human Impact. CATENA supplement* 25: 59-98.

Garay, P. 1986. Informe geológico sobre la sierra de hundimiento de Pecheguer (Alicante). *Jorn. Karst Euskadi* 1: 323-333.

García-Ruiz, F.J., Lasanta, T., Ortigosa, L. & Arnaez, J. 1986. Pipes in cultivated soils of La Rioja: Origin and evolution. *Z. Geomorph. Suppl.* 58: 93-100.

Gärtner, A. 1915. *Die Hygiene des Wassers.*Vieweg, Braunschweig.

Gerba, C.P., Bitton, G. 1984. Microbial pollutants: their survival and transport pattern to groundwater. In Bitton, G. & Gerba, C.P. (eds), *Groundwater Pollution Microbiology*: 65-88. New York: Wiley-Interscience.

Geyh, M.A. & Michel, G. 1979. Hydrochemische und isotopenphysikalische Entwicklung des Grundwassers im Paderborner Aquifer. *GWF Wasser/Abwasser* 120: 576-582. München.

Gierke, J.S. 1990. Modeling the transport of volatile organic chemicals in unsaturated soil and their removal by vapor extraction. Abstract of Ph.D. Dissertation Deo. Civil & Environm. Eng. Mich. Tech, Univ. Houghton.

Gilham, R.W. & O'Hannesin 1994. Enhanced degradation of halogenated aliphatics by zero-valent iron. *Ground Water* 32(6): 958-967.

Gill, D. 1994. Jerusalem's underground water systems, how they met. *Biblical Archaeology Review* 20(4): 21-33.

Gillieson, D.S. 1989. Effects of land use on karst areas in Australia. In Gillieson, D. & Smith D.I. (eds), *Resource Management in Limestone Landscapes: International Perspectives, Spec. Pub. 2, Dept of Geog. & Oceanog., Univ. Coll., Australian Defence Force Academy, Canberra*: 43-60.

Gillieson, D.S. 1996. *Caves: Processes, development, management.* Blackwell, Oxford, 324p.

Gillieson, D. & Ingle-Smith, D. (eds) 1988. Resource management in limestone landscapes. *Proc. Intern. Geographical. Union, Study Group Mans Impact on Karst, Special Publication, 2, Dept of Geography and Oceanography, Australian Defence Force Academy, Canberra,*

Gillieson, D., Cochrane, A. & Houshold, I. 1994a. Rehabilitation of degraded karst ecosystems in southern Australia. *Cave and Karst Science (Trans. B.C.R.A.)* 21(1): 11.

Gillieson, D., Oldfield, F. & Krawiecki, A. 1986. Records of prehistoric soil erosion from rockshelter sites in Papua New Guinea. *Mountain Research and Development* 6(4): 315-324.

Gillieson, D., Wallbrink, P. & Cochrane, A. 1994b. Vegetation change, erosion risk, and land management on the Nullarbor Plain, Australia. *Cave and Karst Science (Trans. B.C.R.A.)* 21(1): 11.

GLA, Geologisches Landesamt Baden-Württemberg 1991. Hydrogeologische Kriterien für die Abgrenzung von Wasserschutzgebieten in Baden-Württemberg. *G.L.A., Informationen* 2/91: 6-21. Freiburg i.Br.

Glass, N.R. 1982. Effects of acid precipitation. *Environ. Sci. Technol.* 16(3): 162A-169A.

Glover, T., Daniel, J. & Lonergan, A. 1995. The use of geochemical methods in the investigation of hazardous waste sites in covered karst terrain. Karst Geo Hazards, In Beck, B.F. (ed.), *Karstgeo Hazards*: 235-240. Rotterdam: Balkema.

Goede, A. 1969. Underground stream capture at Ida Bay, Tasmania, and the relevance of cold climate conditions. *Aust. Geogr. Stud.* 7: 41-48.

Goldie, H. 1987. Human impact on limestone pavement. *Edins* 13: 71-81.

Goldie, H. 1993. The legal protection of limestone pavements in Great Britain. *Environmental Geology* 21: 160-166.

Gomme, J., Shurvell, S., Hennings, S.M. & Clark, L. 1992. Hydrology of pesticides in a Chalk catchment: groundwaters. *J. IWEM* 6: 172-178.

Gospodaric, R. & Habic, P. 1986. Poljes of Hotoussa, Levidi and northern Tripolis. In Morfis, A. & Zojer, H. (eds), *Karst hydrogeology of the central and eastern Peloponnesus (Greece). Steir. Beitr. Z. Hydrogeologie* 37/38: 169-185.

Granat, L. 1972. Deposition of sulfate and acid with precipitation over Northern Europe. Report AC-20, 30, the University of Stockholm, Institute of Meteorology, Stockholm.

Grassi, D. & Tulipano, L. 1983. Connessioni tra assetto morfostrutturale della Murgia (Puglia) e caratteri idrogeologici della falda profonda verificati anche mediante l'analisi della temperatura delle acque sotterranee. *Geol. Appl. e Idrogeol.* 18: 135-154.

Günay, G. & Johnson, A.I. (eds) 1986. *Karst water resources.* IAHS, Publication 161: 642p. Wallingford, UK.

Günay, G. & Johnson, A.I. (eds) 1997. *Karst waters and environmental impacts,* 525 p. Rotterdam: Balkema.

Günay, G., Johnson, A.I. & Back, W. (eds) 1990. *Hydrogeological processes in karst terranes.* IAHS, Publ. 207, 412p. Wallingford, UK.

Günay, G., Simsek, S., Ekmekci, M., Elhatip, H., Yesertener, C., Dilsiz, C. & Cetiner, Z. 1997. Karst hydrogeological and environmental studies of the Pamukkale thermal springs. In Günay, G. & Johnson, I. (eds), *International Symposium and Field Seminar on Karst Water and Environmental Impacts, Antalya, Turkey, 1995. Proceedings*: 29-35. Rotterdam: Balkema.

Gunn, J. 1977. Water pollution in caves. *New Zealand Speleol. Bull.* 5(99): 557-563.

Gunn, J. 1981. Hydrological processes in karst depressions. *Z. Geomorph.* 25: 313-331.

Gunn, J. 1986. Modelling of conduit flow dominated karst aquifers. In Günay, G. & Johnson, A.I. (eds), *Karst water resources.* IAHS, Publication 161: 587-596. Wallingford, UK.

Gunn, J. 1993. The geomorphological impacts of limestone quarrying. In Williams, P.W. (ed.), *Karst terrains: environmental changes and human impact. Catena Supplement* 25: 187-198.

Gunn, J. & Bailey, D. 1993. Limestone quarrying and quarry reclamation in Britain. *Env. Geol.* 21: 167-172.

Gunn, J. & Gagen, P.J. 1987. Limestone quarrying and sinkhole development in the English Peak District. In Beck, B.F. & Wilson W.L. (eds), *Karst hydrogeology: Engineering and environmental applications*: 121-126. Rotterdam: Balkema.

Gunn, J. & Gagen, P.J. 1989. Limestone quarrying as an agency of landform change. In Gillieson, D. & Smith, D.I. (eds), *Resource management in limestone landscapes: international perspectives. Spec. Pub.* 2: 173-181. Dept of Geog. & Oceanog., Univ. Coll., Australian Defence Force Academy, Canberra.

Gutiérrez, J. 1983. *Il Mapa de la Calidad de las aguas de la República de Cuba.* Instituto Nacional de Recursos Hidráulicos, Informe Técnico: 3-83.

Gutiérrez Díaz, J., García, J.M. & Molerio León, L.F. 1982. Vulnerabilidad de los Acuíferos Cársicos a los Procesos de Nitrificación Coloquio Internac. *Hidrol. Cársica de la Región del Caribe*: 523-536. UNESCO, La Habana.

Habic, P. 1987. Use and regulation of karst poljes in Yugoslavia. In *IGU Study Group on Man's Impact in Karst, Proceedings of the 1986 Meeting, Palma de Mallorca, ENDINS* 13: 83-86. Ciutat de Mallorca.

Habic, P., Knez, M., Kogovsek, J., Kranjc, A., Mihevc, A., Slabe, T., Sebela, S. & Zupan, N. 1989. Skocjanske Jame Speleological Revue. *Int. J. Speleol.* 18: 1-2, 1-42.

Hacker, P. 1988. Combined tracer experiments – an important tool for the determination of the sorption capability of a karst aquifer. In Yuan, D. (ed.), *Karst Hydrogeology and Karst Environment Protection. Proceedings 21st IAH Congress, Guilin, China* 2: 962-967. Geological Publishing House, Beijing,.

Hägler, A. 1873. Beiträge zur Entstehungsgeschichte des Typhus und zur Trinkwasserlehre. *Dt. Arch. Klein. Med.* 11: 237-267. Leipzig.

Hall, J.E. 1988. *Sewage sludge in restoration practice*. A seminar on Land Restoration, Investigation and Techniques, 'Ten years of Research – What next?', British Coal Opencast Executive, 1988, University of Newcastle upon Tyne, 124-130.

Hallberg, G.R., Libra, R.D. & Hoyer, B.E. 1985. Nonpoint source contamination of ground water in karst-carbonate aquifers in Iowa. In *Perspectives on Nonpoint Pollution, U.S.E.P.A., Office of Water Regulations and Standards, Washington D.C.*: 109-114.

Harding, K.A. & Ford, D.C. 1993. Impacts of primary deforestation upon limestone slopes in northern Vancouver Island, British Columbia. *Environmental Geology* 21: 137-143.

Hardwick, P. 1995. The impacts of agriculture on limestone caves with special reference to the Castleton catchment, Derbyshire. Manchester Metropolitan University, Unpubl. Ph.D. thesis.

Hardwick, P. & Gunn, J. 1990. Soil erosion in cavernous limestone catchments. In Boardman, J., Foster, I.D.L. & Dearing, J.A. (eds), *Soil erosion on agricultural land*: 301-310. New York: Wiley.

Hardwick, P. & Gunn, J. 1993. The impact of agriculture on limestone caves. In Williams, P.W. (ed.), *Karst Terrains, Environmental Changes, Human Impact. CATENA Supplement* 25: 235-249.

Harker, K. & Ray, J. 1995. Status of Kentucky's regulatory programs that address karst. In Beck B.F. (ed.), *Karst Geohazards. Engineering and environmental problems in karst terrane. Proceedings of the 5th multidisciplinary conference on sinkholes and the environmental impacts of karsts*: 505-509 Rotterdam: Balkema.

Harress, H.M. 1990. Bodenluftabsaugung. In Weber, H. et al. (eds), *Altlasten*. Springer Verlag, Berlin.

Harrison, D.J., Buckley, D. & Marks, R.J. 1992. *Limestone resources and hydrogeology of the Mendip Hills*. Brit. Geol. Surv. Tech. Report SA/92/19. Keyworth, Nottingham.

Hartmann, D. & Michel, P. 1992. Grundwasserschutz in der Schweiz. *gwa* 3: 167-173.

Harward, M.E. & Reisenauer, H.M. 1966. Reactions and movement of inorganic soil sulfur. *Soil Sci.* 101: 326-335.

Heitfeld, H.-H. 1991. Talsperren. In Matthess, G. (ed.), *Lehrbuch der Hydrogeologie* 5: 468p.

Hendry, M.J., Lawrence, J.R., Kirkland, R. & Zannyk, B.N. 1992. Microbial production of carbon dioxide in the unsaturated zone of a meso scale model. In Kharaka & Meast (eds), *Water-rock interaction*: 287-290. Rotterdam: Balkema.

Henry, H., Thorn, R. & Brady, E. 1991. An assessment of the suitability of a range of chemical and biological tracers to monitor the movement of septic tank effluents to groundwater. *Irish Geography* 24(2): 91-105.

Herak, M. 1965. Geologische Übersicht des dinarischen Karstes. *Nase Jame* 7: 5-11.

Herak, M. 1972. Karst of Yugoslavia. In Herak, M. & Stringfield, V.T. (eds), *Karst*: 25-83. Elsevier, Amsterdam.

Hickey, J.J. 1982. *Hydrogeology and results of injection tests at waste-injection test sites in Pinellas County, Florida*. US Geological Survey Water-Supply Paper 2183, 42 p.

Hinchee, R.E., Leeson, A., Semprini, L. & Kee Ong, S. (eds) 1993. *Bioremediation of chlorinated and polycyclic aromatic hydrocarbon compounds,* 864 p. Lewis, Boca Raton.

HMIP-Her Majesty's Inspectorate of Pollution 1989. Waste Management Paper Number 4. The Livensing of Waste Facilities, A revision for Waste Management Paper Number 4 (1976) to provide a technical memorandum on the licensing of waste facilities including a review of relevant legislation, HMSO, London.

Hobbs, S. 1988. Recharge, flow and storage in the saturated zone of the Mendip Limestone Aquifer. Univ. of Bristol, Unpub. PhD thesis.

Hoblea, F. 1995. Problems in relation to the development of ski-resorts on the French mountain Karst. *Acta Carsologica* 24: 267-278. Ljubljana.

Hoenstine, R.W., Lane, E. & Spencer, S.M. 1987. A landfill site in a karst environment, Madison Country, Florida – a case study. In Beck, B.F. & Wilson, W.L. (eds), *Karsthydrogeology, Proc. of the 2nd Multidisciplinary Conference on Sinkholes and the Environmental impacts of karst*: 253-258. Rotterdam: Balkema.

Holler, H. 1976. Gedanken zum Bau des Dobratsch in den östlichen Gailtaler Alpen. *Carinthia* II, 1886/86. Klagenfurt.

Hölting, B., Haertle, T., Hohberg, K.H., Nachtiall, K.H., Villinger, E., Weinzierl, W. & Wrobel, J.P. 1995. Konzept zur Ermittlung der Schutzfunktion der Grundwasserüberdeckung. *Geol.Jb.* C 63: 5-24. Hannover.

Hötzl, H. 1973. Die Hydrogeologie und Hydrochemie des Einzugsgebietes der obersten Donau. *Steir.Beitr. Hydrogeol.* 25: 5-102. Graz.

Hötzl, H. 1989. Schadstoffausbreitung bei Überlagerung eines Karstaquifers mit einem Porengrundwasserleiter. *Oberrhein.geol. Abh.* 35: 17-35. Stuttgart.

Hötzl, H. 1995. Project area Bauschlotter Limestone Platform, National report for Germany. EC-Cost Action 65, EUR 166547 EM, 124-131. Luxembourg.

Hötzl, H. 1996. Grundwasserschutz in Karstgebieten. *Grundwasser* 1(1): 5-11. Heidelberg.

Hötzl, H., Käss, W. & Reichert, B. 1991. Application of microbial tracers in groundwater studies. *Wat.Sci.Tech.* 24(2): 295-300. Great Britain.

Hötzl, H. & Nahold, M. 1992. Rhythmisches Luft-Wasser-Fluten zur CKW-Sanierung im Kluftgrundwasser. Nr. PW 90096. Statuskolloqium Projekt Wasser-Abfall-Boden (PWAB), 18.-19.2.1991, KfK-PWAB 13: 327-338. Kernforschungszentrum Karlsruhe.

Hötzl, H., Nahold, M., Xiang, W. & Bock, P. 1990. CKW-Sanierung – Erfahrungen über einen Schadensfall in Festgesteinen. In Bock, P., Hötzl, H. & Nahold, M. (eds), *Untergrundsanierung mittels Bodenluftabsaugung und In-Situ-Strippen. Schr. Angew.Geol.* 9: 175-186. Karlsruhe

Hötzl, H. & Werner, A. (eds) 1992. *Tracer hydrology,* 464 p. Rotterdam: Balkema.

House, O. 1995. Carbonate rock investigation guidance policy for siting landfills in karst areas in Tennessee. In Beck B.F. (ed.), *Karst Geohazards, Engineering and environmental problems in karst terrane. Proceedings of the 5th multidisciplinary conference on sinkholes and the environmental impacts of karsts*: 511-515. Rotterdam: Balkema.

Houshold, I. 1992. Geomorphology, water quality and cave sediments in the Eastern Passage of Exit Cave and its tributaries. Report to World Heritage Planning Team, Department of Parks, Wildlife & Heritage, Tasmania, 18 p.

Houshold, I. 1997. Karst impacts and environmental rehabilitation of a limestone quarry at Lune River, southern Tasmania. *Proceedings, 11th Australasian Cave and Karst Management Association Converence, Hobart, Tasmania*: 138-175.

Houston, J.F.T. 1982. Rainfall and recharge to a dolomitic aquifer at Kabwe, Zambia. *J. Hydrol.* 59: 173-187.

Howard, K.W.F., Eyles, N. & Livingstone, S. 1996. Municipal landfilling practice and its impact on groundwater resources in and around urban Toronto, Canada. *Hydrogeology Journal* 4(1): 64-79.

Hubbard, D.A. 1993. Status report on the Virginia karst mapping program. In B. Beck (ed.), *Applied karst geology*: 281-284. Rotterdam: Balkema.

Hughes, T.H., Memon, B.A. & Lamoreaux, P.E. 1994. Landfills in karst terrains. *Bull.of the Association of Engineering Geologists* 31(2): 203-208.

Hultberg, H. & Wenblad, A. 1980. Acid groundwater in the southwestern area of Sweden. *Proc. Int. Conf. SNSF Project, Oslo, Norway*: 220-221.

Huntoon, P.W. 1992. Hydrogeologic characteristics and deforestation of the Stone Forest karst aquifers of south China. *Ground Water* 30(2): 167-176.

Huppert, G., Burri, E., Forti, P. & Cigna, A. 1993. Effects of tourist development on caves and karst. *Catena Supplement* 25: 251-268.

Huppert, G.N., Wheeler, B.J., Alexander, E.C. & Adams, R.S. 1989. Agricultural land use and groundwater quality in the Coldriver Cave groundwater basin, upper Iowa river karst region, USA: Part I. In Gillieson, D. & Smith, D.I. (eds), *Resource management in limestone landscapes, International perspectives, Special Publication No.2, Dept. of Geog. & Oceanog., University College, Australian Defence Force Academy, Canberra*: 235-247.

Hutchinson, G.E. 1975. Treatise on Limnology. *Chemistry of Lakes* I(2). New York: Wiley.

IGME 1983. Informe nº 10 sobre el Sistema Acuífero nº 41: calizas y dolomías triásicas de la Sierra Almijara-Sierra de Lújar. Memoria y anejos.

IGME 1986. Las aguas subterráneas en la Comunidad Valenciana, uso calidad y perspectiva de utilización. 298 p.

IGME-IRYDA 1978. Informe técnico nº 7. El Prebético de la provincia de Alicante P.I.H.C.B.S. PNAS., 200 p.

International Geographical Union Study Group 1987. *Mans Impact on karst Proceedings of Meeting in Mallorca, Endins* 13: 48-131. Ciutad de Mallorca.

Issar, A.S. 1990. Water shall flow from the rock. *Hydrogeology and climate in the Lands of the Bible*, 213 p. Springer, New York.

Istituto Nazionale di Geofisica 1990. *Bollettino sismico 1989*. Rome.

James, N.P. & Choquette, P.W. (eds) 1988. *Paleokarst*, 416 p. New York: Springer.

Jankowski, G. 1964. Die Tertiärbecken des südlichen Harzvorlandes und ihre Beziehungen zur Subrosion. *Beih.Geologie* 43: 1-60. Berlin.

Jeanblanc, A. & Schneider, G. 1981. Etude geologique et hydrogeologique du Risoux-Mont d'Or. These 3° cycle, 155 p. Besancon, France.

Jeannin, P.Y., Wildberger, A. & Rossi, P. 1995. Multitracing-Versuche 1992 und 1993 im Karstgebiet der Silberen (Muotatal und Klöntal, Zentralschweiz). *Beiträge zur Hydrogeologie* 46: 43-88. Graz.

Johnson, D.W. & Todd, D.E. 1983. Relationships among iron, aluminum, carbon, and sulfate in a variety of forest soils. *Soil Sci. Soc. Am. J.* 47(4): 792-800.

Johnson, K.S. 1981. Dissolution of salt on the east flank of the Permian Basin in the southwestern USA. *J. Hydrol.* 54: 75-93.

Johnson, K.S. 1987. Development of the Wink sink in west Texas due to salt dissolution and collapse. In Beck, B.F. & Wilson, W.L. (eds), *Karst Hydrogeology: Engineering and Environmental Implications*: 127-136. Balkema, Brookfield, Vermont. Also published in 1989 in *Environ. Geology and Water Sci.* 14: 81-92.

Johnson, K.S. 1992. Evaporite karst in the Permian Blaine Formation and associated strata in western Oklahoma, USA. In Paloc, H. & Back, W. (eds), *Hydrogeology of Selected Karst Regions, Internat. Assoc. Hydrogeol.* 13: 405-420. Heise, Hannover.

Johnson, K.S. 1986. Hydrogeology and recharge of a gypsum-dolomite karst aquifer in SW Oklahoma, USA. *Karst Water Resources. IAHS* 161: 343-357.

Johnson, K.S. & Quinlan, J.F. 1995. Regional mapping of karst terrains to avoid potential environmental problems. *Cave and karst Science* 21(2): 37-39.

Jones, H.C., Noggle, J.C., Young, R.C., Kelly, J.M., Olem, H. & Ruanne, R.J. 1983. Investigations of the cause of fish kills in fish-rearing facilities in Raven Fork watershed. TVA/ONR/WR-83/9. Tenn. Valley Auth., Knoxville, TN.

Jung, W. & Liebisch, K. 1966. Die Grubenhydrogeologie in der Mansfelder Mulde. *Z.angew.Geol.* 12/10: 511-521.

Jung, W. & Spilker, M. 1972. Hydrologische Probleme beim Wasseranstau in der Mansfelder Mulde. *Z.angew.Geol.* 18/1: 17-21.

Jury, A., Winer, A., Spencer, W. & Focht, D. 1987. Transport and transformations of organic chemicals in the soil-air-water ecosystem. *Rev.Env.Cont.Tox.* 99: 120-164. New York.

Kahler, F. 1983. Beobachtungen und Probleme im Thermalgebiet von Warmbad Villach, Neues aus Alt-Villach. *Jb.des Stadtmuseums* 20: 159-213. Villach.

Kammholz, H. 1964. Ingenieurgeologische Situationskarte der Mansfelder Mulde in Maßstab 1: 50 000 mit Erläuterungen. Zentr.Geol.Inst., Berlin.

Käss, W. 1992. Geohydrologische Markierungstechnik. *Lehrbuch der Hydrogeologie* 9: 519 p. Bornträger, Berlin-Stuttgart.

Käss, W., Löhnert, E.P. & Werner, A. 1996. Der jüngste Markierungsversuch im Karst von Paderbron (Nordrhein-Westfalen). *Grundwasser* 1(2): 83-89. Springer, Berlin-Heidelberg.

Kastrinos, J.R. & White, W.B. 1986. Seasonal, hydrogeologic and land-use controls on nitrate contamination of carbonate ground waters. *Proc. Environmental Problems in Karst Terranes and their Solutions Conference (Bowling Green, KY)*: 88-114.

Kenyon, K.M. 1957. *Digging up Jericho,* 272 p. Benn., London.

Keswick, B.H, Wang, D. & Gerba, C.P. 1982. The use of microorganisms as groundwater tracers: a review. *Ground Water* 20(2): 142-149.

Kiel, K. 1958. Wasserversorgung und Wasserhaltung der Mansfelder Mulde unter Berücksichtigung der geologisch-hydrologischen Verhältnisse. *Neu Hütte* 3/10: 577-585.

Kiernan, K. 1981. Man and karst in Tasmania. *Aust. Speleol. Fed. Newsl.* 94: 9-20.

Kiernan, K. 1987a. Soil erosion from hilltribe opium swiddens in the Golden Triangle and the use of karren as an erosion yardstick. In *IGU Study Group on Man's Impact in Karst, Proceedings of the 1986 Meeting, Palma de Mallorca, ENDINS* 13: 59-63. Ciutat de Mallorca.

Kiernan, K. 1987b. Some planning requirements prior to forest industry development of carbonate landscapes. In *IGU Study Group on Man's Impact in Karst, Proceedings of the 1986 Meeting, Palma de Mallorca, ENDINS* 13: 119-125. Ciutat de Mallorca.

Kiernan, K. 1987c. Timber harvesting on karst lands: some operational considerations and procedural requirements. In *IGU Study Group on Man's Impact in Karst, Proceedings of the 1986 Meeting, Palma de Mallorca, ENDINS* 13: 105-109. Ciutat de Mallorca.

Kiernan, K. 1989. Human impacts and management responses in the karsts of Tasmania. In Gillieson, D. & Smith, D.I. (eds), *Resource management in limestone landscapes, International perspectives, Special Publication* 2: 69-92. Dept. of Geog. & Oceanog., University College, Australian Defence Force Academy, Canberra.

Kiernan, K. 1989. Sinkhole hazards in Tasmania. In Beck, B.F. (ed.), *Engineering and environmental impacts of sinkholes and karst*: 123-128. Rotterdam: Balkema.

Kiernan, K. 1993. The Exit Cave Quarry: tracing waterflows and resource policy evaluation. *Helictite* 31: 3-12.

Kiraly, L. 1975. Rapport sur l'etat actuel des connaissances dans le domaine des caracteres physiques des roches karstiques. In Burger, A. & Dubertret, L. (eds), *Hydrology of karstic terrains. Internat. Union Geol. Sci.*, Series B, 3: 53-67.

Klecka, G.M., Davis, J.W., Grey, D.R. & Madsen, S.S. 1990. Natural bioremediation of organic contaminants in ground water: Cliff-Dow superfund site. *Ground Water* 28(4): 534-543.

Klein, M. 1980. Dissolved materials transport, the flushing effect in surface and 'sub-surface' flow. *Earth Surf. Proc. Landfms.* 6: 173-178.

Kleywegt, R.J. & Pike, D.R. 1982. Surface subsidence and sinkholes caused by lowering of the dolomitic water table on the far West Rand gold field of South Africa. *Ann. Geol. Surv. S. Afr.*

Klingebiel, A., Gayet, J. & Maire, R. 1993. Facteurs faciologiques et tectoniques contrôlant la karstification sur la marge nord aquitaine (France): Exemple de l'Oligocène Nord aquitain. *Compte-rendus à l'Académie des sciences* 317(Série II): 523-529. Paris.

Knez, M., Kogovsek, J., Kranjc, A., Mihevc, A., Sebela, S. & Zupan-Hajna, N. 1995. Karst groundwater protection – national report for Slovenia. EC-COST Action 65, Hydrogeological

Aspects of Groundwater Protection in Karstic Areas, Final report, 247-260, European Commission, Luxembourg.

Kölbel, H. 1971. Perm. In *Entwicklungsgeschichte der Erde*: 395-410. Edition Leipzig.

Kogovsek, J. 1982. Vertical percolation in the Planina Cave. *Acta Carsologica Krasoslovni Zbornik* 10: 107-125.

Kogovsek, J. 1987. Natural purification of sanitary sewage during the vertical percolation in Pivka Jama. *Acta Carsologica* 16: 123-139. Ljubljana.

Kogovsek, J. 1991. La qualite de la riviere a perte Pivka dans les annees de 1984 jusqu'au 1990. *Acta Carsologica* 20: 165-186. Ljubljana.

Kogovsek, J. 1992. Consequences of a liquid manure spill into Nanosica Brook. *Ujma* 6: 54-55. Ljubljana.

Kogovsek, J. 1993. Karst waters and how they are endangered. *Nase jame* 35(1): 67-76. Ljubljana.

Kogovsek, J. 1994. Impact of human activity on Skocjanske jame. *Acta Carsologica* 23: 73-80. Ljubljana.

Kogovsek, J. 1995. Some Examples of the Karst Water Pollution on the Slovene Karst. *Acta Carsologica* 24: 303-312. Ljubljana.

Kogovsek, J. 1997. Pollution transport in the vadose zone. In Günay, G. & Johnson, A.I. (eds), *Karst water and environmental impacts. Proc. 5th International Symp. and Field Seminar, Antalya*: 161-166. Rotterdam: Balkema.

Kogovsek, J. & Habic, P. 1980. Study of vertical water percolation in the case of Planina and Postojna caves. *Acta Carsologica Krasoslovni Zbornik* 9: 133-148.

Kovalevsky, V.S. 1976. Basement of natural groundwater regime forecasting. Stroiizdat. M., 204 p.

Kovalevsky, V.S. & Zlobina, V.L. 1988. Location of concealed areas with intensified karst-suffosion processes using helium-survey and hydrobiological data. *Proceedings of the IAH 21st Congress* 21(2): 779-784.

Kovalevsky, V.S. & Zlobina, V.L. 1984. Integration of helium and tritium surveying for mapping of the potential development of karst. Hydrogeology of Karst Terrains. Case Histories., *IAH International Contributions to Hydrogeology* 1: 255-257. Heise, Hannover

Kovalevsky, V.S. & Zlobina, V.L. 1987. Helium survey for delineating areas of karst-suffosion processes caused by high-rate groundwater withdrawal. *Environmental Geology and Water Sciences* 2: 89-94.

Kozary, M.T., Dunlap, J.C. & Humphrey, W.E. 1968. Incidence of saline deposits in geologic time. In *Saline Deposits – International Conference Saline Deposits, Houston, Texas, Symposium, 1962*. Geol. Soc. America, Special Paper 88: 43-57.

Kranjc, A. (ed.) 1995. Man on karst. *Acta Carsologica* 24: 589 p. Ljubljana.

Kranjc, A. & Mihevc, A. 1988. Flood area along the Notranjska Reka river. *Geografski zbornik*, 28: 195-218. Ljubljana.

Kranjc, A. (ed.) 1997. *Tracer hydrology 1997*, 450 p. Rotterdam: Balkema.

Kresic, N. 1991. Quantitative karst hydrogeology with elements of groundwater protection. *Naucna knjiga*, 192 p. Belgrade.

Kresic, N., Papic, P. & Golubovic, R. 1989. The influence of precipitation on the quality of karst groundwater in the industrial zones. *IAHS Pub.* 188: 153-159. Wallingford.

Krothe, N.C. 1990. Delta SUP 15N studies of groundwater nitrate transport through macropores in a mantled karst aquifer. *Transactions American Geophysical Union* 71(28): 876-877.

LaMoreaux, P.E. 1991. Environmental effects to overexploitation in a karst terrane. *23 IAH Congress* I: 103-113. Canary Island.

LaMoreaux, P. 1995. Legal aspects of karst areas and insurability. In Gunay, G. & Johnson, A.I. (eds), *Karst waters and environmental impacts*: 11-18. Rotterdam: Balkema.

LaMoreaux, P.E., Powell, W.J. & LeGrand, H.E. 1997. Environmental and legal aspects of karst areas. *Environmental Geology* 29: 23-36.

LaMoreaux, P. & Warren, W. 1973. Sinkholes. *Geotimes* 18: 3, 15.

LaMoreaux, P., Wilson, B.M. & Memon, B.A. 1984. Guide to the hydrology of carbonate rocks. *Studies and Reports in Hydrology* 41: 345p. UNESCO, Paris.

Lawrence, A.R. & Foster, S.S.D. 1987. The pollution threat from agricultural pesticides and industrial solvents: a comparative review in relation to British aquifers. Hydrogeological Report of the British Geological Survey, 87/2, 30p.

LeGrand, H.E. 1973. Hydrological and econological problems of karst regions. *Science* 279: 859-864.

LeGrand, H.E. 1984. Environmental problems in karst terranes. In Burger, A. & Dubertret, L. (eds), *Hydrogeology of karstic terrains. IAH International Contributions to Hydrogeology* 1: 189-194. Hanover.

LeGrand, H.E. & Stringfield, V.T. 1971. Development and distribution of permeability in carbonate aquifers. *Water Res. Research* 7(5): 1284-1294.

Lerner, D.N., Issar, A.S. & Simmers, I. 1990. Groundwater recharge: A guide to understanding and estimating natural recharge. *IAH International Contrubutions to Hydrogeology* 8. Heise, Hannover.

Lewis, W.C. 1981. Carbon dioxide in Coldwater Cave. *Proc. 8th International Speleological Congress, Kentucky*: 91-92.

LFU-Landesanstalt für Umweltschutz Baden-Württemberg 1995. Handbuch Altlasten und Grundwasserschadensfälle, hydraulische und pneumatische in-situ-Sanierungstechniken. *Materialien zur Altlastensanierung* 16: 390 p. Eigenverlag der LFU, Karlsruhe.

Libra, R.D., Hallberg, G.R. & Hoyer, B.E. 1987. Impacts of agricultural chemicals on groundwater quality in Iowa. In Fairchild, D.M. (ed.), *Ground Water Quality and Agricultural Practices*: 185-217. Lewis Publishers, Chelsea.

Libra, R.D., Hallberg, G.R., Hoyer, B.E. & Johnson, L.G. 1986. Agricultural impacts on groundwater quality: the Big Spring basin study, Iowa. In *Agricultural Impacts on Groundwater, National Water Well Association, Worthington OH*: 253-273.

Lichon, M. 1993. Human impacts on processes in karst terranes, with special reference to Tasmania. *Cave Science* 20(2): 55-60.

Li Zhenshuan 1996. Environment divided zones with chemical thermodynamic and protection programs in the karst water system. *30th Intern. Geol. Congress*, Abstracts, 3,16-5-20, 277, Beijing.

Llamas, M.R., Back, W. & Margat, J. 1992. Groundwater use: equilibrium between social benefits and potential environmental cost. *Applied Hydrogeology* 2: 3-14.

Löhnert, E.P. 1992. Tracing of the Paderborn karst aquifer system (Westphalia, Germany). A critical review. In Hötzl & Werner (eds), *Tracer Hydrology*: 243-250. Rotterdam: Balkema.

Lorah, M.M. & Herman, J.S. 1988. The chemical evolution of a travertine depositing stream, geochemical processes and mass transfer reactions. *Water Resources Research* 24(9): 1541-1552.

Lorenz, S. 1962. Wassereinbrüche im Mansfeleder. Kupferschieferbergbau. *Z.angew.Geol.* 8/6: 310-316.

Lozek, V. 1990. Karst and human impact. *Stuida Crasologica* 3: 61-67.

Lushichik, A. 1982. Formation of hydrochemical groundwater regime of karstifying carbonaceous deposits within the limits of irrigated land masses of the Flat Crimea. In *Impact of Agricultural Activities on Groundwater (IAH International Symposium, Prague, Czechoslovakia), IAH Memoires* 16(3): 307-315. Novinar Publishing House, Prague.

Mackay, D.M. & Cherry, J.A. 1989. Groundwater contamination: Pump- and -treat remediation. *Envir. Sci. and Engineering* 23: 630-637.

Malanchuk, L.J. & Turner, R.S. 1987. *Effects on Aquatic Systems*. Chapter 8 in NAPAP (The National Acid Precipitation Assessment Program) Interim Assessment, Volume IV: Effects of Acidic Deposition, 8-1 to 8-81.

Mangin, A. 1975. Contribution à l'étude hydrodynamique des aquifères karstiques. Laboratoire Sout. du CNRS, Thèse Doct.

Mangin, A. 1981b. Apports des analyses corrélatoire et spectrale croisées dans la connaissance des systèmes hydrologiques. *C.R. Acad. Sc. Paris* 293: 1011-1014.

Mangin, A. 1981a. Utilisation des analyses corrélatoire et spectrale dans l'approche des systèmes hydrologiques. *C.R. Acad. Sc. Paris* 293: 401-409.

Mangin, A. 1984. Pour une meilleure connaissance des systèmes hydrologiques à partir des analyses corrélatoire et spectrale. *Journal of Hydrology* 67: 25-43.

Mangin, A. & D'Hulst, D. 1995. Fréquentation des grottes touristiques et conservation. Méthode d´approche pour en étudier les effets et proposer une réglementation. *Simposio Internazionale Grotte turistiche e Monitoraggio Ambientale, Stazione Scient, Bossea, Italia*: 117-145.

Mangin, A. & Pulido-Bosch, A. 1983. Aplicación de los análisis de correlación y espectral en el estudio de acuíferos kársticos. *Tecniterrae* 51: 53-65.

Marker, M.E. & Gamble, F.M. 1987. Karst in southern Africa. In International Geographical Union Study Group: Mans Impact on karst. *Proceedings of Meeting in Mallorca, Endins* 13: 93-98. Ciutat de Mallorca.

Marsden, R.W. & Lucas, J.R. 1973. Specialized underground extraction systems. In Cummins, A.B. & Given, I.A. (eds), *SME Mining Engineering Handbook, Soc. Min. Engrs. of Amer., Inst. Min., Metall., Petrol. Engrs.* 2(21): 1-118.

Marsily, G. de 1978. De l'identification des systèmes hydrogéologiques. Univ. Paris VI, Thèse Doct. Sci. Nat., 215 p.

Martínez-Alfaro, P.E., Montero, E. & López-Camacho, B. 1992. The impact of the overexploitation of the Campo de Montiel aquifer on the Lagunas de Ruidera ecosystem. *IAH, Selected Papers* 3: 87-91. Heise, Hannover.

Meja, B., Ros, M., Dular. M., Rejic, M. & Ponikvar-Zorko, P. 1983. Onesnasevanje Notranjske Reke.- Med. simp. 'Zagita Krasa ob 160-letnici tur. razv. skocjanskih jam', 48-51, Secana.

Mercer, J.W. & Cohen, R.Y. 1990. A review of immiscible fluids in the subsurface. *3rd of Contamination Hydrogeology* 6: 107-163.

Metcalf & Eddy Inc. 1991. *Wastewater engineering: treatment, disposal and reuse.* 3rd edn. (revised by Tchobanoglous, G. & Burton, F.L). McGraw-Hill, New York.

Michel, G. 1988. Conflict between groundwater exploitation and limestone quarrying in the karst region of Warstein (Federal Republic of Germany). *Proc. Karst Hydrogeology and Environment Protection Conf.*: 178-182. Guilin, China.

Mihevc, A. 1984. Nova spoznanja o Kani jami. *Naje Jame* 26: 11-20. Ljubljana.

Mijatovic, B.F. 1975. Exploitation rationnelle des eaux karstiques. In *Hydr. Terrains Karstiques'. AIH*: 123-135.

Mijatovic, B.F. 1984a. Captage par galerie dans un aquifère karstique de la côte Dalmate: Rimski Bunar, Trogir (Yougoslavie). In Burger, A. & Dubertret, L. (eds), *Hydrogeology of karstic terrains, case histories, IAH International Contributions to Hydrogeology* 1: 152-155.

Mijatovic, B.F. (ed.) 1984b. *Hydrogeology of the Dinaric karst. IAH Intern.Contribution to Hydrogeology* 4: 255 p. Heise, Hannover.

Mijatovic, B.F. 1984c. Karst poljes in Dinarides. In Mijatovic, B.F. (ed.), *Hydrogeology of the Dinaric Karst, IAH International Contributions to Hydrogeology* 4: 87-109. Heise, Hanover.

Milanovic, P.T. 1981. Karst Hydrogeology. *Water Res. Publ.*, 434 p. Littleton, Colorado.

Milanovic, P.T. 1984. Some methods of hydrogeologic exploration and water regulation in the Dinaric karst with special reference to their application in eastern Herzegovina. In Mijatovic, B.F. (ed.), *Hydrogeology of the Dinaric Karst, IAH Intern. Contributions to Hydrogeology* 4: 160-200. Heise, Hannover

Milde, K., Milde, G., Ahlsdorf, B., Litz, N., Muller-Wegener, U. & Stock, R. 1988. Protection of highly permeable aquifers against contamination by xenobiotics. In Yuan, D. (ed.), *Karst Hydrogeology and Karst Environment Protection (Proceedings 21st IAH Congress, Guilin, China)* 1: 194-201. Geological Publishing House, Beijing.

Miller, J.C. & Eingold, J.C. 1987. Vulnerability of a karst terraine to groundwater contamination by the agricultural chemical ethylene dibromide, northwest Florida (USA). In *IAH Symposium on Ground-water Protection Areas, Karlovy Vary, Czechoslovakia, 1986, IAH Memoires* 19: 2, 240-253. Novinar Publishing House, Prague.

Mishu, L.P., Godfrey, J.D. & Mishu, J.R. 1997. Foundation remedies for residential construction over karst limestone in Nashville, Tennessee. In Beck, B.F. & Stephenson, J.B. (eds), *The Engineering Geology and Hydrogeology of Karst Terranes*: 319-328. Rotterdam: Balkema.

Molerio León, L.F. 1975. Esquema geoespeleológico preliminar de cuba. Memoria Explicativa del Mapa de las Regiones Cársicas de Cuba a escala 1:1 000 000.

Molina, M. & McDonald, F. 1987. Sinkhole management and flooding in Jamaica. In Beck, B.F. & Wilson, W.L. (eds), *Karst hydrogeology: Engineering and environmental applications*: 293-298. Rotterdam: Balkema.

Moore, B. 1995. Landspreading animal wastes. In *The Role of Groundwater in Sustainable Development, Proceedings of IAH Irish Group 15th Annual Groundwater Seminar, Portlaoise*.

Moore, H. & Amari, D. 1987. Sinkholes and gabions – a solution to the solution problem. In Beck, B.F. & Wilson, W.L. (eds), *Karst hydrogeology: Engineering and environmental applications*: 305-310. Rotterdam: Balkema.

Moore, H.L. 1984. Geotechnical considerations in the location, design and construction of highway in karst terrain – 'The Pellissippi Parkway extension', Knox-Blount Countries, Tennessee. In Beck, B.F. (ed.), *Proceedings of the 1st multidisciplinary conference on sinkholes*: 385-390. Rotterdam: Balkema.

Moore, H.L. 1995. Field trip through east Tennessee karst with emphasis on practical problems. In Beck, B.F. (ed.), *Proceedings of the 5th multidisciplinary conference on sinkholes*: 549-557. Rotterdam: Balkema.

Morfis, A. & Zojer, H. (eds) 1986. Karst hydrogeology of the Central and Eastern Peloponnesus (Greece). *Steir.Beitr.z.Hydrogeologie* 37/38: 1-301. Graz.

Morse, J.W. 1983. The kinetics of calcium carbonate dissolution and precipitation. In Reeder, R.J. (ed.), *Mineralogy and Chemistry: Reviews in Mineralogy* 11. Mineralogical Society of America.

Motyka, J. 1988. Triassic carbonate sediments of Olkusz-Zawiercie ore-bearing district as an aquifer. *Sci. Bull. of Acad.of Mining and Metallurgy*, 157, *Geology Bull.* 36: 106 p.. Kraków.

Motyka, J. & Wilk, Z. 1984. Hydraulic structure of karst-fissured Triassic rocks in the vicinity of Olkusz (Poland). *Kras i speleologia* 5(14): 11-24. Katowice.

Nahold, M.E. 1996. Zur Sanierung von CKW-Kontaminationen in komplexen Grundwasserleitern. *Schr. Angew. Geologie* 32. Universität Karlsruhe, 310 p.

Nahold, M. & Hötzl, H. 1993. Effects of different hydropneumatic in situ remediation systems. In Hinchee, R.E., Leeson, A., Semprini, L. & Kee Ong, S. (eds), *Bioremediation of chlorinated an polycyclic aromatic hydrocarbon compounds*: 114-154. Lewis, Boca Raton.

NAPAP-National Acid Precipitation Assessment Program 1991. Integrated Assessment Report 1990. Office of the Director, 520 p., Washington, D.C.

Naughton, M. 1983. Pollution at Teesan Springs, County Sligo – a case study. *Irish Journal of Environmental Science* 2(2): 52-56.

Newton, J.G. 1984. Review of induced sinkhole development. In Beck, B.F. (ed.), *Proceedings of the 1st multidisciplinary conference on sinkholes,* 429 p. Rotterdam: Balkema.

Newton, J.G. 1987. Development of sinkholes resulting from man´s activities in the eastern United States. US Geological Survey, Cir. 968, 54 p.

Nicod, J. 1991. Natural hazards and engineering impacts in the karsts of Mediterranean France. In Sauro, U., Bondesan, A. & Meneghel, M. (eds), *Proceeding of the International Conference on environmental changes in karst areas. Department of Geography, University of Padua*, p. 9-16.

Nicod, J. 1995. Artificial Drainage of the Poljes and Karst Depressions in the South-Eastern France. *Acta Carsologica* 24: 413-427. Ljubljana.

Nyer, E.K. & Morello, B. 1993. Trichloroethylene treatment and remediation. *GWMR spring*: 98-103.

Oden, S. 1968. The acidification of air and precipitation and its consequences in the natural environment. *Ecology Committee Bulletin* 1. Swedish National Science Research Council, Stockholm.

Overrein, L.N., Seip, H.M. & Tollan, A. 1980. Acid precipitation – effects on forests and fish. Final report of the SNSF Project 1972-80, SNSF Research Report No. 19, SNSF Project, Oslo.

ÖVGW 1995. Schutz- und Schongebiete.- Regeln der Österr. Vereinig. für das Gas- und Wasserfach, Richtlinie W 72, 45 p., Vienna.

ÖWWV 1984. Leitlinien für die Nutzung und den Schutz von Karstwasservorkommen für Trinkwasserzwecke. – Österr. Wasserwirtschaftsverband, *Regelblatt 201*, 55 p., Vienna.

Pace, P. 1986. Gli acquedotti di Roma, Sp. ed. 16th International Water Supply Congress and Exhibition, Azt. Studio S. Eligio, Roma.

Padilla, A., Pulido-Bosch, A. & Mangin, A. 1994. Relative importance of baseflow and quickflow from hydrographs of karst spring. *Ground Water* 32(2): 267-277.

Papic, P., Kresic, N. & Golubovic, R. 1991. Acid rains and their influence on the quality of Petnica karstic spring water. Zbor. radova Odbora za kras i speleologiju., knj. IV. Posebna. izdanja. Srpske akademije nauka i umetnosti., knj. DCXIV. *Odeljenje prirodno-matematickih nauka* 67: 95-105. Belgrade.

Pasquarell, G.C. & Boyer, D.G. 1995. Agricultural impacts on bacterial water quality in karst groundwater. *Journal of Environmental Quality* 24: 959-969.

Pasquarell, G.C. & Boyer, D.G. 1996. Herbicides in karst groundwater in southeast West Virginia. *Journal of Environmental Quality* 25: 755-765.

Pavlin, B. & Fritz, F. 1978. La protection du système des sources karstiques de Golubinka contre la contamination par la mer. *SIAMOS* 1: 227-235. Granada.

Pederson, T.A. & Curtis, J.T. 1991. Soil Vapor Extraction Technology – Reference Handbook, EPA/540/2-91/003, Cincinnati.

Peterlin, S. (ed.) 1972. The Green Book on the threat to the environment in Slovenia, 255 p. Ljubljana.

Pevalek, J. 1938. Biodinamika Plitvickih Jezera i njena zastita, Zastite Prirode 1, and Der Travertin und die Pltitvicer Seen. *Verhand.Int.Ver.Limnol.* 7: 165-181. Belgrad.

Polsak, A. 1963. Les rapports hydrogeologique des lacs de Plitvice. In Guide pour Voyage. *Assoc.Intern.Hydrogeol.Congr.* 77-81. Belgrade.

Poltnig, W., Probst, G. & Zojer, H. 1994. Untersuchung zur Speicherung und zum Schutz von Karstwässern der Villacher Alpe (Kärnten). *Mitt.Österr.Geol.Ges.* 87: 75-90. Wien.

Potié, L. & Tardieu, B. 1977. Aménagement et captage sous-arins dans les formations calcaires. *Karst Hydrogeology (Congress of Alabama, 1975)*: 39-56.

Prinz, H. 1980. Erscheinungsformen des tiefen Salinekarstes an der Trasse der DB-Neubaustrecke Hannover-Würzburg in Osthessen. *Rock Mech. Suppl.* 10: 23-33.

Prinz, H. 1982. *Abriß der Ingenieurgeologie,* 419 p. Enke Verlag Stuttgart.

Prohic, E. 1989. Pollution assessment in carbonate terranes. In LaMoreaux, P. (ed.), *Hydrology of limestone terranes: Annotated bibliography of carbonate rocks No. 4, I.A.H. International Contributions to Hydrogeology* 10: 61 –82. Heise, Hanover.

Pulido-Bosch, A. 1985. L'exploitation minière de l'eau dans l'aquifère de la Sierra de Crevillente et ses alentours (Alicante, Espagne). *Hydrogeology in the Service of Man. 18th Cong., IAH, Cambridge*: 142-149.

Pulido-Bosch, A. 1986. Reflexiones sobre Hidrogeología Kárstica basadas en ejemplos de las Cordilleras Béticas. In *El Karst en Euskadi* 2: 31-50.

Pulido-Bosch, A. 1991. The overexploitation of some karstic aquifers in the province of Alicante (Spain). *XXIII IAH Congress* 1: 65-72. Canary Islands.

Pulido-Bosch, A. (ed.) 1993. *Some Spanish Karstic Aquifers.* Univ. Granada, 310 p.

Pulido-Bosch, A. 1995. Los acuíferos kársticos españoles. Investigación y Ciencia. (in print).

Pulido-Bosch, A., Castillo, A. & Padilla, A. (eds) 1989. La sobreexplotación de acuíferos. *Temas Geológico-Mineros* 10: 687 p. IGME, Madrid.

Pulido-Bosch, A., Martin-Rosales, W., López-Chicano, M., Rodríguez-Navarro, C.M. & Vallejos, A. 1997. Human impact in a tourist karstic cave (Aracena, Spain). *Environmental Geol.* 31: 142-149.

Pulido-Bosch, A., Morell, I. & Andreu, J.M. 1995. Hydrogeochemical effects of the groundwater mining of the Sierra de Crevillente (Alicante, Spain). *Environmental Geology* 26: 232-239.

Pulido-Bosch, A., Sánchez, F, Navarrete, F. & Martínez-Vidal, J.L. 1992. Groundwater problems in a semiarid area. *Environmental Geology* 20(3): 195-205.

Quinlan, J.F. 1974. Origin, distribution, and detection of development of two types of sinkholes in anthropogenic karst, South Africa. In *Proceedings of the 4th Conference of Karst Geology and Hydrology, West Virginia Geological and Economic Survey*, 161 p.

Quinlan, J.F. 1983. Groundwater pollution by sewage, creamery waste, and heavy metals in the Horse Cave area, Kentucky. In Dougherty, P.H. (ed.), *Environmental Karst, Geo. Speleo. Pubs.* 52. Cincinnati.

Quinlan, J.F. 1988. Protocol for reliable monitoring of groundwater quality in karst terranes. In Yuan, D. (ed.) *Karst Hydrogeology and karst environment protection, Proceedings 21st IAH Congress, Guilin, China* 2: 888-893. Geological Publishing House, Beijing.

Quinlan, J.F. 1989. Groundwater monitoring in karst terranes: recommended protocol and implicit Assumptions. US Environmental Protection Agency, Envir. Monitor. Systems Laboratory, Las Vegas, Nev. EPA/600/X-89/050, 88 p.

Quinlan, J.F. 1990. Special problems of ground-water monitoring in karst terranes. In Nielson, D.M. & Johnson, A.I. (eds), *Ground Water and Vadose Zone Monitoring, American Society for Testing and Materials, Philadelphia*: 275-304.

Quinlan, J.F. & Alexander, E.C. 1987. How often should samples be taken at relevant locations for reliable monitoring of pollutants from an agricultural, waste disposal or spill site in a karst terrane? A first approximation. In Beck, B.F. & Wilson, W.L. (eds), *Karst hydrogeology: Engineering and environmental applications*: 277-286. Rotterdam: Balkema.

Quinlan, J.F. & Ewers 1985. Ground water flow in limestone terranes: Strategy rationale and procedure for reliable, efficient monitoring of ground water quality in karst areas. *Proc. of the 5th Nat.Symp. and Expo. on Aquifer Restor. and Groundwater monitoring. National Water Well Association, Woshington, OH*: 197-234.

Quinlan, J.F., Ewers, R.O. & Field, M.S. 1986. How to use groundwater tracing to 'prove' that leakage of harmful materials from a site in a Karst Terrane will not occur. In *Proceedings Environmental Problems in Karst Terranes and their Solutions Conference, National Water Well Association*: 289-301.

Quinlan, J.F. & Ray, J.A. 1981. *Groundwater basins in the Mammoth Cave Region, Kentucky*. Occ. Publ., 1, Friends of the Karst, Mammoth Cave.

Quinlan, J.F. & Ray, J.A. 1991. Groundwater remediation may be achievable in some karst aquifers that contaminated, but it ranges from unlikely to impossible in most. I. Implication of long-term tracer tests. *Proceedings of the 3rd Conference Hydrogeology, ecology, monitoring and management of groundwater in karst terrains. USEPS and NGWA, Nashville, Tennessee*: 553-558.

Quinlan, J.F., Ray, J.A. & Schindel, G. 1995. Intrinsic limitations of standard criteria and methods for delineation of groundwater-source protection areas (springhead and wellhead protection areas) in carbonate terranes: Critical review, technically – sound resolution of limitations, and case study in a Kentucky karst. In Beck, B.F. (ed.), *Karst Geohazards. Engineering and environmental problems in karst terrane. Proceedings of the 5th multidisciplinary conference on sinkholes and the environmental impacts of karsts*: 525-537. Rotterdam: Balkema.

Quinlan, J.F. & Rowe, D.R. 1977. Hydrology and water quality in the central Kentucky karst: Phase I. Univ. Kentucky Water Resources Research Inst., Research Report, 101, 93p.

Quinlan, J.F., Smart, P.L., Schindel, G.M., Alexander, E.C.Jr., Edwards, A.J. & Smith, A.R. 1992. Recommended administrative/regulatory definition of karst aquifer, principles for classification of carbonate aquifers, practical evaluation of vulnerability of karst aquifers, and determination of optimum sampling frequency at springs. In Quinlan, J.F. (ed.), *Hydrogeology, ecology, montioring and managment of groundwater in karst terranes Conference (3rd, Nashville, Tenn.) Proceedings*: 573-635. National Ground Water Association, Dublin, Ohio.

Rathfelder, K., Lang, J.R. & Abriola, L.M. 1995. Soil vapor extraction and bioventing. US Nat. Report to Intern. Union of Geodesy and Geophysics, Review of geophysics, supplement, p. 1067-1081.

Reade, J. 1983. *Assyrian sculpture*. British Museum Publications, 72 p., London.

Recker, S. 1991. Petroleum hydrocarbon remediation of the subcutaneous zone of a karst aquifer, Lexington, Kentucky. In *Proceeding of the Third Conference on hydrogeology, ecology, montioring and managment of groundwater in karst terranes*: 447-473. U.S.EPA and NGWA, Nashville, Tennessee.

Reddy, M.M. 1977. Crystallisation of calcium carbonate in the presence of trace concentrations of phosphorus containing anions. 1. Inhibition of phosphate and glycerophosphate ions at pH 8.8 and 25°C. *J. Crystal Growth* 41: 287-295.

Reeder, P.P. & Crawford, N.C. 1989. Potential groundwater contamination of an urban karst aquifer: Bowling Green, Kentucky. In Beck, B.F. (ed.), *Proceedings of the 3rd multidisciplinary conference on sinkholes*: 197-206. Rotterdam: Balkema.

Reeder, P.P. & Day, M.J. 1993. Seasonality of chloride and nitrate contamination in the Wouthwestern Wisconsin karst. In Beck, B.F. (ed.), *Proceedings of the 4th Multidisciplinary Conference on sinkholes and the environmental impacts of karsts*: 53-61. Rotterdam: Balkema.

Reichert, B. 1991. Anwendung natürlicher und künstlicher Tracer zur Abschätzung des Gefährdungspotentials bei der Wassergewinnung durch Uferfiltration. *Schriftenr. Angew. Geologie Karlsruhe* 13: 226 p. Karlsruhe.

Reiss, F., Ramspacher, P. & Zojer, H. 1986. Upper Ladon river system. In Morfis, A. & Zojer, H. (eds), *Karst hydrogeology of the central and eastern Pelopennesus (Greece). Steir. Beitr. z. Hydrogeologie* 37/38: 115-127. Graz.

Reuter, F. 1962. Ingenieurgeologische Beurteilung und Klassifikation von Auslaugungserscheinungen. *Freib. Forsch. H.*, C 127: 1-47.

Reuter, F., Molek, H. & Kockert, W. 1972. Exkursionsführer zu ausgewählten Objekten des Salzund Gipskarstes im Subherzynischen Becken, in der Mansfelder Mulde und im Südharz-Gebiet. Bergakademie Freiberg.

Reuter, F. & Tolmacev, V.V. 1990. Bauen und Bergbau in Senkungs- und Erdfallgebieten. *Schriftenreihe f. Geolog. Wissenschaft* 28: 176 p.

Ripp, B.J. & Baker, J.A. 1997. Urbanization in karst sinkhole terrain – a St. Louis perspective. In Beck, B.F. & Stephenson, J.B. (eds), *The engineering geology and hydrogeology of karst terranes*: 293-298. Rotterdam: Balkema.

Rismal, M., Lovgin, N. & Cvitanovi, I. 1994. Notranjska Reka kot moeni vir za preskrbo pitno vodo Krasa in obalne regije – primer integralnega gospodarjenja z vodami. Elaborat, FAGG, Institut za zdravstveno hidrotehniko, 1-13, P1.1-P2.14, Ljubljana.

Röckel, T. & Hötzl, H. 1986. Polje of Stymfalia. In Morfis, A. & Zojer, H. (eds), *Karst hydrogeology of the Central and Eastern Peloponnesus (Greece). Steir.Beitr.z. Hydrogeologie* 37/38: 127-130. Graz.

Rodríguez-Estrella, T. & Gómez de las Heras, J. 1986. Principales características de los acuíferos kársticos de la provincia de Murcia. *Jorn. Karst Euskadi* 1: 187-203. San Sebastián.

Roglic, J. 1951. Unsko-koranska zaravan i Plitvicka jezera. Geomorfoloska promatranja. *Geograf.Clasnik* 13: 49-68.

Roglic, J. 1981. Les barrages de tuf calcaire aux lacs de Pltvice. Actes du Coll. de la formations carbonates externes, tufs et travertines. *Assoc.Francaise de Karstologie* 3: 37-48.

Rojsek, D. 1990. Human impact on Skocjanske jame system. *Studia Carsologica* 2: 120-132. Brno.

Rojsek, D. 1994. Inventarisation of the Natural Heritage. *Acta Carsologica* 23: 111-121. Ljubljana.

Rojsek, D. 1995. Inventory of the Skocjan world heritage site. *Acta Carsologica* 24: 475-485. Ljubljana.

Rojsek, D. 1996. Velika voda – Reka – A karst river. *Acta Carsologica* 25: 193-206. Ljubljana.

Romero, P., Borrego, J.J., De Vicente, A., Moriñigo, A., Martínez-Manzanares, E., Arrabal, F., Florido, J.A., Avilés, M., Cornax, R., Codina, J.C. & Arcos, M.L. 1991. Estudio microbiológico y químico de las aguas de la Cueva de Nerja (Málaga). In Marín, F. & Carrasco, F. (eds), *Investigación biológica y edafológica de la Cueva de Nerja Trabajos sobre la Cueva de Nerja* 2: 45-109.

Rosenzweig, A. 1972. Study of the differences in effects of forest and other vegetative covers on water yield (Final report, Project A-10-FS-13). State of Israel Ministry of Agriculture, Soil Conservation and Drainage Division Research Unit, Research Report No. 33, Tel Aviv.

Rosenzweig, A. 1973. Determination of evapotranspiration for short time intervals in a karst benchmark basin. State of Israel Ministry of Agriculture, Soil Conservation and Drainage Division Research Unit, Research Report No. 37, Tel Aviv.

Roth, R. & Peterson, M. 1994. Remediation of gasoline-contaminated soils and groundwater using soil vapor extraction and airsparging. *Proc. 6th Annual New England Environmental EXPO, Boston, Massachusetts.*

Rovey, C.W. & Cherkauer, D.S. 1995. Scale dependence of hydraulic conductivity measurements. *Ground Water* 33: 769-780.

Rozkowski, A., Kowalczyk, A., Motyka, J. & Rubin, K. (eds) 1996. *Karst-fractured aquifers – vulnerability and sustainability.* University of Silesia, Katowice-Ustrom, Poland, 309 p.

Ruiz-Sánchez, I., Marín-Girón, F., Ojeda, F., Marín-Olalla, F., Berros, J. & Marín-Olalla, E. 1991. Estudio macroscópico 'in situ' y microscópico-ecológico de pequeñas zonas de flora verde (algas verdes y verdeazuladas) del interior de la Cueva de Nerja. In Marín, F. & Carrasco, F. (eds), *Investigación Biológica y edafológica de la Cueva de Nerja, Trabajos sobre la Cueva de Nerja* 2: 113-125.

Sabatini, D.A. & Austin, T.A. 1991. Characteristics of Rhodamine WT and Fluorescein as adsorbing ground-water tracers. *Ground Water* 29(3): 341-349.

Saleem, Z.A. 1977. Road salts and quality of groundwater from a dolomite aquifer in the Chicago area. In Dilamarter, R.R. & Csallany, S.C. (eds), *Western Kentucky University, Bowling Green KY*: 364-368.

Sanz de Galdeano, C. 1986. Structure et stratigraphie du secteur oriental de la Sierra Almijara (Zone Alpujárride, Cordilléres Bétiques). *Estudios Geol.* 42: 281-289.

Sasowsky, I. & White, W. 1993. Geochemistry of the Obey River Basin, northcentral Tennessee: a case of acid mine water in a karst drainage system. *Journal of Hydrology* 146: 29-48.

Sauro, U., Bondesan, A. & Meneghel, M. (eds) 1991. *Proceeding of the International Conference on environmental changes in karst areas. Department of Geography, University of Padua,* 414 p.

Sauter, M. 1992. Assessment of hydraulic conductivity in a karst aquifer at local and regional scale. In Quinlan, J.F. & Stanley, A. (eds), *Hydrology, ecology, monitoring and management of karst terranes. Conference Proceedings*: 39-57. National Ground Water Association, Dublin, Ohio.

Schleyer, R., Renner, I. & Mühlhausen, D. 1991. Beeinflussung der Grundwasserqualität durch luftgetragene organische Schadstoffe. *WaBoLu-Hefte* 5/1991, 96 p.

Schmidl, A. 1851. Unterirdischer Lauf des Recca-Flusses. *Jahrbuch geol. Reichanst.* 22: 184. Wien.

Schrale, G., Smith, P.C. & Emmett, A.J. 1984. Limiting factors for land treatment of cheese factory waste in the Mount Gambier area, South Australia. *Water (Official Journal of the Australian Water and Wastewater Association)* 11(4): 10-14.

Schwartz, C. & Drewes, D. 1995. A partnership approach in marketing land use regulations for karst areas . In Beck, B.F. (ed.), *Karst geohazards, engineering and environmental problems in karst terrane. Proceedings of the 5th multidisciplinary conference on sinkholes and the environmental impacts of karsts*: 521-524. Rotterdam: Balkema.

Schwille, F. 1971. Die Migration von Mineralöl in porösen Medien. *Gwf-wasser/abwasser* 112, part 1, 6/1971, 307-311, part 2, 7/1971, 331-339, part 3, 9/1971, 465-472.

Schwille, F. 1984. Migration of organic fluids immiscible with water in the unsaturated zone. In Yaron, B., Dagan, G. & Goldshmid, J. (eds), *Pollutants in porous media*: 27-48. Springer, Berlin.

Schwille, F. 1988. *Dense chlorinated solvents in porous and fractured media,* 146 p. Lewis, Boca Raton.

Scott, J.R. 1994. Catchment water quality deterioration as a result of water-level recovery in abandoned gold mines on the Eastern and Central Witwatersrand. Water Research Commission Report: K5/864, Provisional final report, August 1994.

Seip, H.M. 1980. Acidification of freshwater – sources and mechanisms. In Drablos, D. & Tollan, A. (eds), *Ecological impact of acid precipitation. Johs. Grefslie Trykkeri A/S, Mysen, Norway*: 350-351.

Sendlein, L.V.A. & Palmquist, R.C. 1977. Strategic placement of waste disposal sites in karst regions. In Tolson, J.S. & Doyle, F.L. (eds), *Karst Hydrogeology*: 323-336.

Settergren, C.D. 1977. Early conflicts over effluent irrigation in Missouri's Ozarks. In Dilamarter, R.R. & Csallany, S.C. (eds), *Hydrologic problems in karst regions, Western Kentucky University, Bowling Green KY*: 405-410.

SGOP 1991. Estudio hidrogeológico de las Sierras Tejeda, Almijara y Guájares (Málaga y Granada).

Simmers, I., Villarroya, F. & Rebollo, L.F. (eds) 1992. Selected Papers on Aquifer overexploitation. *IAH Selected Papers* 3: 391 p. Heise, Hannover.

Simmleit, N. & Herrmann, R. 1987a. The behaviour of hydrophobic organic micropollutants in different karst water systems, I: Transport of micropollutants and contaminant balances during the melting of snow. *Water, Air and Soil Pollution* 34: 79-95.

Simmleit, N. & Herrmann, R. 1987b. The behaviour of hydrophobic organic micropollutants in different karst water systems, II: Filtration capacity of karst systems and pollutant sinks. *Water, Air and Soil Pollution* 34: 97-109.

Simonic, M. 1993. Interpretation of aquifer pollution in the Zuurbekom South-eastern subcompartment. Technical report GH 3800, Directorate Geohydrology, Department of Water Affairs, R.S.A.

Simpson, T.W. & Cunningham, R.L. 1982. The occurrence of flow channels in soils. *Journal of Environmental Quality* 11(1): 29-30.

Simsek, S. 1990. Karst hot water aquifers in Turkey. In Günay, G., Johnson, A.I. & Back, W. (eds), *Hydrogeological processes in karst terranes. IAHS Publ.* 217: 29-35.

Singh, B.R. 1984. Sulfate sorption by acid soils: 2. Sulfate adsorption isotherms with and without organic matter and oxides of aluminum and iron. *Soil Sci.* 138: 94-297.

Slack, L.J. 1977. Hydro-environmental effects of sprayed sewage effluent, Tallahassee, Florida. In Tolson, J.S. & Doyle, F.L. (eds), *Karst Hydrogeology, IAH Memoirs* 12: 309-322. University of Alabama Press, Huntsville.

Slifer, D.W. & Erchul, R.A. 1989. Sinkhole dumps and the risk to groundwater in Virginia's karst areas. In Beck, B.F. (ed.), *Engineering and environmental impacts of sinkholes and karst, Proc. of the 3rd Multidisciplinary Conference on sinkholes and the environmental impacts of karst*: 207-212. Rotterdam: Balkema.

Smart, P.L., Edwards, A.J. & Hobbs, S.L. 1991. Heterogeneity in carbonate aquifers: Effects of scale, fissuration, lithology and karstification. In *Proceedings of the Third Conference on Hydrogeology, ecology, monitoring and management of groundwater in karst terranes, National Water Well Association*: 373-388.

Smart, P.L. & Friederich, H. 1986. Water movement and storage in the unsaturated zone of a maturely karstified carbonate aquifer, Mendip Hills, England. In *Proceedings Environmental Problems in Karst Terranes and their Solutions Conference, National Water Well Association*: 59-87.

Smart, P.L. & Hobbs, S.L. 1986. Characterisation of carbonate aquifers. A conceptual base. In *Proceedings environmental problems in karst terranes and their solutions Conference, National Water Well Association*: 1-14.

Smith, D.I. 1993. The nature of karst aquifers and their susceptibility to pollution. In Williams, P.W. (ed.), *Karst Terrains: Environmental Changes and Human Impact. Catena Supplement* 25: 41-58.

Smith, D.I. & Finlayson, B.L. 1988. Water in Australia: Its role in environmental degradation, in Land, Water and People: Geographical Essays. In Heathcote, R.L. & Mabbutt, J.A. (eds), *Australian Resource Management, Academy of Social Sciences*: 7-48. Allen and Unwin, Sydney.

Smith, P.C. & Schrale, G. 1982. Proposed rehabilitation of an aquifer contaminated with cheese factory wastes. *Water (Official Journal of the Australian Water and Wastewater Association)* 9(1): 21-24.

Smith-Carington, A.K., Bridge, L.R., Robertson, A.S. & Foster, S.S.D. 1983. The nitrate pollution problem in groundwater supplies from Jurassic limestones in central Lincolnshire. Institute of Geological Sciences Report, 83/3, 22 p., H.M.S.O., London.

Smolen, M.D., Humenik, F.J., Brichford, S.L., Spooner, J., Bennett, T.B., Lanier, A.L., Coffey, S.W. & Adler, K.J. 1989. NWQEP 1988 Annual Report: Status of agricultural nonpoint source projects. Biological and Agricultural Engineering Dept., N.C. Agricultural Extension Service, North Carolina State University, Raleigh, N.C.

Sociedad Excursionista de Málaga 1985. *La Cueva de Nerja*. Sociedad grupo de espeleólogos granadinos, 87 p.

Stanton, W. 1966. The Impact of Quarrying on the Mendip Hills, Somerset. *Proc. Univ.Bristol Spel. Soc.* 11: 54-62.

Stanton, W. 1977. A view of the Hills. In Barrington, N. & Stanton, W. (eds), *Mendip – the complete caves and a view of the hills*: 193-234. Cheddar Valley Press, Somerset.

Stanton, W. 1989. Bleak prospects for limestone. *New Scientist,* May 1989: 56-60.

Stanton, W. 1990. Hard limestone: too valuable to quarry. *Minerals Planning* 43: 3-9.

Starr, R.C. & Cherry, J.A. 1994. In situ remediation of contaminated ground water. The funnel-and-gate system. *Ground Water* 32(3): 465-476.

Stefka, L. 1990. Some problems of environmental conservation in the protected landscape region 'Moravian Karst'. *Studia Carsologica* 3: 116-119.

Stephenson, J.B. & Beck, B.F. 1995. Management of the discharge quality of highway runoff in karst areas to control impacts to ground-water – a review of relevant literature. In Beck, B.F. (ed.), *Proceedings of the 5th Multidisciplinary Conference on sinkholes and the environmental impacts of karsts, Galtinburg, Tennessee*: 297-321. Rotterdam: Balkema.

Stephenson, J.B., Zhou, W.F., Beck, B.F. & Green, T.S. 1997. Highway stormwater runoff in karst areas – Preliminary results of baseline monitoring and design of a treatment system for a sinkhole in Knoxville, Tennessee. In Beck, B.F. & Stephenson, J.B. (eds), *The Engineering Geology and Hydrogeology of Karst Terranes*: 173-181. Rotterdam: Balkema.

Stille, H. 1903. Geologisch-hydrologische Verhältnisse im Ursprungsgebiet der Paderquellen zu Paderborn. *Abh.Königl.Preuß.Geol.Landesanst.u.Bergak.*, *N.F.* 38: 1-129. Berlin.

Stini, J. 1937. Zur Geologie der Umgebung von Warmbad Villach. *Jb.Geol.B.-A.* 87. Wien.

Stumm, W. & Morgan, J.J. 1981. *Aquatic Chemistry.* 2nd ed. John Wiley & Sons, NY

TA-Abfall 1991. Technische Anleitung zur Lagerung, chemisch-physikalischen, biologischen Behandlung, Verbrennung und Ablagerung von besonders überwachungsbedürftigen Abfällen. Bundesministerium für Umwelt, Naturschutz und Reaktorsicherheit. Gesamtfassung der Zweiten Allgemeinen Verwaltungsvorschrift zum Abfallgesetz vom 10.3.1991, Bonn.

Tadolini, T. & Tulipano, L. 1979. The evolution of fresh water-salt water equilibrium in connection with drafts from the coastal carbonate and karst aquifer of the Salentine Peninsula (Southern Italy). *Proc. 6th Salt Water Intrusion Meeting*: 69-86. Hannover, Germany.

TA-Siedlungsbfall 1993. Technische Anleitung Siedlungsabfall. Bundesministerium für Umwelt, Naturschutz und Reaktorsicherheit (1993). *Dritte allgemeine Verwaltungsvorschrift zum Abfallgesetz. Bundesanzeiger* 45, 99a, 14.5.1993. Bonn.

Tennyson, L.C. & Settergren, C.D. 1977. Subsurface water behavior and sewage effluent irrigation in the Missouri Ozarks. In Dilamarter, R.R. & Csallany, S.C. (eds), *Hydrologic problems in karst regions, Western Kentucky University, Bowling Green KY*: 411-418.

Thomson, K.C. 1997. Field trip through the karst areas of southern Creene and northern Christian Counties, Missouri, emphasizing the relationships between surface features and springs. In

Beck, B.F. & Stephenson, J.B. (eds), *The Engineering Geology and Hydrogeology of Karst Terranes*: 475-506. Rotterdam: Balkema.

Thorn, R.H. & Coxon, C.E. 1992. Hydrogeological aspects of bacterial contamination of some western Ireland karstic limestone aquifers. *Environ. Geol. Water Sci.* 20(1): 65-72.

Troester, J.M. & White, W.B. 1984. Seasonal fluctuations in the carbon dioxide partial pressure in a cave atmosphere. *Water Resources Research* 20: 153-156.

Trombe, F. 1952. *Traité de Speleologie*. Payot. París.

Tucker, N.L. 1982. Nonpoint agricultural pollution in a karst aquifer: Lost River groundwater drainage basin, Warren county, Kentucky. Unpubl. MS Thesis, Western Kentucky University, Bowling Green, Kentucky.

Tulipano, L. 1988. Temperature logs interpretation for identification of preferential flow pathways in the coastal carbonatic and karstic aquifer of the Salento Peninsula (Southern Italy). *Proc. 21st I.A.H. Congress, Karst Hydrogeology and Karst Environment Protection, Guilin, China*: 956-961.

Tulipano, L. & Fidelibus, M.D. 1988. Temperature of ground waters in coastal aquifers: some aspects concerning saltwater intrusion. *Proc. 10th Salt Water Intrusion Meeting, 308-316.* Gent, Belgium.

Tulipano, L. & Fidelibus, M.D. 1995. National report for Italy. In EC-COST action 65 (1995). Hydrogeological aspects of groundwater protection in karstic areas. Final report. European Commission, EUR 16547EN, 171-202, Luxembourg.

Tyc, A. 1989. Contemporary karst processes in the zone of influence of the lead and zinc mines in the Olkusz Region. *Kras i speleologia* 6(15): 23-38. Katowice.

Tyc, A. 1990. Sink forms in the karst of the Olkusz district caused by the mining and water pumping activities. *Sci. Bull. of Acad. of Mining and Metallurgy, 1368, Sozology and Sozotechnics* 32: 100-112. Kraków.

Tyc, A. 1994. Anthropogenic Impact on Karst Processes in the Olkusz-Zawiercie Karst Area (Silesian-Cracow Upland). University of A. Mickiewicz, Poznañ PhD Thesis, 183 p.

US Environmental Protection Agency (USEPA) 1987. National pesticide survey health advisory communication: atrazine. Office of Drinking Water, USEPA, Washington D.C.

US Environmental Protection Agency (USEPA) 1991. Solid Waste Disposal Facility Criteria, Final Rule, Federal Register., 56, 40 CFR Parts 257 and 258, 51016-51119, US Government Printing Office, Washington, DC.

US Environmental Protection Agency (USEPA) 1992. RCRA Ground-Water Monitoring. Draft Technical Guidance, D.C., EPA 530-R-93-001, 243 p. (NTIS PB 93-139 350), Office of Solid Waste, Washington.

Urich, P.B. 1991. Stress on tropical karst resources exploited for the cultivation of wet rice. In *Proceedings of the International Conference on Environmental Changes in Karst Areas, IGU/UIS, Italy, 15-27 Sept. 1991, Quaderni del Dipartimento di Geografia* 13: 39-48. Universita di Padova.

Urich, P.B. 1993. Stress on tropical karst cultivated with wet rice: Bohol, Phillippines. *Environmental Geology* 21: 129-136.

Van Miegroet, H. & Cole, D.W. 1984. The impact of nitrification on soil acidification and cation leaching in red alder ecosystem. *J. Environ. Qual.* 13: 586-90.

Vegter, J.R. & Foster, M.B.J. 1992. The hydrogeology of dolomitic formations in the Southern and Western Transvaal. In Back, W., Herman, J.S. & Paloc, H. (eds), *Hydrogeology of Selected Karst Regions*: 355-376. Heise, Hannover.

Vera, J.A., Ruiz-Ortiz, P.A., Garcia-Hernandez, M. & Molina, J.M. 1987. Paleokarst and related pelagic sediments in the Jurassic of the sub-Betic zone, Southern Spain. In James, N.P. & Choquette, P.W. (eds), *Paleokarst*: 364-384. Springer-Verlag, Berlin.

Villinger, E. 1977. Über Potentialverteilung und Strömungssysteme im Karstwasser der Schwäbischen Alb (Oberer Jura, SW-Deutschland). *Geol.Jb*, C 18, 93 p. Hannover.

Voigt, H.-J. 1989. *Hydrogeochemie, Eine Einführung in die Beschaffenheitsentwicklung des Grundwassers*, 310 p. VEB Verlag, Leipzig.

Volker, A. & Henry, J.C.(eds) 1988. Side effects of water resources management. *IAHS Publ.* 172, 269 p.

Vollenweider, R. 1982. *Eutrophication of Waters- Monitoring Assessment and Control.* OECD, Paris.

Vrba, J. & Romijn, E. (eds) 1986. Impact of agricultural activities on ground water. *International Contributions to Hydrogeology* 5. Heise, Hannover.

Vrba, J. & Svoma, J. (eds) 1982. Impact of agricultural activities on Groundwater. *IAH International Symposium, Prague, Czechoslovakia, IAH Memoires* 16. Novinar Publishing House, Prague.

Wagner, G. 1954. Der Karst als Musterbeispiel der Verkarstung. *Aus der Heimat* 62(9/10): 193-212. Rau, Öhringen.

Walters, R.F. 1978. Land Subsidence in Central Kansas Related to Salt Dissolution. *Kansas Geol. Survey, Bull.* 214.

Walton, D.G. & Levin, M. 1993. The identification and verification of polluted areas in the dolomitic aquifer of the PWV area. Report GEA-1045, Earth and Environmental Department, Atomic Energy Corporation of South Africa, Pretoria.

Warner, F. 1992. Risk analysis, perception and management: Introduction. Royal Society (London), Study Group, 12 p.

Werner, E. 1983. Effects of highways on karst springs – an example from Pocahontas Country, West Virginia. In Dougherty, P.H. (ed.), *Environmental Karst*: 3-13. GeoSpeleo Publ., Cincinnati, Ohio.

Wheeler, B.J., Alexander, E.C., Adams, R.S. & Huppert, G.N. 1989. Agricultural land use and groundwater quality in the Coldwater Cave groundwater basin, upper Iowa river karst region, USA: Part II. In Gillieson, D. & Smith, D.I. (eds), *Resource management in limestone landscapes, International perspectives, Special Publication No.2, Dept. of Geog. & Oceanog., University College, Australian Defence Force Academy, Canberra*: 249-260.

Whelan, M.P., Voudrias, E.A. & Pearce, A. 1994. DNAPL pool dissolution in saturated porous media. *Proc. Developement and Preliminary Results. Journal of Contaminant Hydrology* 15: 223-237. Elsevier Science, Amsterdam.

Whitaker, F.F., Smart, P.L., Vahrenkamp, V.C., Nicholson, H. & Wogelius, R.A. 1994. Dolomitization by near-normal seawater ? Field evidence from the Bahamas. *Spec. Publs. Int. Ass. Sediment.* 21: 111-132.

White, W.B. 1988. *Geomorphology and hydrology of karst terrains,* 464p. Oxford Univ. Press, Oxford.

Wicks, C. & Groves, C. 1993. Acidic mine drainage in carbonate terrains: geochemical processes and rates of calcite dissolution. *Journal of Hydrology* 146: 13-27.

Wiersma, J.H., Stieglitz, R.D., Dewayne, L.C. & Metzler, G.M. 1986 Characterisation of the shallow groundwater system in an area with thin soils and sinkholes. *Environmental Geology and Water Sciences* 8(1-2): 99-104.

Williams, P.W. 1983. The role of the subcutaneous zone in karst hydrology. *J. Hydrol.* 61: 45-67.

Williams, P.W. (ed.) 1993. Karst Terrains: Environmental Changes and Human Impact. *Catena Supplement* 25: 268 p.

Work Group I 1983. Impact Statement; Summary. United States-Canada Memorandum of Intent (concerning transboundary air pollution). Final report, January 1983, (1-1)-(1-24).

Yates, M.V. 1985. Septic tank density and ground-water contamination. *Ground Water* 23(5): 586-591.

Yesertener, C. & Elhatip, H. 1997. Evaluation of groundwater flow by means of dye-tracing techniques, Pamukkale Thermal Springs, Wertern Turkey. *Hydrogeology Journal* 5(4): 51-59.

Yuan, D. 1981. *A brief introduction to China's research in karst.* Inst. Karst Geol., Guilin, Guangxi, China.

Yuan, D. 1983. *Problems of environmental protection of karst areas.* Inst. Karst Geol., Guilin, Guangxi, China.

Yuan, D. (ed.) 1988a. Karst hydrogeology and karst environmental protection. *Proc. 21st Congress I.A.H.,* 1261p. Guilin.

Yuan, D. 1988b. Karst Environmental Systems. In Gillieson, D. & Ingle Smith, D. (eds), *Resource Management in Limestone Landscapes. Proc. Intern. Geographical. Union Study Group, Mans Impact on Karst. Dept of Geography and Oceanography, Australian Defence Force Academy, Special Publication* 2: 149-163.Canberra.

Yuan, D. 1991. *Karst of China,* 224 p. Geol.Publ.House, Beijing.

Yuan, D. 1992. Karst and karst water in China. In Back, W., Herman, J.S. & Paloc, H. (eds), *Hydrogeology of Selected Karst Regions*: 315-338. Heise, Hannover.

Yuan, D. 1993. Environmental change and human impact on karst in southern China. In Williams, P.W. (ed.), *Karst Terrains: Environmental Changes and Human Impact. Catena, Supplement* 25: 99-107.

Zhang Zeng Xiang 1996. Some environmental problems in karst region of southern China. *30th Intern.Geol. Congress, Abstracts* 3, 16-12-20, 320. Beijing.

Zötl, J. 1974. *Karsthydrogeologie,* 291 p. Springer, Wien-New York.

Zojer, H. 1980. Beitrag zur Kenntnis der Thermalwässer von Warmbad-Villach. *Steir. Beitr. z. Hydrogeologie* 32. Graz.

Zojer, H. & Zötl, J. 1993. Warmbad Villach. In Zötl, J. & Groldbrunner, J.E. (eds), *Die Mineral- und Heilwässer Österreichs*: 130-137. Springer, Wien- New York.

Zupan, M. & Kolbezen, M. 1976. Hydrochemische Untersuchungen im Einzugsgebiet der Ljubljanica. In Gospodaric, R. & Zötl, J.G.(eds), *Markierung unterirdischer Wässer, Untersuchungen in Slowenien 1972-1975. Steir.Beitr.z.Hydrogeologie* 28: 63 78. Graz.

Subject index

Location index

319